Nucleic acid and protein sequence analysis
a practical approach

TITLES PUBLISHED IN
THE
PRACTICAL APPROACH
SERIES

Nucleic acid and protein sequence analysis
a practical approach

Edited by
M J Bishop
University of Cambridge, Computer Laboratory,
Corn Exchange Street, Cambridge CB2 3QG, UK

C J Rawlings
Biomedical Computing Unit, Imperial Cancer Research Fund,
PO Box 123, Lincoln's Inn Fields, London WC2A 3PX, UK

IRL PRESS
Oxford · Washington DC

IRL Press Limited
P.O. Box 1,
Eynsham,
Oxford OX8 1JJ,
England

British Library Cataloguing in Publication Data

Nucleic acid and protein sequence analysis: a practical approach.—(The
  Practical Approach Series)
  1. Deoxyribonucleic acid—Data processing
  2. Nucleotide sequence—Data processing
  I. Bishop,M.J. (Martin J.) II. Rawlings,C.J. III. Series
  574.87'3282 QP624

ISBN 1-85221-007-9 (hardbound)
ISBN 1-85221-006-0 (softbound)

Printed in England by Information Printing, Oxford.

Preface

Both molecular biology and computer science are rapidly evolving disciplines. Macromolecular sequences of DNA, RNA and protein are the encoded form of the genetic information of living organisms and as such are particularly adapted to manipulation by computer. However, the use of computers for sequence analysis has developed into a complex field in the last few years and may be a bewildering enterprise to the molecular biologist with little or no previous computing experience. This book is designed as a practical aid to biologists wishing to use computers for the acquisition, storage, or analysis of nucleic acid or protein sequences. Introductory chapters deal with computer systems, routine analysis and commercial software. Molecular sequence databases are then described, as well as various online services by which they may be accessed. The computer is of great value in the experimental determination of sequences and restriction map analysis and DNA sequencing are both described, including automation. Specialized topics are the prediction of regions of DNA sequences which code for protein, secondary structure prediction of RNA and protein structure prediction. There are details of sequence comparison methods, including database searching, and the inference of evolutionary relationships from sequence data.

M.J.Bishop and C.J.Rawlings

Contents

8. COMPUTER HANDLING OF DNA SEQUENCING PROJECTS 173
R.Staden

11. SECONDARY STRUCTURE PREDICTION OF RNA   259
M.Gouy

Contributors

M.J.Bishop
University of Cambridge, Computer Laboratory, Corn Exchange Street, Cambridge CB2 3QG, UK

Sir Walter Bodmer
ICRF Laboratories, PO Box 123, Lincoln's Inn Fields, London WC2A 3PX, UK

J.F.Collins
Department of Molecular Biology, University of Edinburgh, King's Buildings, Mayfield Road, Edinburgh EH9 3JR, UK

A.F.W.Coulson
Department of Molecular Biology, University of Edinburgh, King's Buildings, Mayfield Road, Edinburgh EH9 3JR, UK

J.K.Elder
Department of Biochemistry, Oxford University, South Parks Road, Oxford OX1 3QU, UK

A.E.Friday
University of Cambridge, Department of Zoology, Downing Street, Cambridge CB2 3EJ, UK

A.Frischauf
European Molecular Biology Laboratory, Meyerhofstrasse 1, D-6900 Heidelberg, FRG

M.Ginsburg
Imperial Cancer Research Fund, Clare Hall Laboratories, Blanch Lane, South Mimms, Potters Bar, Herts EN6 3LD, UK

M.Gouy
Laboratoire de Biometrie, Universite Lyon I, 43, Bd du 11 novembre 1918, F-69622 Villeurbanne Cedex, France

P.Hoyle
8 Walpole Gardens, Chiswick, London W4 4HG, UK

H.Lehrach
European Molecular Biology Laboratory, Meyerhofstrasse 1, D-6900 Heidelberg, FRG

C.J.Rawlings
Biomedical Computing Unit, Imperial Cancer Research Fund, PO Box 123, Lincoln's Inn Fields, London WC2A 3PX, UK

E.M.Southern
Department of Biochemistry, Oxford University, South Parks Road, Oxford OX1 3QU, UK

R.Staden
MRC Laboratory of Molecular Biology, Hills Road, Cambridge CB2 2QH, UK

P.A.Stockwell
Department of Biochemistry, University of Otago, Dunedin, PO Box 56, New Zealand

G.D.Stormo
Department of Molecular, Cellular and Developmental Biology, University of Colorado, Boulder, CO 80309, USA

W.R.Taylor
Laboratory of Molecular Biology, Department of Crystallography, Birkbeck College, Malet Street, London WC1E 7HX, UK

E.A.Thompson
Department of Statistics GN22, University of Washington, Seattle, WA 98195, USA

R.Wakeford
The European Biotechnology Information Project, The British Library, 9 Kean Street, London WC2B 4AT, UK

G.Zehetner
European Molecular Biology Laboratory, Meyerhofstrasse 1, D-6900 Heidelberg, FRG

Introduction

SIR WALTER BODMER

The development of recombinant DNA techniques, encompassing the ability to clone, sequence, manipulate, analyse and express DNA sequences from any organism, has revolutionized our ability to analyse and manipulate fundamental genetic and cellular processes, at all levels. These techniques have become an integral part of most biological, agricultural and medical research and, apart from their contribution to fundamental understanding of biological processes, have spawned a major new industry − the biotechnology industry.

Even before the development of DNA cloning and sequencing techniques, computer storage, analysis and comparison of amino acid sequences of proteins was well developed, especially for the analysis of evolutionary relationships and rates of change. Computers have also been used for many years for the analysis of protein structure from X-ray crystallographic data, and for the analysis of population genetic models which provide the quantitative basis for understanding the evolutionary process. But the advent of recombinant DNA techniques brought the need for computer analysis right into the laboratory. Even the simple assembly of DNA sequences, analysis of overlaps, and search for the restriction enzyme sequences, is hard to do without the use of computers. As the bank of sequences has grown, so computers have become an essential tool for working with DNA in the laboratory.

The rapid development of laboratory techniques, together with the large number of research groups involved in their use, has generated a huge and burgeoning body of data that requires a new generation of computer techniques for their analysis. Searching a large DNA data base for similar sequences is a complicated problem. Levels of similarity and sequence patterns, not just sequence identity, need to be taken into account. And multiple comparisons dictate the need for highly efficient algorithms.

The linguistics of the DNA sequence need to be understood and worked out by analysis of available sequences. Even the apparently simple task of searching for expressed sequences is complicated and eukaryotic organisms by the existence of short exons and the lack of consistency in the types of control sequences found at the five prime end of an expressed sequence. Evolutionary comparisons can become more and more sophisticated as data builds up not only on exons, but also on introns and flanking regions, and with genetic changes now known commonly to include insertions, deletions, transpositions, gene conversion and other related mechanisms. Beyond all this lies the ultimate challenge of predicting the tertiary structure of proteins from their primary amino acid sequences, obtained from the DNA sequence. Energy minimization approaches have not succeeded, and so far the best results have been obtained from empirical approaches using sequence data from proteins with known structures. As more structures are worked out and key features of the relationship between the linear sequence and the three-dimensional structure are established, it should gradually become possible to use highly organized empirical approaches, coupled with theoretical backing,

1

to work out structure prediction procedures. Somewhat simpler will be the problem of predicting the structure derived from a sequence that is very similar to one whose structure is already determined. Since the eukaryotic genes occur in related gene families, this possibility will become increasingly important as more structures are defined.

However, it is not only protein structure that is important. Tertiary structures of DNA, and more especially RNA sequences are becoming of increasing interest, particularly following the demonstration that the RNA sequence corresponding to an intron can itself function as an enzyme. It seems possible that the complicated patterns of sequences found at the five prime ends of genes, where they are controlled, need to be thought of not so much in simple linguistic terms, but rather in terms of the tertiary structures that can be formed by the sequences. For all these developments, many of which are described and their future hinted at in this book, computers with increasing power are an essential tool.

The human genome contains about 3×10^9 base pairs, of which however, perhaps only $5-10\%$ maybe functionally significant. These sequences are probably organized into some $5000-15\ 000$ different genetic functions, many involving clusters of related genes with similar structures. Knowledge of the total human genome mapping sequence, with all its products defined, would be a revolutionary step forward in our ability to analyse the genetic contribution to all aspects of normal and abnormal human variation. The comparison of this sequence with that of sequences from other species will be of enormous interest in unravelling the evolutionary relationships and hierarchies between gene products, and will also help reveal the control language by which complex patterns of differential gene expression are achieved. Establishing the total human genome sequence is surely a major challenge that must be taken up worldwide. It will of course be paralleled by much sequence information from other species. Such endeavours again require further developments in computing. First and most obviously comes the need for automating all aspects of the DNA cloning and sequencing process. Secondly, there is the need not only for comparison of DNA sequences and structure prediction, but simply for handling very large data-bases on the DNA sequences and inter-relating these with information on the proteins and their functions, and on the phenotypes that they determine. Furthermore, if such a project is to be successful it must be carried out worldwide and networking the sources of information and approaches for their analysis will be needed for this. Thus, computers will play an ever increasing role in the revolution that is being forged by the application of recombinant DNA techniques.

The chapters in this book represent an important step forward in bringing together a wide variety of computer applications for the analysis of nucleic acid and protein sequences in a practical and useful way. It should therefore be widely appreciated by the many research workers that depend on the techniques of molecular genetics and molecular biology.

CHAPTER 1

Introduction to computer hardware and systems software

MARTIN J. BISHOP

1. INTRODUCTION

It is natural that computers are of great importance in the acquisition and analysis of biological macromolecular sequences. Computers are our tools for encoding and manipulating information of all kinds. DNA encodes the genetic information of the cell, while RNA and protein are involved in implementing the development and functioning of the cell. DNA sequences are linear or circular unbranching polymeric chains which may total more than 10^9 nucleotides per cell. It is the order of the four possible nucleotides along the chain which enables information to be encoded. Sequences amounting to more than 5×10^6 nucleotides have already been determined and the pace at which DNA is sequenced is unlikely to diminish in the near future. Computer methods are, therefore, already proving essential for the storage and analysis of nucleic acid and protein sequences.

The use of computers to analyse DNA sequences became important with the development of practicable methods of DNA sequencing in the mid-seventies (1). By 1980 a number of reviews had appeared (2,3). In the short time since, a considerable literature on the subject has accumulated, largely published as three collections of papers in Nucleic Acids Research (4−6). There are now further reviews (7−12) and a comprehensive bibliography (13).

Computer technology (14) is evolving very rapidly at the present time so that it is difficult to make decisions about the purchase of suitable equipment which will be cost effective for a range of tasks. The choice will be dominated by the availability of software without which even the most sophisticated hardware will be useless. By hardware we mean the physical computer and its peripheral devices. Software is a term for the programs which the computer executes as a series of instructions which act on input and transform it to output: the results of running the program. Computers need an operating system which is a special program to control what the computer does. Users' programs are executed under the supervision of the operating system which decides what resources to allocate to the program. Simple operating systems allow only one task to be performed at a time, while more complex operating systems allow many tasks and are therefore described as *multi-tasking*. Still more complex operating systems allow more than one user to access the computer at a given time and are therefore described as *multi-user*. Tasks running in the background are called *batch processing* or *jobs*.

The machine instructions which the computer executes are relatively simple and primitive operations. A complex computer program may consist of millions of instructions. A programming language allows the programmer to use less primitive constructs which may be later expanded into machine instructions. A compiler is a computer program which converts the text of programs written in a programming language (the *source code*) into a set of instructions (the *object code*) which can be executed by the computer. The part of the hardware which executes instructions is called the central processing unit (*CPU*). Different computers may have different CPUs and therefore execute different machine codes. The rules of a programming language are fixed and can be mapped into a variety of machine codes. Common programming languages used for applications in molecular biology include BASIC, C, FORTRAN and Pascal. It is possible to compile the text of a computer program on different machines by using the appropriate language compiler for the machine in question. A computer program which can be moved to a variety of computers with different CPUs is said to be portable. Program portability is extremely important because the considerable human effort required to develop software is not wasted when it is required to move software onto a new machine (15). However, it is hardly ever possible to move computer programs from machine to machine without some changes. Programs should be designed so as to isolate the machine dependent sections (16). In the majority of cases, portability remains an ideal rather than a reality.

To the user of a computer the most important issue is the provision of an appropriate *man—machine interface* (MMI). This is an area of rapid change at the present day and there are no standards because the concepts are innovative. The earliest means of direct interaction with computers involved teletype devices. There followed the use of visual display units acting as *dumb terminals* allowing interaction with a single input line and *graphics terminals* allowing the drawing of pictorial vector representations. Full screen interaction on video terminals allows selection from menus using the cursor keys and is the prevalent MMI at the present day. The use of colour may also be very effective for some applications. Becoming popular is the extremely easily used *WIMP* interface (Windows, Icons, Mice, Pull-down menus) based on ideas developed at Xerox Palo Alto Research Center (17). The mouse is controlled by hand and has a number of buttons. Movements of the mouse are interpreted as movements of a cursor on the screen. Actions of the computer are controlled by actions of the mouse such as *dragging* or *clicking* and it becomes possible to run sophisticated applications with little or no typing ability. Interaction with the machine is speeded and the rate of errors is reduced because multiple choices may be rapidly presented and selected. Multiple windows allow the user to perform a number of related tasks simultaneously, simulating sheets of paper or pages of a book. The interface is successful because it models traditional human ways of working on a desk top.

In recent years, computers have been classified as *micro, mini* or *main-frame*. The advent of inexpensive microprocessor based computers (*workstations*) with advanced interactive graphics running applications which formerly required the use of a mainframe has revolutionized this position (18−19). It is predicted that the new pattern will be a network of personal workstations linked with servers for file storage, printers and other specialized services as well as high performance *supercomputer* processing nodes. The most important applications of supercomputers in molecular biology are for searching molecular sequence databases and for studies of macromolecular structure and dynamics (20,21). Molecular sequence analysis as described in this book is a first step towards a deeper understanding of biological function. A long range objective is the prediction of the folded conformation of proteins and RNA molecules given their primary sequences. The function of biologically important molecules depends on their conformation in three dimensions and on the dynamics of their interactions. The tools of site-directed mutagenesis and expression of cloned genes enable RNA or protein molecules to be engineered to aid these functional studies. Computational methods will continue to play an essential role in biotechnology (22).

2. THE COMPUTER

The choice of a computer for applications in molecular biology will be largely dominated by the availability of software. The software should exist and it is most unwise to purchase hardware based on a promise of future software. Both hardware and software are rapidly evolving and specific recommendations made in a book of this sort are likely to be out of date almost immediately. However, we can outline the guiding principles, illustrate with examples, and recommend consultation with experts for further advice.

The three major hardware considerations are the type of CPU, the amount of random access memory (*RAM*), and the access speed and size of mass storage devices. A comprehensive workstation for molecular biology should have a 32 bit processor, at least 1 Mbyte of RAM and 20 Mbyte of mass storage in the form of hard disk (23). There should be a graphics capability as prediction of sequence function and comparisons of sequences need to be visualized. Experimental determination of sequences typically needs a digitizer for sizing restriction fragments or reading DNA sequencing gels from autoradiograms, though this technology may be replaced by automated methods using densitometry. A printer will be required to produce a copy of results on paper. There should also be a means of communicating with other computers to exchange information with colleagues and to access molecular sequence databases.

The limitations on the size of a problem which can be tackled by a computer are related to its memory capacity and speed of execution. The smallest unit of computer memory is termed a bit (which has two states: one or off). The smallest unit which can be conveniently accessed is called a byte (often 8 bits). The power of the central processing unit of a computer is related to the largest width in bits of its registers and the speed of execution of its instructions. A register is a storage location in the CPU. Microprocessor CPUs commonly have 8-bit, 16-bit or 32-bit registers. The CPU is connected to the RAM and both of these are connected to peripherals by a *data bus*. The data bus of a microcomputer is commonly either 8-bit or 16-bit depending on the number of parallel conductors. The destination of information on the data bus is defined using the *address bus* which is commonly 16-bit or 24-bit. The width of the address bus limits the amount of RAM which can be readily addressed. Please refer to ref. 11 for further details and diagrams of microprocessor configurations.

The most common 8-bit microprocessor chips are the Mos Technology 6502 and the Zilog Z80. The 6502 has one 8-bit general register, two 8-bit index registers, an 8-bit stack pointer, a 16-bit program counter and two interrupt levels. Bytes in the lowest page of memory can be paired for use in indexing 64 Kbytes. The stack is an area of memory organized such that items are added to or removed from the top of the stack. The stack is used for temporary storage of data and addresses when programs are running. The program counter keeps track of the location in memory containing the next instruction to be obeyed. The interrupt facility allows signals from outside to interrupt the sequence of instructions in the CPU. The 6502 is the CPU of popular machines such as the Apple II, the Commodore Pet and the Acorn BBC B. The Z80 has eight 8-bit general registers (six of which may be paired to give three 16-bit registers), a 16-bit stack pointer and a 16-bit program counter. The Z80 has been used in a variety of machines with the CP/M operating systems such as the Osborne, Kaypro and North Star. (CP/M stands for Control Program for Microprocessors.) The 6502 and Z80 machines are now inexpensive and are often used for applications such as interfacing laboratory equipment or terminal emulation. They are not sufficiently powerful for the wide range of DNA or protein sequence analysis tasks.

The most population 16-bit microprocessor chips in present use are the Intel 8086 and 8088. The 8088 can run the same software as the 8086 but has an 8-bit rather than a 16-bit data bus. The Intel 80286 is a more powerful 16-bit microprocessor which can run in 8086-compatible 'real address' mode. The 8086 processor can address 1 Mbyte of memory in 64 Kbyte segments. It has therefore replaced the Z80 as the most popular chip for personal computers. The 8086 has four 16-bit general registers, four 16-bit index registers and four 16-bit segment registers. There are two interrupt levels. The 8086 and 8088 chips are used as the CPUs of a variety of machines such as the Sirius, Apricot and the IBM PC and its compatibles include the Compaq and the Olivetti M24. The machines use the MSDOS and PCDOS operating systems (MSDOS stands for MicroSoft Disk Operating System). The Apricot Xen and the IBM PC AT use the 80286 chip. There is a 80287 math-coprocessor to perform high-speed arithmetic, logarithmic functions, and trigonometric operations with great accuracy. In the future, machines will be based on the 80386 chip which has 32-bit registers. 16-bit machines are suitable for many DNA or protein sequence analysis tasks, but 32-bit machines are more powerful and are easier to program.

A number of microprocessor chips with 32-bit registers are now in commercial use, notably the Motorola 68000 family, the National Semiconductor 32016 and the Digital MicroVAX chips. These chips are at the heart of the present generation of supermicro workstations. The 68000 has eight 32-bit data registers, seven 32-bit index registers, two 32-bit stack pointers, one 24-bit program counter, and one 16-bit status register. There are seven interrupt levels. Most operations can be performed on 8-, 16-, or 32-bit values. The only important omission from full 32-bit capability is the lack of 32-bit multiply and divide, though these are implemented in the 68020. The 68000 and 68010 have a 16-bit data bus. A version of the 68000 with an 8-bit data bus is named the 68008. The 68020 has a 32-bit data bus and a coprocessor interface allowing the addition of chips to provide fast floating point arithmetic. Important 68000 based systems include the Apple Macintosh and the Sun-3 (68020) both of which have the WIMP type of operating system. The MicroVAX chip has sixteen 32-bit general purpose registers and eighteen 32-bit privileged registers. The VAX-11 supermini computers have 304 instructions. Of these, 175 are implemented in the MicroVAX chip, 70 in a companion floating point chip and 59 via emulation. Performance of MicroVAX II computers is similar to that of the VAX 11/780 (as far as the single user is concerned).

The machine is proving extremely popular for scientific applications and has virtual memory management. Virtual memory means that more than the available RAM may be used by a program by transferring infrequently used pages to disk.

In the future, microprocessors will be faster and less expensive. One approach to this is the simplification of the instruction set in the RISC (Reduced Instruction Set Computer) architecture. The IBM PC RT workstation uses such a chip and has 32-bit registers, 40-bit addressing and virtual memory management.

Another trend is towards the building of machines optimized for particular applications. For example, the Inmos Transputer is a computer on a chip and by connecting transputers in a suitable configuration it may prove possible to build a machine for very rapid lookup of a query DNA sequence in a large database.

3. THE PERIPHERALS

3.1 Storage

3.1.1 *Diskette drives*

The most common storage medium for small computers at the present time is a flexible diskette made of plastic coated with a magnetic oxide and contained in a square jacket. Initially, 8 inch floppy disks were favoured but now the most common size is 5.25 inch. These are in turn being replaced by 3.5 inch disks in a rigid plastic jacket protecting the disk from mechanical damage.

Floppy disks may be either single-sided or double-sided which means that data are recorded on one or both surfaces. There are different recording densities known as single density, double density or even quadruple density. The number of sectors and tracks recorded on the disk may vary, as may the method of locating these. On a hard-sectored disk the sectors are marked by a series of holes while on a soft-sectored disk there is a single hole to mark the origin. Most machines now use soft-sectored disks. Circuitry called the disk-controller is used to move the heads and control reading and writing on the disk.

Different operating systems use different directory and file formats even when the physical recording of the data is the same. It is hardly ever possible to read disks from a different machine and operating system unless a special conversion program is available. It may be possible to move data between machines using the serial ports and it is worth investigating the problem in advance of the necessity to perform such a transfer (see Section 4.1).

Backup copies of floppy disks should always be kept to avoid loss of data. In spite of such warnings, failure to do so is common until bitter experience convinces us of this wisdom.

3.1.2 *Hard disk drives*

A hard disk holds much more data than a diskette and the drive has faster access times for reading and writing information. The term Winchester disk is often used as a synonym for hard disk. The term fixed disk may also be used when the hard disk is built into the disk drive. Hard disks are precision made from metal with a magnetic oxide coating and a number of disks may be mounted on a single spindle. The hard disk has a disk-controller and associated software. Capacities of hard disks for small computers are commonly in the range 10−40 Mbytes.

Failure of a hard disk can lead to disastrous loss of data and regular backup is essential. The smaller hard disks may be backed up on floppy disks. This would be too tedious for larger disks which are backed up on magnetic tape (see Section 3.1.4).

3.1.3 *Optical disks*

There is the promise that plastic disks written with laser light will provide inexpensive storage of very large amounts of data in the near future. These are expected initially to be of write once/read type for archiving. Such technology may become important for distribution and storage of molecular sequence databases (see Chapter 3).

3.1.4 *Magnetic tape*

Magnetic tape is extremely important for backing up or transferring large files between different types of computers. Large computers are always provided with 0.5 inch reel to reel tape drives. Three recording densities in common use are 800, 1600 or 6250 bytes per inch (bpi). Data written in card image format (80 byte records padded with trailing spaces if necessary) at moderate block sizes (800 bytes per block) in the ASCII code at 1600 bpi will guarantee portability. A difficulty is that 0.5 inch tape drives are more expensive than small computers so that drives are found only at the larger installations. The problem of communicating the data to microcomputers remains (see Section 4.1).

Cartridge tape drives are less expensive and so are often provided with smaller computers. Unfortunately, there are no widely accepted standards for recording on cartridge tape so that cartridge drives from different manufacturers cannot read or write the tapes of others. Nevertheless, cartridge tapes are important for distribution and backup of software for the more powerful microcomputers. More sophisticated models allow automatic retrieval of individual files from tape to disk on demand (see Chapter 3).

Audio cassette recorders can be used for recording computer data and are sometimes supplied with inexpensive microcomputers. They are best avoided as being slow and unreliable.

3.2 Terminals

The cathode ray tube is still the most used device for visual displays. For monochrome, phosphors are commonly green, amber, or white. Some visual displays called storage tubes keep an image on the screen which is not refreshed. Others have to be continually refreshed so that the picture has to be rewritten on the screen many times per second. In a domestic television the screen is divided into a number of horizontal lines (often 625) and the electron beam scans across each in turn in a zig-zag pattern. The technique is known as raster. Some computer graphics systems do not move the electron beam in a zig-zag pattern but move the beam along the lines required to generate the picture and are known as vector displays.

There are various ways of dividing up the screen and of storing the data to be displayed. These can differ vastly in the resolution which they provide. A low resolution scheme might divide the screen into 80 columns and 25 lines and use one byte in memory to contain the symbol to be displayed in a location. This method gives a memory mapped display and is limited to one of 256 symbols to be displayed at each location. To obtain medium or high resolution the screen is divided into a large number of tiny locations called pixels. There might be 1000 × 1000 of these in the display. The status of each pixel could be represented by a single bit in memory and the display is described as bit-mapped. This assumes that a pixel is to be either black or white, but if a byte were used per pixel then 256 grey tones could be indicated. Similarly, a full colour display may be built up if 3 bytes per pixel are used. The bit-mapped display is important because it can reproduce the effects obtained with paper and ink and is coming into widespread use.

The simplest type of video-terminal is known as a 'dumb terminal' or 'glass teletype' and allows editing only on the current input line. It is now more common to use a terminal which allows interaction with the full screen. There is an ANSI standard (X3.41-1974 and X3.64-1977) defining escape and control sequences in terminals. A well known example of a terminal conforming to the standard is the Digital VT100. For graphics devices there is a standard called GKS (Graphical Kernel System) which defines a large number of conceptual graphical operations. These are implemented as bindings for various languages (a subroutine library) and by device drivers for graphics terminals or other devices. An often used protocol for driving a graphics terminal is the Tektronix 4010 plotstream which, as well as drawing vectors, allows the user to input the position of the cross hairs on the screen. Some software requires a terminal which allows more than one plane for the display. Typically, menus and textual information are displayed in one plane and graphical information in another. An example of such a terminal is the Selanar Hirez which is Ansi and Tektronix 4010 compatible. Screens for WIMP machines like the Sun-3 are bit-mapped to support work in multiple windows and to give high resolution. A graphics processor may be added to enhance graphics performance and a graphics buffer for high speed hidden surface removal in 3-D applications.

3.3 **Printers**

3.3.1 *Daisy wheel printer*

The daisy wheel type of printer works much like a typewriter. It carries a set of characters around the circumference of a wheel which is rotated to the desired letter and then strikes a coated plastic or inked fabric ribbon. Such printers are slow (up to 50 characters per second) but produce high quality type. They cannot be used to print graphics output. They are being replaced for many applications by the more versatile dot-matrix or laser printers (see below).

3.3.2 *Line printers*

These are low quality printers which print a line at a time. In one design a chain of raised characters of type rotates at high speed and hammers force the type against the paper when the characters to be printed are at the correct positions. Such printers are fast (several thousand lines per minute) but produce low quality output only suitable for the proof stage of documents or for program listings. Their cost means that they are only provided at centralized facilities.

3.3.3 *Dot-matrix printers*

Dot-matrix printers work by selecting a set of pins to form the character to be printed from a matrix. The selected pins are then driven against the ribbon and paper by small hammers. Dot-matrix printers have the advantage of being faster than daisy wheel printers but the quality of the print is not so high. Dot-matrix printers can also be used for printing a graphics image. This is because the printer can be made to print dot by dot as well as character by character. With suitable software, the matrix of pixels on the graphics screen can be mapped to a matrix of dots on the printer.

3.3.4 *Laser printers*

Laser printers are based on photocopier technology and have a photosensitive drum. An image is projected onto the drum using a beam of laser light controlled by a computer. Ink particles are attracted to the parts of the drum which have been charged by the laser light. The ink particles are then transferred to paper to which they are fused using heat and pressure.

3.4 **Digitizers**

A digitizer is a device for entering a series of coordinates from a drawing, photograph or other material. The most common application in DNA sequence analysis is the entry of data from autoradiograms of restriction enzyme digest gels or DNA sequencing gels. As these have to be illuminated from below it is appropriate to have a digitizer with a translucent tablet or to be able to work without a tablet and to place the photographic film on a light box. An example of a suitable digitizer is the Science Accessories Corporation GrafBar GP-7. [Science Accessories Corporation, Southport, CT, USA, and P.M.S. (Instruments) Ltd., Waldeck House, Reform Road, Maidenhead, Berkshire SL6 8BX, UK.] This has a pair of microphones which receive ultra-sound from a spark produced at the stylus. The Cartesian coordinates of the stylus are computed by triangulation with an inbuilt microprocessor.

3.5 **Speech synthesis**

The goal of being able to talk to computers in continuous speech has not yet been reached though progress is being made. It is practical to use voice synthesis and one application is to have the machine read back in voice form a DNA sequence which has been entered by another method.

4. COMPUTER COMMUNICATIONS

The subject of computer communications is very complex and difficult (24). There are an infinite number of ways of communicating (whether by human language or between computers) and a good many of these have been tried. It is convenient here to deal with aspects of the subject under three headings: asynchronous communications, local area networks and wide area networks.

4.1 **Asynchronous communications**

Digital information has to be encoded for transmission through a communications medium (usually wire, but optical fibre or radio frequency or microwave links may be used). The rate of transmission may be measured in bits per second (bps) which will be the same as baud rate if one symbol corresponds to one bit. Commonly, both parties send and receive information at the same time which is known as full duplex operation. This requires three wires: transmit, receive and signal ground and the method of coding is part of an international standard called V24 (corresponding to RS232C in the US). Transmission speeds are from an agreed set between 75 and 19 200 bps (commonly 300, 1200 or 9600). The receiver has to be able to decode a serial stream of bits as bytes or characters, that is to determine the correct reading frame. There are two methods of determining the correct frame either asynchronous or synchronous. In asynchronous framing there is no common clock and characters can be sent or received at any time. The asynchronous framing technique puts the bits in a character into a frame of a start bit and stop bits. This method is typically used to connect terminals to computer hosts. In synchronous framing there is a common clock and characters are sent continuously at a specified rate. This method is used for high speed computer to computer communications as in local or wide area networks.

Asynchronous communications are suitable for the intermittent transfer of data which occurs when a user interacts with a computer from a terminal. They are not appropriate for file transfer between computers. However, because the V24 serial interface is ubiquitous this is commonly used for transfer of small files, particularly between microcomputers or between microcomputers and larger hosts. Software for this purpose is obviously machine dependent though a common protocol and set of programs in the public domain developed at Columbia University, New York and known as Kermit has gained widespread acceptance in the academic community (25). It is necessary to have a Kermit program running on both machines and packets sent to the receiver are returned to the transmitter for verification. If an error is detected the packet is retransmitted. This is not efficient but is often useful for moving small amounts of data between very different machines. More satisfactory methods are magnetic tape (Section 3.1.4) or FTP (Section 4.3).

4.2 Local area networks

Local area networks are privately owned networks that provide reliable high-speed communications in a limited geographical area such as a laboratory, complex of buildings or, at most, a University campus (26,27). The maximum cable length is a few kilometres. The data transmission rate is in the range $1-20$ Mbps, much higher than asynchronous communications. There are two sorts of signalling, base band and broad band. Broad band signalling can make use of the technology developed for cable television and the very broad bandwidths mean that the same physical cable can carry a number of separate networks by frequency division multiplexing.

The most popular base band local area network is the Ethernet system (28) developed by Xerox and launched commercially by Digital, Intel and Xerox (29) and can be used to connect machines from a large number of manufacturers. However, it is the communication protocol that determines whether two machines can communicate. Microcomputer manufacturers have often developed base band local area networks specific to their own products (which reduces costs). Some of these do not work very well and cynics have noted that the transmission rate when a floppy disk is carried by hand from one machine to a machine in the next office approaches 0.5 Mbps. Local area networks permit sharing of expensive resources such as laser printers and plotters.

4.3 Wide area networks

The term wide area network is applied to a network that covers a large area such as a whole country. It may even be world-wide where a multinational organisation owns the network. The IBM network BITNET (USA) provides connections to Canada (under the name NORTHNET), Europe (under the name EARN), the Middle East, Japan, South East Asia and Australia. Other examples of networks are TRANSPAC, DATA-PAC, TELENET and IPSS in Western Europe and the US, ARPANET and TYMNET in the US and AGRENET and JANET in the UK. More details of some of these are given below and in Chapter 5.

In most countries it is illegal for individuals or companies to transmit data outside their own property. There exists, however, a system of wires (or other media) to carry telephone conversations. In Europe the licensees running such systems are often government bodies called PTTs (Postal, Telegraph and Telephone authorities), and the network they manage is called the PSTN (Public Switched Telephone Network). The present PSTN was designed to carry speech and has a bandwidth of only 3 KHz. A device called a MODEM (MOdulator and DEModultor) is needed to convert digital signals to audio signals and back again. The PTTs work through a standards organisation called the CCITT (International Consultative Committee for Telegraphs and Telephones) which has specified the V24 interface (see Section 4.1) for connecting asynchronous terminals to modems. A microcomputer with an RS232 port and a modem enables one to connect to remote hosts and transfer files over the PSTN. Users can work from an asynchronous terminal at 300/300 or 1200(receive)/75(transmit) bps using an inexpensive modem. A more expensive modem enables one to use a 1200/1200 terminal [for example: WS3000, Miracle Technology Ltd., St. Peters Street, Ipswich, Suffolk, UK; and Smartmodern 1200, Hayes Microcomputer Products Inc., Norcross, GA, USA (split baud rates are not used in USA)].

PTTs are replacing analog telephone equipment by digital technology which has the advantage of lower noise, allowing digital multiplexing on trunk networks and allowing the inexpensive VLSI (very large scale integration) microprocessor technology of computers to be used for switching equipment. The digital circuit of System X has a 64 KHz bandwidth and will eventually be used for every telephone in the UK. Handsets will convert analog speech signals to digital signals using pulse code modulation. In Britain, British Telecom operate a packet switched network called PSS (Packet Switch Stream) which forms part of an international system known as IPSS. The local packet switching exchanges (PSEs) are linked by a synchronous network with some very fast links (Megastream).

The CCITT have defined a standard interface known as X21 for connecting communications equipment to the digital switched network. The access protocol adopted by the CCITT for use between a host (DTE data terminal equipment) and a node (DCE data communications equipment) on the network is called X25. X25 is defined to use X21 at its lowest level. On PSS, the local PSE can be accessed either via the PSTN or via a leased asynchronous line to the site.

Higher level protocols have to be implemented in hosts for host – host communication over the network and might involve file transfer or terminal access (which have very different attributes). A *conceptual file store* is a definition of how files are maintained on any computer. A file transfer protocol (FTP) has a conceptual file store which is mapped into the actual file store by the implementation. The requirement is for high throughput of data while moderate delays are acceptable. Terminal access, on the other hand, requires bursts of activity with low delay. An intermediate processor to interface asynchronous terminals with single character interaction to synchronous packet networks is called a PAD (packet assembler and disassembler) on X25 networks. A *conceptual terminal* is a definition of the characteristics of any terminal. A conceptual terminal is described by a protocol (X3) with a number of parameters the values of which are stored in the PAD in a table for each terminal connected. A second protocol (X28) defines the commands which can be used from the terminal to alter parameter values and to connect and disconnect calls. A third protocol (X29) defines the control packet format to be used between the PAD and the remote host. The three terminal related protocols are often collectively referred to as *Triple X*. Because X29 uses X25 directly its use is limited to X25 networks.

In the United States different systems are in use. The best known example is the ARPA (Advanced Research Projects Agency) network. The mail facility on this network is a major source of traffic and enables communication between researchers in similar subjects at remote locations. The host to host protocol implemented in hosts on the ARPA network is called TCP (Transmission Control Protocol) and terminal access is provided by a terminal interface processor (TIP). A protocol called TELNET allows the host and the TIP to negotiate on the features which will be supported.

Whilst a broad parallel may be drawn between the X25 - PAD - X3/X28/X29 mechanism on PSS and the TCP - TIP - TELNET mechanism on ARPANET, they are quite different in detail. Connecting different networks together introduces new problems and the need for a gateway protocol bridging the two. The gateway also needs to take account of differences in addressing and facilities available in the two networks. The UCL gateway allows users of JANET (the Joint Academic network) in the UK to exchange electronic mail with users of networks in the US. [Users of JANET in the UK should apply to Irene Hassel, Liaison, UCL Gateway Service Project, Computing Service, University College, London, UK.]

5. EXAMPLE SYSTEMS

5.1 **IBM Personal Computer AT**

The IBM Personal Computer AT is a member of the IBM PC family of microcomputers which share a common software base (30). It typically supports only a single use and costs $4000. The IBM PC AT uses the Intel 80286 processor and will run most of the wide range of software available for the less powerful IBM Personal Computer. The Enhanced model AT has 512 Kbyte of memory, one 1.2 Mbyte diskette drive and one 20 Mbyte fixed disk. The memory can be upgraded to 640 Kbyte which is the maximum supported by the operating system DOS 3.0. The fixed disk operates at 5 Mbps with 40 milliseconds (ms) average access time. The 1.2 Mbyte high capacity diskette drive stores data on 5.25 inch double-sided floppy disks. The drive is also able to read disks written in single- or double-sided formats by the IBM PC. The high capacity drive cannot write these formats, but a suitable drive may be added to do so. Within the system unit are eight slots for the attachment of cards. One slot contains the disk controller and another contains a card with a serial port for asynchronous communications and a parallel port for the attachment of a printer. There is a choice of four different video units, the Monochrome Display (80 characters in 25 lines), the Colour Display (640 × 200 resolution, 16 colours), the Enhanced Colour Display (640 × 350 resolution, 16 simultaneous colours from a palette of 64) and the Professional Graphics Display (640 × 480 resolution, 256 simultaneous colours from a palette of 4096). Each display needs a supporting adaptor card. A number of printers are available. The IBM Proprinter is of the dot-matrix type for text or graphics at near letter quality or data processing quality. The IBM Wheelprinter is a letter quality impact printer. The IBM Quietwriter Printer is fast and almost silent, producing letter quality printing by a technology involving resistive thermal transfer.

The most used operating system for the IBM PC AT is DOS 3.0, though Xenix (a flavour of Unix) is also available. PCDOS is related to MSDOS which is similar to the earlier CP/M and is composed of four main segments: the BIOS (Basic Input/Output System), the DOS (Disk Operating System), the Command Processor and the Program Segment. Like CP/M, MSDOS was originally a very simple operating system with commands like COPY, DEL (delete), DIR (directory), FORMAT, RENAME, TYPE used for manipulating files. It has been enhanced by the provision of a Unix-like tree structured file system, Unix-like redirection of input and output, and a Unix-like pipe facility (see Section 5.3). MSDOS is a single user, single task operating system, except that printing can be run in the background while other work proceeds.

The most popular applications for the IBM PC AT include the work processing program DisplayWrite 3, the database program dBaseIII and Lotus 1-2-3 which includes a spread sheet, graphical and database facilities. Programming languages available for the IBM PC include BASIC, FORTRAN, Pascal, C, and APL. A popular FORTRAN compiler is IBM PC Professional FORTRAN by the Ryan-McFarlan Corporation which requires the 80287 math-coprocessor. There is a wide choice of C compilers, notable examples being Microsoft C and Lattice C. Programming under DOS is not without its problems. The architecture of the 8086 processor leads to Small, Medium, Compact and Large memory models and this is carried over to the 80286 when running in 'real mode'. Restrictions on addressing and the size of program segments makes programming less easy. Many successful applications have been developed on the IBM PC, but scientific programmers should be aware that programs developed on mainframe computers may be difficult to move to the IBM PC AT under DOS because of the problems mentioned above. (These limitations do not exist on the IBM PC RT under Unix.) A wide range of IBM graphics software is available to support the variety of displays. The Graphics Development Toolkit contains a standard Virtual Device Interface, a set of device drivers, and a set of high level language bindings for BASIC, FORTRAN and Pascal. The Graphical Kernel System (GKS) is based on the standard and is a powerful tool giving graphics program portability while using as much as possible of the inherent capabilities of a device through enquiry routines. There are bindings for FORTRAN, BASIC and Lattice C. The Graphical File System is provided for the compact storage of graphic information. The Plotting System provides a high level subroutine library for bar charts, scatter diagrams and text charts. Finally, there is a Graphics Terminal Emulator which permits the IBM PC to act as a Tektronix 4010 or 4014 terminal when attached to remote hosts. There are a number of commercial nucleic acid and protein sequence analysis packages for the IBM PC including DNASTAR, MICROGENIE, DNASIS and IBI-Pustell (Chapter 3). There is also a wide variety of academic research software which is made available free of charge.

5.2 **MicroVAX II**

The MicroVax II is the smallest member of the VAX family of computers which range up to the 8800 rated at 11 mips (million instructions per second). A typical configuration costs \$30 000 and supports eight users. By clustering VAX computers even more powerful installations may be created. The operating system VMS is a multi-user, multi-tasking, virtual memory operating system allowing a virtual address space of 4 Gbytes. Swapping pages of memory to disk enables more than the physical RAM to be addressed from a users program. An alternative choice is Ultrix-32M, Digital's version of the Unix operating system derived from 4.2BSD. The MicroVax II 630QB has a MicroVax II processor chip and a companion floating point chip. The memory can be expanded up to 9 Mbyte. Memory is connected to a special memory bus having a peak bandwith of 9 Mbyte/sec. There is a 12 slot backplane and cards communicate via the 22-bit Q-bus having a peak bandwith of 3 Mbyte/sec. The mass storage devices include a diskette drive, fixed disk and a cartridge tape drive. The RX50 dual diskette drive holds two 5.25 inch diskettes each storing 400 Kbyte. The RD53 is a 71 Mbyte 5.25 inch fixed disk operating at 5 Mbps with an average access time of 38 ms. The TK50 cartridge tape has a capacity of 95 Mbyte. Alternatively, a 456 Mbyte RA81 disk and a 0.5 inch TSV05 tape drive recording at 1600 bpi can be added. There is a serial port and the DHV11 multiplexer provides a further eight asynchronous ports for terminal lines. Synchronous interfaces for Ethernet (DEQNA) or X25 can be added. Video terminals for the MicroVax are the VT220 (with VT100, VT52 and Ansi capability) and the VT240 (which adds to the VT220 ReGIS and Tektronix 4010/4014 graphics). A wide range of printers is available including the LN03 laser printer.

The VMS operating system (31) is a collection of software that organises the processor and peripherals into a high-performance system. The operating system has jobs running as independent activities on the system. These include the *Job Controller* which initiates and terminates user processes and manages *spooling*, the *Operator Communications Manager* which handles messages queued to the system operators, the *Swapper* which controls the swapping of the *working set* for a process in and out of memory, and the *Error Logger* which collects all hardware and software errors detected by the system. Digital Command Language (DCL) is the name of the command language used to communicate with VAX/VMS. It is a large and comprehensive language, more complex than DOS or Unix. The most popular editor is EDT which is a full screen editor operated from a special keypad on the VT220 keyboard. A rich variety of programming languages are available including Ada, APL, BASIC, C, COBOL, FORTRAN, Pascal and PL/I. Software development tools such as language sensitive editors and source level debuggers are available for the majority of these languages. A version of GKS is available for graphics programming. VAX/VMS is very important as an environment for molecular biology software at the present time. Major packages include the IntelliGenetics software, the Batelle-Northwest software, the Staden software and the UWGCG software of Devereux, Haeberli and Smithies. The EMBL and NBRF database groups use VAX/VMS.

5.3 **Sun-3**

Sun computers are high performance graphics workstations in both monochrome and colour. Each screen supports one user for graphics, though other terminals may be added. The cost is around $25 000. The processor is the Motorola 68020 and the operating system is Unix 4.2BSD, which is multi-user, multi-tasking and offers a 256 Mbyte virtual address space. The deskside Sunstation 3/160M-4 has a 68020 processor running at 16.7 MHz and a 68881 floating point coprocessor. There are 4 Mbytes of main memory (expandable up to 16 Mbytes). An optional floating point accelerator increases floating-point performance by a factor of 4 to 500 kflops and uses the Weitek chip set. A memory management unit facilitates multiple processes. There is a 12 slot card cage and boards communicate via two buses. A high speed synchronous bus provides access to main memory at 15 Mbytes/sec. A 32-bit VME bus is used for peripheral data at 5 Mbytes/sec. Two 71 Mbyte 5.25 inch hard disks and a 60 Mbyte 0.25 inch streaming cartridge tape drive on a SCSI bus can be mounted in the system pedestal. The disk operates at 5 Mbps with an average access time of 35 ms. Alternatively, a 380 Mbyte disk and a 0.5 inch tape drive can be accommodated in a separate rack. The disk operates at 1.8 Mbytes/sec with an average access time of 18 ms. There is a 19 inch bit-mapped monochrome display monitor with a resolution of 1152 × 900. The display is flicker free and responds at high speed. There is a 3-button mouse. There are two serial ports and an ethernet connection. X25 hardware and software is available. Sun hardware adheres to an open systems philosophy: the architecture is based on industry standards allowing connection to equipment from other manufacturers. The Sun LaserWriter provides hard copy for both text and graphics. It is based on a Canon LBP-CX laser-xerographic engine and is controlled by a 68000 processor with 1.5 Mbyte. Software called TranScript running in the Sun converts troff or plot files into source in the PostScript language. The PostScript language interpreter is held in ROM (read only memory) in the LaserWriter. The PostScript machine receives commands over the serial interface and interprets these as images for the laser printing marking engine.

The operating system of the Sun-3 is Unix (32). Unix consists of a kernel, one or more shells, and a large collection of utilities (many of which may be ignored by the molecular biologist). The kernel manages the hardware and provides fundamental services such as filing and input and ouput. The shell is the interface between the kernel and the user providing a command language and a command interpreter. Utilities are programs such as editors, compilers, type-setting software or indeed any application.

On a WIMP machine the underlying operating system need not be obtrusive. Unix is remarkable because it may be run on all three of the example systems described here, and many more besides. Unix is a portable operating system and enables users to move quite effortlessly to entirely different hardware. The tools provided by Unix are liked by programmers but there is little conciliation to novice users who find Unix terse, cryptic, inconsistent in the specification of options and capable of destroying their files without a murmur. It does have an elegant underlying design born out of power derived from simplicity rather than complexity (33). The structure in which Unix keeps files is called a tree. Branches split into sub-branches and a hierarchical organisation of information is easily achieved. Peripherals on the Unix system are all treated with files. If you wish to send the output from a program to a file rather than to your terminal you just redirect it. Similarly, you can redirect input from wherever you want. This idea is extended to pipes which link programs in tandem. Output from the previous program becomes input to the next.

The great strength of the Sun is the SunWindows system which implements a WIMP user interface. The high resolution bit-mapped display and the mouse pointing device provide an extremely versatile medium for presenting and manipulating information. Multiple tasks can be managed with ease in a number of windows displayed on the screen. If the screen becomes cluttered, windows which are temporarily not needed can be closed down to a small pictorial representation called an icon and opened again when needed. Pop-up menus can be displayed with the mouse and an item rapidly selected. When not in use they do not occupy space on the screen. The SunWindows system is a versatile system from which an infinite variety of user interfaces may be constructed. The programming languages C, FORTRAN and Pascal are supplied and GKS is available for graphics programming. There are a number of other standard tools to help the programmer. *Shelltool* consists of a single window that emulates a terminal running a shell. Simultaneous remote login to a number of hosts can be managed in such windows. *Graphicstool* allows an application which has been written with no knowledge of the SunWindows system to be run. The Graphicstool window consists of two subwindows; an upper window contains scrolling text and a lower window receives the graphical output. *Icontool* enables programmers to use a bit-map editor to construct their own icons. *Fonttool* is another bit-map editor that enables users to create their own language fonts. *Dbxtool* is a an outstandingly powerful debugger for diagnosing run-time errors in programs and uses multiple subwindows. A wealth of third party applications, compilers and tools are available through the Catalyst referal program. The notable tool for molecular biologists is the IntelliGenetics Bion software.

6. REFERENCES

1. Staden,R. (1977) *Nucleic Acids Res.*, **4**, 4037.
2. Gingeras,T.R. and Roberts,R.J. (1980) *Science*, **209**, 1322.
3. Queen,C.L. and Korn,L.J. (1980) *Methods in Enzymology.* Vol. **65**, 595.
4. Soll,D. and Roberts,R.J. (eds.) (1982) *The Applications of Computers to Research on Nucleic Acids.* IRL Press, Oxford.
5. Soll,D. and Roberts,R.J. (eds.) (1984) *The Applications of Computers to Research on Nucleic Acids II.* IRL Press, Oxford.
6. Soll,D. and Roberts,R.J. (eds.) (1986) *The Applications of Computers to Research on Nucleic Acids III.* IRL Press, Oxford.
7. Bishop,M.J. (1984) *Bioessays*, **1**, 29.
8. Bishop,M.J. (1984) *Bioessays*, **1**, 75.
9. Bishop,M.J. (1984) *Bioessays*, **1**, 126.
10. Korn,L.J. and Queen,C. (1984) *DNA*, **3**, 421.
11. Smith,R.J. (1985) in *Microcomputers in Biology — a Practical Approach.* Ireland,C.R. and Long,S.P. (eds.), IRL Press, Oxford, p. 151.
12. Friedland,P. and Kedes,L.H. (1985) *Comm. A.C.M.*, **28**, 1164.
13. Rawlings,C. (1986) *Software Directory for Molecular Biology.* Macmillan, London.
14. Curran,S. and Curnow,R. (1983) *The Penguin Computing Book.* Penguin Books.
15. Shooman,M.L. (1983) *Software Engineering.* McGraw-Hill, Maidenhead.

16. Wallis,P.J.L. (1982) *Portable Programming*. Macmillan, London.
17. Thacker,C.P., McCreight,B.W., Lampson,B.W., Sproull,R.F. and Boggs,D.R. (1982) in *Computer Structures: Principles and Examples*, Siewiorek,D.P., Bell,C.G. and Newell,A. (eds.), McGraw Hill, New York, NY, pp. 549−572.
18. Joy,W. and Gage,J. (1985) *Science, 228*, 467.
19. Crecine,J.P. (1986) *Science, 231*, 935.
20. Hillman,D.E. (ed.) (1984) *Advanced Computing in the Life Sciences*. Proceedings of the workshop on the applications of supercomputers in life sciences. Dec. 10−12, 1984, Airlie, VA.
21. Forty,A.J. (1985) *Future Facilities for Advanced Research Computing*. Science and Engineering Research Council, Swindon.
22. Blundell,T.L., Sternberg,M.J.E. and Cooper,S. (1985) *Biomolecular Modelling in the EEC*. Commission of the European Communities, Brussels.
23. Bishop,M.J. (1985) *Bioessays, 2*, 218.
24. Cole,R. (1981) *Computer Communications*. Macmillan, London.
25. Da Cruz,F. and Catchings,W. (1984) *Byte,* **June**, 255, **July**, 143.
26. Anon (1982) *Introduction to Local Area Networks*. Digital Equipment Corporation.
27. Gee,K.C.E. (1983) *Introduction to Local Area Computer Networks*. Macmillan, London.
28. Metcalfe,R. and Boggs,D. (1976) *Comm. A.C.M., 19*, 395.
29. Anon (1980) *The Ethernet: a Local Area Network Data Link Layer and Physical Layer Specification*. Digital, Intel and Xerox.
30. Barnetson,P. (ed.) (1985) *The Research and Academic Users Guide to the IBM Personal Computer*. IBM United Kingdom Limited.
31. Anon (1982) *VAX Technical Summary*. Digital Equipment Corporation.
32. Courington,B. (1985) *The Unix System: a Sun Technical Report*. Sun Microsystems Inc.
33. Brown,P. (1984) *Starting with Unix*. Addison-Wesley, Reading, MA.

CHAPTER 2

DNA sequence analysis software

P.A. STOCKWELL

1. INTRODUCTION

This chapter examines some of the processes and manipulations of DNA sequence data which are now regarded as standard in the sense that they can be expected to be widely implemented and readily available to molecular geneticists. These techniques may be applied to newly determined sequences, to sequences taken from databases or to data manually entered from printed publications.

There are many different programs for DNA sequence analysis and many of these tend to do the same things. Section 2 of this chapter looks at some of the points to be considered in choosing programs. Sequence data, like all data in computers, are stored in files. Section 3 looks at some of the different types of sequence files which may be encountered. Altering or editing the sequence data in files is examined in Section 4, along with the potential for sequence editing to allow simulation of spliced DNA constructs. The remaining Sections 5 – 10 look at typical manipulations of sequence data.

2. THE RANGE OF DIFFERENT PROGRAMS

Chapter 1 outlines the wide range of computer systems that are available and the extent to which the performance and resources which they provide may differ. The same diversity occurs within sequence analysis software so it is impossible to provide a complete guide to sequence manipulation. Rather, I have attempted to outline the concepts and show examples of practical programs which apply the principles. You may find that you have a different range of programs available, but you should find the same functionality. If a given function does not exist in your set of programs then it should be possible to obtain the necessary software.

2.1 Sources of software

Chapter 3 examines some commercially available sequence analysis packages but there are many other sources of sequence software, some of which are available at little or no cost. Published lists (1,2) of sequence software, including free packages, make it a relatively simple matter to find an appropriate program to perform a given function on your computer system. In deciding whether to buy commercial packages or to obtain free software you should bear in mind that if you pay for software then there is some implied guarantee that it has been thoroughly tested and should be reliable. There should therefore be some form of support or recourse in the event of problems. Free software may be every bit as reliable as commercial equivalents and may be equally as well supported, but authors cannot provide a formal guarantee of quality and support for products which return no profit.

A good criterion for choosing a given program or package is that your colleagues are using it and finding it to be effective on computers which sufficiently resemble your own. You may be able to copy the software directly from them, but before you do so, you must ensure that it is not protected by copyright or license agreements. It is a sad comment on attitudes to computers and their software that many people, who would be outraged at a theft from themselves, will happily make and use illegal copies of copyright programs without a thought for the financial loss to the author.

If you find yourself obliged to look for software, rather than accessing an on-line system (as described in Chapter 5) or using facilities provided by the Computer Services section of your organization, you should consider a further point: complete packages, whether commercially marketed or not, should provide a unified system of files and conventions. They should also provide a consistent environment for user interaction. Your own 'mix-and-match' selection of programs or pieces of other packages will probably not achieve this type of unity.

2.2 Quality of software

There are many criteria for the quality of computer software but here I am concerned with four major areas. As you use various sequence software products you will observe quite a wide degree of variation in these points.

2.2.1 *User friendliness*

How easily can a novice user learn and use this program/package? All normal human computer users periodically make mistakes, irrespective of their experience. What happens when the user makes a wrong response — does the program crash with '?FORTRAN-F-error 6 in line ???' or 'PA1050 error @36694A9F', or does it give a helpful error message and reprompt the user?

To what extent does the program help the user's response by offering sensible defaults or examples of the type of response necessary? For example, if the program requires a YES/NO answer does it say 'TYPE Y FOR YES' and fail to mention that only an uppercase Y will be accepted as YES, all else as NO, or does it say 'Y/N? [Y] >' and accept responses in either case? These are not trivial points, they are most important in preventing user frustration and in avoiding undermining the confidence of users.

2.2.2 *Program reliability and robustness*

Program reliability and robustness is a measure of the extent to which programs fail unexpectedly when faced with certain conditions. For example, what happens when the program encounters a larger sequence than was expected by the author? Does the program carry on without warning to produce incorrect results which you may or may not notice later? Does it crash the program or even the whole computer system, or does it at least warn the user that an excessively long sequence has been found and either stop or continue as appropriate?

Some programs fail unexpectedly with certain sets of data, sometimes through unpredicted system requirements or though undetected or unresolved bugs. There is a further possibility that an algorithm applied by the program may be incorrectly coded, or even incorrectly understood by the author of the program. Program reliability or robustness depends on whether the program has been tested with a wide enough variety of data to give us confidence that it will work reliably and effectively on our data.

2.2.3 *Program limits*

Most sequence software has some upper limit on the sequence length which can be handled. This may vary widely, from programs running on large mainframes with capabilities exceeding hundreds of kilobases of sequence down to microcomputer systems where programs may handle less than 10 kb. You should know what the limit is for any program you use. Does the program mention any limit in its heading or must you read all the fine print in the manual to check on this? Does a large utility package have different limits for different functions, e.g. edit up to 30 kb but translate only 24.5 kb? If it does have this problem, does it warn you? These points are real problems which have been observed with a variety of existing sequence software.

Whether you do your sequence analysis on a micro or a mainframe will depend on many factors including convenience and budgetary considerations. If you use a microcomputer system then you will probably have to accept short sequence limits which will force you to examine long sequences in a piece-meal fashion. These limitations should diminish in future as microcomputers become increasingly powerful.

2.2.4 *Documentation and help*

No matter how carefully programs are written or how user friendly they are, there is still some need for written documentation. Aside from providing a user guide, documentation is necessary to explain particularly abstruse aspects of the algorithms, to provide notes on complex aspects of program commands and to provide an ultimate specification for the program.

There is an increasing tendency to provide documentation on-line in the form of interactive HELP information. If this is well done, it can replace much conventional documentation although not all, since calling on the program's HELP facility generally obliterates the current screen contents and can leave the user wondering to what to apply his new-found knowledge. Poorly conceived on-line help can be an infuriating nuisance.

It is more difficult to provide on-line documentation for microcomputer systems since microcomputer programs may have to run with limited disk resources.

You should ensure that you obtain the necessary documentation for the program/packages which you intend to run. You will find it particularly useful if the documentation contains printed command summaries since they can prevent much furious scrabbling through turgid manuals.

3. SEQUENCE FILES

The same spirit of inventiveness which has generated many different computer systems and many different sequence data handling packages has been applied to sequence data files. I shall not discuss all the file types in use but shall consider a few examples.

3.1 **Sequential text files**

These are normal computer files which consist of lines of text (or sequence data), which

```
ATGGCTGTTTATTTTGTAACTGGCAAATTAGGCTCTGGAAAGACGCTCGTTAGCGTTGGT
AAGATTCAGGATAAAATTGTAGCTGGGTGCAAAATAGCAACTAATCTTGATTTAAGGCTT
CAAAACCTCCCGCAAGTCGGGAGGTTCGCTAAAACGCCTCGCGTTCTTAGAATACCGGAT
AAGCCTTCTATATCTGATTTGCTTGCTATTGGGCGCGGTAATGATTCCTACGATGAAAAT
AAAAACGGCTTGCTTGTTCTCGATGAGTGCGGTACTTGGTTTAATACCCGTTCTTGGAAT
GATAAGGAAAGACAGCCGATTATTGATTGGTTTCTACATGCTCGTAAATTAGGATGGGAT
ATTATTTTTCTTGTTCAGGACTTATCTATTGTTGATAAACAGGCGCGTTCTGCATTAGCT
GAAAATGTTGTTTATTGTCGTCGTCTGGACAGAATTACTTTGCCTTTTGTCGGTACTTTA
TATTCTCTTATTACTGGCTCGAAAATGCCTCTGCCTAAATTACATGTTGGCGTTGTTAAA
TATGGCGATTCTCAATTAAGCCCTACTGTTGAGCGTTGGCTTTATACTGGTAAGAATTTG
TATAACGCATATGATACTAAACAGGCTTTTTCCAGTAATTATGATTCCGGTGTTTATTCT
TATTTAACGCCTTATTTATCACACGGTCGGTATTTCAAACCATTAAATTTAGGTCAGAAG
ATGAAATTAACTAAAATATATTTGAAAAAGTTTTCTCGCGTTCTTTGTCTTGCGATTGGA
TTTGCATCAGCATTTACATATAGTTATATAACCCAACCTAAGCCGGAGGTTAAAAAGGTA
GTCTCTCAGACCTATGATTTTGATAAATTCACTATTGACTCTTCTCAGCGTCTTAATCTA
AGCTATCGCTATGTTTTCAAGGATTCTAAGGGAAAATTAATTAATAGCGACGATTTACAG
AAGCAAGGTTATTCACTCACATATATTGATTTATGTACTGTTTCCATTAAAAAAGGTAAT
TCAAATGAAATTGTTAAATGTAATTAA@
```

Figure 1. Gene 1 from bacteriophage f1 (22) written as a sequence data file in the format of Staden. The sequence is 1047 bases in length.

are read from top to bottom. They are available on all computers from all language systems so that sequence data files have largely been implemented as sequential text files. Sequential text files can be printed or listed directly for examination. A variety of different formats have been adopted for the way in which the data are contained in sequential text files.

3.1.1 *Staden type files*

The format which was used by Staden in his well known suite of sequence programs (3), consisted of the sequence written out in 60 character lines and terminated with a @ character. There was no provision for name, comment or topology information. Any name or identifier for the sequence (e.g. LAMBDA, or PX174) had to be specified in the file name. *Figure 1* shows a region of DNA sequence as it appears in a Staden file.

3.1.2 *MOLGEN/SEQ type files*

The format adopted for the series of programs originating from the Stanford molecular genetics project (MOLGEN) and used with the SEQ (4) and PEP programs which are now offered as part of the IntelliGenetics package is illustrated in *Figure 2*. This format allows a comment field in which a detailed description of the sequence can be written. The comments are followed on the next line by a sequence name or identifier, followed on subsequent lines by the sequence itself, and terminated with a topology character (to distinguish linear from circular sequences). The original MOLGEN format allows multiple sequences in one file, each separated by an END line (not illustrated). Most programs which read this file format (e.g. Wilbur and Lipman homology searching programs) (5) only accept one sequence per file, as illustrated here. The END line then becomes optional.

3.1.3 *University of Wisconsin Genetics Computer Group*

Devereux *et al.* (6) have extended the idea of annotating the sequence within its file

```
; This is the sequence of Gene I from bacteriophage fl as determined by
; Hill and Petersen, J. Virology (1982), 44, 32-46.
; The gene 1 protein product is involved in phage maturation.
;
; Edited by SSEDIT on 16-Jan-1986 at 14:52
F1-GENE1
ATGGCTGTTTATTTTGTAACTGGCAAATTAGGCTCTGGAAAGACGCTCGTTAGCGTTGGTAAGATTCAGG
ATAAAATTGTAGCTGGGTGCAAAATAGCAACTAATCTTGATTTAAGGCTTCAAAACCTCCCGCAAGTCGG
GAGGTTCGCTAAAACGCCTCGCGTTCTTAGAATACCGGATAAGCCTTCTATATCTGATTTGCTTGCTATT
GGGCGCGGTAATGATTCCTACGATGAAAATAAAAACGGCTTGCTTGTTCTCGATGAGTGCGGTACTTGGT
TTAATACCCGTTCTTGGAATGATAAGGAAAGACAGCCGATTATTGATTGGTTTCTACATGCTCGTAAATT
AGGATGGGATATTATTTTTCTTGTTCAGGACTTATCTATTGTTGATAAACAGGCGCGTTCTGCATTAGCT
GAAAATGTTGTTTATTGTCGTCGTCTGGACAGAATTACTTTGCCTTTTGTCGGTACTTTATATTCTCTTA
TTACTGGCTCGAAAATGCCTCTGCCTAAATTACATGTTGGCGTTGTTAAATATGGCGATTCTCAATTAAG
CCCTACTGTTGAGCGTTGGCTTTATACTGGTAAGAATTTGTATAACGCATATGATACTAAACAGGCTTTT
TCCAGTAATTATGATTCCGGTGTTTATTCTTATTTAACGCCTTATTTATCACACGGTCGGTATTTCAAAC
CATTAAATTTAGGTCAGAAGATGAAATTAACTAAAATATATTTGAAAAAGTTTTCTCGCGTTCTTTGTCT
TGCGATTGGATTTGCATCAGCATTTACATATAGTTATATAACCCAACCTAAGCCGGAGGTTAAAAAGGTA
GTCTCTCAGACCTATGATTTTGATAAATTCACTATTGACTCTTCTCAGCGTCTTAATCTAAGCTATCGCT
ATGTTTTCAAGGATTCTAAGGGAAAATTAATTAATAGCGACGATTTACAGAAGCAAGGTTATTCACTCAC
ATATATTGATTTATGTACTGTTTCCATTAAAAAAAGGTAATTCAAATGAAATTGTTAAATGTAATTAA1
```

Figure 2. Gene 1 from bacteriophage fl (22) written as a sequence data file in the MOLGEN/SEQ format. The lines at the top of the file beginning with semicolons are comments, including the line to the effect that the file has been edited/created by SSEDIT. The first line without a semicolon is the sequence identifier, followed by the sequence. The 1 at the end indicates that this sequence is linear — a 2 would indicate a circular sequence — (the total genome of fl is circular, however the sequence of an isolated gene is linear).

so that the sequence data in a sequential file may be included with a potentially large amount of commenting. The comments are even accepted when intercalated between lines of sequence information. This strategy can allow large amounts of pertinent information about the biological function of the sequence, position numbers, its restriction map, sites of genes, etc. to accompany the sequence albeit at some cost in disk space. This approach to sequence files is similar to that employed in sequence databases (see Chapter 4).

3.2 **Direct access files**

Direct access (or random access) files consist of a number of components (called records) which can be read or written in any order at any time. They cannot usually be printed directly in the same manner as sequential files and are not as uniformly implemented across computer and language systems. This has limited their use for sequence data storage despite some potential advantages. The main example in which direct access files have been used for sequence data storage is Stockwell's LDNA program and its file structure (7) in which files are used to permit very long sequences to be handled and stored on computers with limited memory. The file becomes, in effect, an extension of the system memory. The LDNA file structure has the potential to permit comment fields to accompany the sequence but this feature has not been used in current versions so that, like Staden-type files, any name or identifier must be contained in the file name itself.

3.3 **Sequence symbols**

DNA sequences consist of the letters A,C,G and T (or U for RNA) representing the

Table 1. IUB Group Codes for Incompletely Specified Bases

Code	Bases coded		Matching bases	Complement
G	=		G,R,K,S,B,V,D,N,-	C
A	=		A,R,M,W,H,V,D,N,-	T
T	=		T,Y,K,W,H,B,D,N,-	A
C	=		C,Y,M,S,H,B,V,N,-	G
R	= A,G	(Purine)	G,A,R,V,D,N,-	Y
Y	= C,T	(Pyrimidine)	T,C,Y,H,B,N,-	R
M	= A,C		A,C,M,H,V,N,-	K
K	= G,T		G,T,K,B,D,N,-	M
S	= G,C		G,C,S,B,V,N,-	S
W	= A,T		A,T,W,H,D,N,-	W
H	= A,T,C	(not G)	A,T,C,H,Y,M,W,N,-	D
B	= G,T,C	(not A)	G,T,C,B,Y,K,S,N,-	V
V	= G,A,C	(not T)	G,A,C,V,R,M,S,N,-	B
D	= G,A,T	(not C)	G,A,T,D,R,K,W,N,-	H
N (-)	= G,A,T,C		G,A,T,C,R,Y,M,K,S,W,H,B, V,D,N,-	N(-)

A list of the IUB standard symbols for incompletely specified bases and their complements from (8,9). The use of a dash (-) for any base is not strictly part of the standard but is a recognized alternative.

four bases. At times it is convenient to use additional symbols to specify a sequence position in a less rigid manner, for example you may wish to indicate that a certain sequence position is occupied by either A or T, but not G or C. There is a now an IUB standard (8,9) for incompletely specified bases and this is shown in *Table 1*, along with the complementary symbols and the base symbols which might match each for search purposes. The standard regards upper and lower case characters as being equivalent.

A specialized use of additional symbols is the set of uncertainty codes defined by Staden (10) for use while reading DNA sequencing gels and this process is described in Chapter 8.

There is as yet no standard series of codes for chemically modified (e.g. methylated) or unusual bases in DNA sequences.

You will find that not all sequence programs or packages are capable of interpreting incompletely specified base codes. Some which do may differ from the standard (which has only recently been implemented). If programs do offer such codes they can be most useful when you wish to specify sub-sequences for searching (see Section 7).

4. SEQUENCE EDITING

When you start using a computer for molecular genetics you will need to enter new sequences into files or edit files which have already been created. These functions are performed with a sequence editor: a program which reads sequence files, modifies them according to your instructions and writes the modified sequence to a new file. Existing sequence editors vary widely in their manner of operation and convenience of use and they can be classified as line-orientated, screen-orientated or hybrid. You should also be aware of two other aspects of sequence editors: choosing and learning an editor and security against loss or corruption of the sequence information that you are editing. Finally I shall examine the way in which sophisticated sequence editors enable simulation of physical experiments, e.g. in the construction of artificial plasmids by combining parts of different sequence files together.

4.1 **Line editors**

Older editors tended to be line-orientated in the sense that they were designed to be used at printing terminals (e.g. a TELETYPE) and the printed editor responses were limited to save time and paper. Line editors accept lines of edit specifications or commands which are then applied as a batch operation to the sequence. Once you have made a series of changes you may elect to have the sequence listed at the terminal, or the program may do this automatically. This latter method is illustrated in *Figure 3* which shows a short editing session using Staden's SEQEDT. This editor provides only three commands: Find (a sequence position), Delete (*n* characters) and Insert (a string). You type a combination of these commands as a single command string (which may be more than one line) and it is applied as a single operation, producing the output file.

There is quite a variety of other line-orientated sequence editors; they may offer different commands including search functions but their overall concept is very similar to that illustrated for SEQEDT. The problem with most line editors is that you may find them rather obscure in operation. You will undoubtedly need a printed listing of the initial sequence for reference before starting any appreciable editing, and you may find that the lack of immediate feedback about the changes you have made causes confusion.

4.2 **Screen-orientated editors**

Screen-orientated editors are designed to make full use of the features available on modern VDU terminals and microcomputers. The editor responds immediately to any valid keystroke or combination of keystrokes that you make by altering the display accordingly. As you insert, delete or overwrite sequence characters — at a clearly recognizable position (e.g. at the cursor position on the display) — the changes are made immediately. You can move the cursor (the active editing position) by simple keystrokes (e.g. by using the arrow keys provided on most modern keyboards).

Figure 4 shows the terminal during an editing session with Stockwell's SSEDIT full screen sequence editor. The screen is immediately updated as each of the characters is inserted at the cursor position, and it is very simple to realize if you have made the wrong change and it is equally simple to correct the error. The SSEDIT editor also provides a search function which was used in this case to locate the cursor on the required position in the sequence.

You will probably find that screen-orientated editors are the most straightforward to use since the immediate feedback from the editor makes the effects of changes very easy to see. If you have any experience with modern, high-quality computer text editors you will appreciate the directness and simplicity of sequence editing with a good screen-orientated editor.

DNA sequence analysis software

```
$seqedt

          PROGRAM TO EDIT SEQUENCE DATA STORED ON DISK

          COMMANDS ARE ENTERED FROM KEYBOARD, UPTO 80 PER LINE
          MAXIMUM OF 1000 EDIT STRING CHARACTERS PER EDIT
          COMMANDS ARE: I=INSERT, F=FIND, D=DELETE.
          ALL COMMANDS ARE PRECEDED AND FOLLOWED BY /
          EDITS ARE FINISHED BY TYPING "//", "@"

          TO EDIT AN OLD FILE TYPE Y

Y
          INPUT FILE

          PLEASE TYPE NAME OF FILE 1

f1gene1.dat
          OUTPUT FILE

          PLEASE TYPE NAME OF FILE 2

f1gene1.new
          TYPE EDITS NOW


f/557/i/cct//@

                10        20        30        40        50        60
          ATGGCTGTTT ATTTTGTAAC TGGCAAATTA GGCTCTGGAA AGACGCTCGT TAGCGTTGGT
                70        80        90       100       110       120
          AAGATTCAGG ATAAAATTGT AGCTGGGTGC AAAATAGCAA CTAATCTTGA TTTAAGGCTT
               130       140       150       160       170       180
          CAAAACCTCC CGCAAGTCGG GAGGTTCGCT AAAACGCCTC GCGTTCTTAG AATACCGGAT
               190       200       210       220       230       240
          AAGCCTTCTA TATCTGATTT GCTTGCTATT GGGCGCGGTA ATGATTCCTA CGATGAAAAT
               250       260       270       280       290       300
          AAAAACGGCT TGCTTGTTCT CGATGAGTGC GGTACTTGGT TTAATACCCG TTCTTGGAAT
               310       320       330       340       350       360
          GATAAGGAAA GACAGCCGAT TATTGATTGG TTTCTACATG CTCGTAAATT AGGATGGGAT
               370       380       390       400       410       420
          ATTATTTTTC TTGTTCAGGA CTTATCTATT GTTGATAAAC AGGCGCGTTC TGCATTAGCT
               430       440       450       460       470       480
          GAAAATGTTG TTTATTGTCG TCGTCTGGAC AGAATTACTT TGCCTTTTGT CGGTACTTTA
               490       500       510       520       530       540
          TATTCTCTTA TTACTGGCTC GAAAATGCCT CTGCCTAAAT TACATGTTGG CGTTGTTAAA
               550       560       570       580       590       600
          TATGGCGATT CTCAATCCTT AAGCCCTACT GTTGAGCGTT GGCTTTATAC TGGTAAGAAT
               610       620       630       640       650       660
          TTGTATAACG CATATGATAC TAAACAGGCT TTTTCCAGTA ATTATGATTC CGGTGTTTAT
               670       680       690       700       710       720
          TCTTATTTAACGC CTTATTTATC ACACGGTCGG TATTTCAAAC CATTAAATTT AGGTCAG
               730       740       750       760       770       780
          AAGATGAAATTAA CTAAAATATA TTTGAAAAAG TTTTCTCGCG TTCTTTGTCT TGCCGATT
               790       800       810       820       830       840
          GGATTTGCAT CAGCATTTAC ATATAGTTAT ATAACCCAAC CTAAGCCGGA GGTTAAAAAG
               850       860       870       880       890       900
          GTAGTCTCTC AGACCTATGA TTTTGATAAA TTCACTATTG ACTCTTCTCA GCGTCTTAAT
               910       920       930       940       950       960
          CTAAGCTATC GCTATGTTTT CAAGGATTCT AAGGGAAAAT TAATTAATAG CGACGATTTA
               970       980       990      1000      1010      1020
          CAGAAGCAAG GTTATTCACT CACATATATT GATTTATGTA CTGTTTCCAT TAAAAAAGGT
              1030      1040      1050      1060      1070      1080
          AATTCAAATG AAATTGTTAA ATGTAATTAA
FORTRAN STOP
$
```

Figure 3. Staden's SEQEDT in use to edit a sequence file. The sequence CCT has been inserted at position 557 of f1 Gene 1 and the new sequence written to a file FIGENE1.NEW. The figure has been slightly compressed by removal of some blank lines from the original program output. The underscored characters are those typed by the user and the $ characters are the system command prompt.

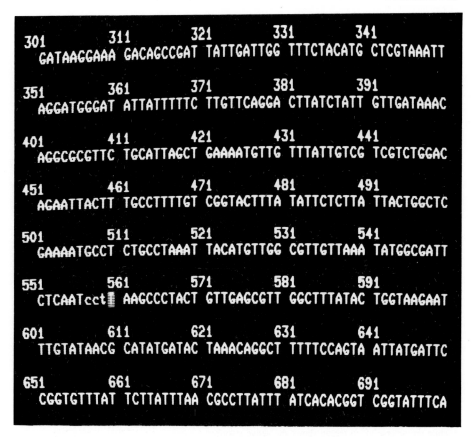

Figure 4. The SSEDIT full screen sequence editor in use. The sequence CCT has just been inserted at position 557 of gene 1 from phage f1. The inserted characters, which are inserted at the cursor position (now on 560) are in lower case for clarity, SSEDIT will convert them to uppercase on writing the sequence to the output file.

4.3 Hybrid editors

I apply this term to sequence editors which are basically line-orientated editors in their command system but which provide an automated screen update of the relevant region of sequence as you make each change. This differs from true screen-orientated editors in that you must make line commands to move through the sequence, or to delete or insert characters, but the editor makes enough use of screen features to display sequence changes and the active editing position. The distinction is made clearer by following the example of the use of Stockwell's LDNA program (7) in *Figure 5*. As the legend states, line commands were used to prepare the editor for the insertion of three bases at the position shown in reverse video on the screen. The bases will be inserted by pressing the <RETURN> key and the display will then be updated to show the changes.

You may find that good hybrid editors are nearly as transparent and reassuring in operation as a true screen-orientated editor. You still need to translate your intentions into a series of editor commands to use them, but with a good program this is not a difficult process. There is something of a continuum between screen-orientated and hybrid editors, since some editors will not fit clearly into one or the other category.

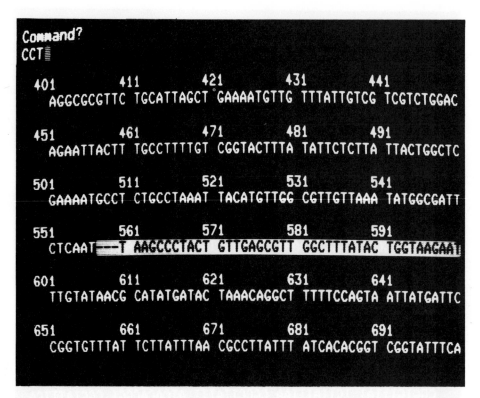

Figure 5. The LDNA editor in use. This hybrid line/screen editor is about to insert CCT into position 557 of gene 1 from phage f1. The 3 dashes have been inserted by prior commands ?EDIT 557 and ?INSERT 3 while the reverse video region shows the active position.

4.4 Choosing and learning a sequence editor

You may be in the fortunate position to be able to choose from a number of different sequence editors, in which case the best advice is to try all of them and assess which you regard as being the most appropriate to use. You should also consider the availability of long-term support and updates for a given editor. Finally, make sure that your chosen editor creates or edits sequence files in the correct format for performing other manipulations, or at least make sure that you can convert the format if necessary (see Section 9). If your preferred editor is part of a package, then it will already create and edit files in the correct format for the rest of the package.

You must expect to spend some time getting used to the sequence editor which you elect to use. Try creating new sequence files and editing copies of old ones to gain confidence and experience in the editing process. Investigate any other features which your editor provides (e.g. translation, insertion of files, searching, etc.) so that you will be able to make use of them when you need them later. If you are not very adventuresome you may prefer to take the approach that you would rather leave the awkward operations to those who understand or enjoy them rather than bother yourself. This approach is fine so long as these people will always be around when you want them. In the long run, you will be better off making the initial effort yourself.

4.5 **Security against data loss with editors**

There are two sides to this question. The first is that of security against data loss on computers in its own right, the second that of what intrinsic security is provided against accidental corruption by the editor itself.

4.5.1 *Preventing data loss*

Computers are complex electro-mechanical devices and as a result they tend to fail occasionally. In addition computers are controlled by human beings who have a tendency to make mistakes. These factors in association with 'acts of God' such as fire, flood damage, etc. can result in loss or corruption of your data, whether it is on a mainframe system, or on your own personal microcomputer. It is necessary for you to ensure that some precautions are taken to reduce the risk of permanent loss of data. As you spend more time editing and analysing sequence data, and as your operations become more dependent on the use of computers and the availability of the data, you become more vulnerable to serious disruption in the event of data loss whether it is through mechanical failure, fire or user error.

Large mainframe installations may already offer an archiving service for files which are held in long-term storage off the site. You should always be able to recover lost data with such systems, although you may have to spend some time in the event of problems. If such services are not available, for example, you are running your own microcomputer based system, then backup storage of data is your own concern. The rule is simple: always maintain an up-to-date copy of any important information on a separate disk from the main working copy and preferably store this disk in a different location. If your computer system is unreliable, or if the information is especially valuable then you should keep more than one backup copy.

4.5.2 *Editor data backup*

Sequence editors vary widely in their ability to provide an automatic backup of edited sequence files. Backup files can be particularly important when you are editing a sequence file and storing the new sequence in a file of the same name. Some computer operating systems immediately overwrite the old version of a file when creating a new file of the same name, while others store some (or all) old versions of a file until specifically told to delete them. When you use a system which will not save old versions, it can be very helpful if the editor itself provides some means of saving the old file from immediate deletion, e.g. by renaming the old file in some appropriate way. This gives at least one chance of recovering from an unfortunate editing mistake. Stockwell's program SSEDIT and LDNA both follow a widespread practice where editing a file called, say, MYSEQ.DAT will produce two files MYSEQ.DAT containing the new edited version and MYSEQ.BAK containing the original sequence.

If neither the system nor the editor itself provide automated backup then you must take extra care when working on important files. For example you could copy the file to a temporary backup version before editing. You will find it safer to use editors which provide an effective display of the final state of the edited sequence prior to the end of editing sessions since you will have time to abort the process if something has gone seriously wrong.

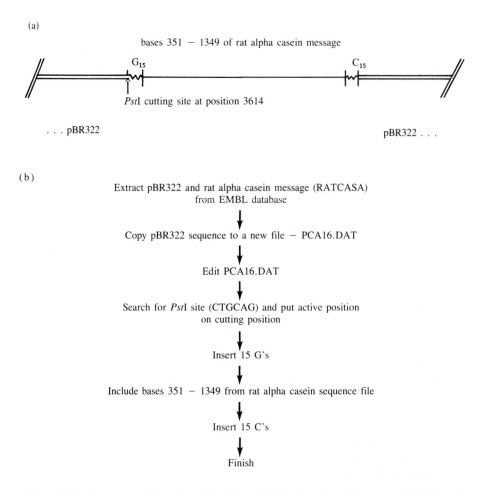

Figure 6. (a) The structure of the *p*Cα16 clone from (11). The insert has a 15 long poly-G leader and a 15 long poly-C tail. **(b)** The steps used to create this sequence on the computer using the include function of a powerful sequence editor — in this case the SSEDIT program described in Section 4.2. The completed sequence was subjected to computer restriction site analysis as described in Section 7.

4.6 Simulation of artificial sequence constructs

More sophisticated sequence editors allow you to copy parts of sequence files into other sequences. This is useful on occasions when you may wish to simulate an artificially constructed sequence in which part of one sequence is ligated into another. (In the absence of an INCLUDE or INSERT function you would have to type in the sequence of the inserted fragment manually.)

I have chosen an actual example to illustrate this method. A laboratory wished to use a series of clones of casein cDNA fragments inserted into the vector pBR322. When they performed molecular weight analysis of restriction fragments they found a discrepancy with the published description of the clone (11). To investigate this problem we reconstructed the clone according to the description and subjected it to restriction site analysis as described in Section 7.1. *Figure 6a* shows the published structure of the clone *pC*α16 which comprises most of a cDNA rat alpha casein message inserted into the single *Pst*I site of pBR322. *Figure 6b* shows the steps used to recreate this sequence on the computer. From the results obtained, it appeared that the published structural description of the clone was inaccurate. Incidentally in this case we were lucky that the sequences of both pBR322 and the rat alpha casein cDNA were available directly from the EMBL database (see Chapter 4). If this had not been so then it would have been necessary to create the sequence files in advance by typing in the sequence (using the editor) from printed listings.

If your sequence editor can perform this type of construction you will find that agreement between computer-generated predictions and experimental data provide compelling evidence for your interpretation of molecular events.

5. SEQUENCE MANIPULATION

In this section I shall describe the basic manipulations that can be performed on sequence data in the computer. These include the interconversion of RNA and DNA sequences, complementation, translation and searching for open reading frames.

5.1 Transcription − RNA to DNA

Nucleic acid sequences can be expressed as DNA or RNA sequences, since the only difference on the computer is the interchanging of T and U. However many sequence utilities have been designed to read DNA sequences only so that you may have to convert RNA sequences to DNA for computer analysis. Some sequence editors and general purpose packages provide this function, but you may not have access to such programs. There are alternative solutions: the programming required to read a sequence file and substitute T for U is trivial, and you or a co-erced colleague could write a suitable program. Alternatively, sequences stored in sequential text files (see Section 3) can always be modified with a normal text editor. You should be able to use an automated search and replace function to complete this task rapidly, but make sure that you have a satisfactory backup copy of the file in case of problems.

If you require the reverse process, that of converting DNA to RNA sequences, then the same method will suffice.

5.2 Complementing sequences

DNA molecules consist of two base-paired strands each of opposite polarity. It is conventional to store the sequence of one strand only and in the direction 5′ to 3′ (the direction of transcription and of protein translation). Determining the other strand (the − strand) requires that the sequence be reversed and each base replaced by its pairing complement. These two processes are generally applied simultaneously since a reversed uncomplemented sequence is not biologically meaningful. Good sequence editors, sequence utilities or packages generally provide a complement function, which will produce the complementary strand from a DNA sequence.

You will appreciate that many of the analytical techniques (e.g. homology searching, open reading frame scans, etc.) must be applied to both strands of a sequence to give a complete picture. Some of the programs concerned will generate the complement

```
$ analyseq

  ANALYSEQ

  type y for embl format n
  sequence file name=casa.dat
 sequence length=  1349
 TYPE CARRIAGE RETURN FOR MENUS NOW AND ALSO AFTER OTHER OPTIONS
 NOTE THAT MENUS HAVE NEGATIVE NUMBERS
 ............


 ............
 translation menu
 15 = translate and list
 36 = translate and list in six phases
 19 = translate and write protein sequence to disk
 17 = count codons, bases, amino acids etc
 52 = write codon table to disk
  option=36
 line length(multiple of 3,def=60)=
 first sequence number =1
 last sequence number =1349
```

```
     I  F  L  I  I  S  Q  L  L  S  P  Y  S  W  V  Q  D  L  S  N
    S  S  *  S  S  P  S  F  S  H  P  T  L  G  F  K  I  L  A  T
   L  L  D  H  L  P  A  S  L  T  L  L  L  G  S  R  S  *  Q  P
 ATCTTCTTGATCATCTCCCAGCTTCTCTCACCCTACTCTTGGGTTCAAGATCTTAGCAAC
      10        20        30        40        50        60
 TAGAAGAACTAGTAGAGGGTCGAAGAGAGTGGGATGAGAACCCAAGTTCTAGAATCGTTG
   D  E  Q  D  D  G  L  K  E  *  G  V  R  P  N  L  I  K  A  V
    R  R  S  *  R  G  A  E  R  V  R  S  K  P  E  L  D  *  C  G
     K  K  I  M  E  W  S  R  E  G  *  E  Q  T  *  S  R  L  L  W

     H  E  T  S  Y  P  H  L  P  R  G  C  C  S  C  S  A  *  S  S
    M  K  L  L  I  L  T  C  L  V  A  A  A  L  A  L  P  R  A  H
   *  N  F  L  S  S  P  A  S  W  L  L  L  L  L  C  L  E  L  I
 CATGAAACTTCTTATCCTCACCTGCCTCGTGGCTGCTGCTCTTGCTCTGCCTAGAGCTCA
      70        80        90       100       110       120
 GTACTTTGAAGAATAGGAGTGGACGGAGCACCGACGACGAGAACGAGACGGATCTCGAGT
   M  F  S  R  I  R  V  Q  R  T  A  A  A  R  A  R  G  L  A  *
    H  F  K  K  D  E  G  A  E  H  S  S  S  K  S  Q  R  S  S  M
     S  V  E  *  G  *  R  G  R  P  Q  Q  E  Q  E  A  *  L  E  D

     S  *  K  C  S  Q  Q  S  N  S  A  R  E  *  Q  Q  *  G  T  G
    R  R  N  A  V  S  S  Q  T  Q  Q  E  N  S  S  S  E  E  Q  E
   V  E  M  Q  S  A  V  K  L  S  K  R  I  A  A  V  R  N  R  K
 TCGTAGAAATGCAGTCAGCAGTCAAACTCAGCAAGAGAATAGCAGCAGTGAGGAACAGGA
      130       140       150       160       170       180
 AGCATCTTTACGTCAGTCGTCAGTTTGAGTCGTTCTCTTATCGTCGTCACTCCTTGTCCT
   R  L  F  A  T  L  L  *  V  *  C  S  F  L  L  L  S  S  C  S
    T  S  I  C  D  A  T  L  S  L  L  L  I  A  A  T  L  F  L  F
     Y  F  H  L  *  C  D  F  E  A  L  S  Y  C  C  H  P  V  P  F

     N  C  *  T  T  K  V  S  Q  S  *  *  G  V  R  Q  Q  P  E  Q
    I  V  K  Q  P  K  Y  L  S  L  N  E  E  F  V  N  N  L  N  R
   L  L  N  N  Q  S  I  S  V  L  M  R  S  S  S  T  T  *  T  D
 AATTGTTAAACAACCAAAGTATCTCAGTCTTAATGAGGAGTTCGTCAACAACCTGAACAG
      190       200       210       220       230       240
 TTAACAATTTGTTGGTTTCATAGAGTCAGAATTACTCCTCAAGCAGTTGTTGGACTTGTC
   I  T  L  C  G  F  Y  R  L  R  L  S  S  N  T  L  L  R  F  L
    N  N  F  L  W  L  I  E  T  K  I  L  L  E  D  V  V  Q  V  S
     Q  *  V  V  L  T  D  *  D  *  H  P  T  R  *  C  G  S  C  V

 ............
```

Figure 7. Part of a run of Staden's package ANALYSEQ showing the translation of the rat alpha casein gene (11) in 6 phases (3 for each strand). The translation is to single letter amino acid codes. Commands typed by the user are underscored.

automatically, but many do not. You must check whether the program you wish to use will examine both strands, and if they do not, then you must complement the sequence by some suitable means and run your program on both strands separately.

5.3 Sequence translation

Sequence translation is the formation of a protein sequence by replacing each triplet codon of the nucleic acid sequence with its corresponding amino acid residue. Sequence translation is a standard operation which is widely available in sequence packages and utilities. There are quite a few options to translation since each strand of the nucleic acid sequence can be translated in three phases, with the output being in one- or three-letter amino acid codes. If you wish to use the derived protein sequence for some subsequent analysis (as described in Chapter 12) then you will need to write it to a disk file in the appropriate format. Alternatively you may wish to have the nucleic acid listed in parallel with its translation in one or more phases. *Figure 7* shows part of a translation listing, produced on this occasion by Staden's ANALYSEQ package (12). As in the SSEDIT program described in Section 4.2, translation can be made into a sequence editor function.

Most DNA sequences require the use of the standard genetic code for translation, but sequences of mitochondrial origin require different codes. You may also wish to use a 'special' code if you intend to predict the results of experiments with mutant 'suppressor' bacterial strains in which some normal terminator codons actually code for an amino acid. Some, but not all, sequence translation utilities provide the option of alternate genetic codes. Those which do may require you to specify a table of codon translations in the form of a data file for any other than the standard code.

5.4 Finding open reading frames

Finding regions of sequence which actually code for protein genes is a complex process and is covered in detail in Chapter 10. The process of translating a sequence and identifying the regions which contain no terminator codons is an important first step in analysing nucleic acid sequences. As for translation, you should scan for open reading frames on both strands for completeness and you may need to use alternative genetic codes if the origin of the DNA sequence indicates this.

Figure 8 shows the output from Stockwell's program ORF when scanning for open reading frames in the rat alpha casein cDNA sequence. This program only scans one strand, but does so in each of the three phases. Similar programs are available in various packages or utilities.

6. SEQUENCE COMPOSITION

Once DNA sequence information is stored in computer files, it is a relatively trivial operation to perform various composition analyses simply by scanning through the sequence. In fact, computers are ideal for this tedious type of work.

6.1 Base composition

This is a simple count of the frequency of each base in the sequence. From this it is possible to work out the composition of the complementary strand, the molecular weights

```
                         PROGRAM ORF
                         ===========

                        Version 1.00

                 Open Reading Frame searching program

            Written by Peter A. Stockwell, Otago University, NZ.

             COPYRIGHT (c) 1985 by MRC (NZ) and P.A. Stockwell

                    Run on 22-Jan-1986 at 15:41

              Sequence file = casa.dat, Sorting: by length
              Finding ORF's >=  30 to a maximum number of 200
```

Orf No.	Start	Stop	Phase	Length (Bases)	First ATG	GTG	Peptide length total	initiated
1	11	913	2	903	62	89	301	284 A
2	463	849	1	387	0	0	129	0 -
3	285	659	3	375	372	321	125	113 G
4	747	962	3	216	795	0	72	56 A
5	66	233	3	168	129	168	56	35 A
6	1054	1188	1	135	0	0	45	0 -
7	1142	1258	2	117	0	0	39	0 -
8	917	1030	2	114	0	962	38	23 G
9	1	111	1	111	0	0	37	0 -
10	966	1055	3	90	993	0	30	21 A
11	304	390	1	87	0	0	29	0 -
12	961	1038	1	78	1012	0	26	9 A
13	1059	1133	3	75	0	0	25	0 -
14	394	459	1	66	430	0	22	10 A
15	1237	1299	1	63	0	1294	21	2 G
16	1290	1349	3	60	0	0	20	0 -
17	217	267	1	51	0	0	17	0 -
18	3	53	3	51	0	0	17	0 -
19	1262	1303	2	42	1271	1292	14	11 A
20	1307	1348	2	42	0	0	14	0 -
21	1170	1211	3	42	0	0	14	0 -
22	1034	1072	2	39	0	0	13	0 -
23	663	701	3	39	0	0	13	0 -
24	1227	1265	3	39	0	0	13	0 -
25	1076	1111	2	36	0	0	12	0 -
26	127	159	1	33	0	0	11	0 -
27	925	957	1	33	0	0	11	0 -
28	865	894	1	30	0	0	10	0 -

Figure 8. The output from ORF when scanning for open reading frames in the rat alpha casein cDNA. The final column shows whether the first possible initiator encountered was ATG or GTG (which can act as an initiator in prokaryotes).

of either strand or both strands. It is also possible to estimate parameters such as the melting point for the DNA based on the GC/AT ratio. Many sequence utilities or packages contain this type of function and you should have no difficulty in applying this type of analysis to your sequence data.

A further elaboration of composition counting is the process of taking a moving average of the composition along the molecule and plotting its asymmetry to show regions of high GC content, etc. A moving average is determined by summing the composition in a window of specified length, then moving the window one base forwards and summing the composition again, until the entire sequence has been scanned.

6.2 Dinucleotide frequency

It is possible to sum the occurrences of each of the 16 dinucleotides in a sequence. *Figure 9* shows Staden's ANALYSEQ (12) in use for this purpose.

Obviously we could examine the frequency of occurrence of all possible oligonucleotides in a sequence and if we did so we would find any long internal repeated sequences. In practice it is simpler to locate internal repeats by homology searching methods (as described in Section 8 and in Chapter 13) so that the only other useful oligonucleotide to examine is the triplet codon.

```
   ...
menus are
-1 = general
-2 = screen control
-3 = gene search by content
-4 = gene search by signal
-5 = statistical analysis of content
-6 = translation and codons
-7 = structures and repeats
23 = HELP
 8 = STOP
 option=-5
statistical analysis of content
 7 = count dinucleotide frequencies
17 = count codons, bases, amino acids etc
52 = write codon table to disk
 2 = plot base composition
26 = plot local deviations in base composition
28 = plot local deviations in dinucleotide composition
29 = plot local deviations in trinucleotide composition
27 = plot negentropy
40 = calculate codon pressure
 option=7
```

	T		C		A		G	
	obs	expected	obs	expected	obs	expected	obs	expected
T	7.12	6.47	8.16	6.81	4.60	7.53	5.56	4.62
C	9.12	6.81	6.75	7.17	10.09	7.93	0.82	4.87
A	5.93	7.53	6.08	7.93	9.05	8.76	8.53	5.38
G	3.26	4.62	5.86	4.87	5.79	5.38	3.26	3.30

```
.....
```

Figure 9. Staden's ANALYSEQ used for scanning the sequence of rat alpha casein cDNA for dinucleotide frequency. The expected frequencies are calculated from the base composition. The user responses are underscored.

6.3 **Codon frequency**

If you count trinucleotide occurrence in known protein-coding regions of a sequence then you have established the codon frequency for those genes. The problems of identifying protein-coding regions are not so straightforward (see Chapter 10) but once identifiable genes have been located, the codon frequency within them can be used as a tool to help locate further genes (13). If you examine the statistical content menu of ANALYSEQ (shown in *Figure 9*) you will see that codon frequencies can be determined by this program, and that they can be written to disk files for subsequent use.

7. MAPPING

Mapping the positions of features on a DNA sequence is an important step in investigating structure and function. The process of mapping restriction endonuclease cutting sites on the sequence serves a further important function: it can provide evidence for accuracy of sequencing when the computer-predicted restriction map is in accord with experimental observations.

Restriction mapping is a relatively straightforward process and I shall examine it in some detail. Mapping other features, such as promoter sequences and ribosomal binding sites in DNA sequences is less reliable and is covered elsewhere (Chapter 10) so that I make only a brief comment here.

7.1 Restriction endonuclease mapping

Most sequence utility programs provide search functions to scan for sub-sequences and most extend the search process to scan for restriction enzyme recognition sites (termed restriction sites) by enzyme name. The strategy is fairly simple, the program will search for the given enzyme name through a file which contains the enzymes and their recognition or cutting sites in a format the program recognizes. Once the recognition site is found it will be used to search through the DNA sequence.

7.1.1 Restriction site files

You will probably find that your sequence utility program or package comes complete with a file of restriction enzyme recognition sites. There are several points which you should consider when you carry out computer searching for restriction sites.

(i)   *Isochizomers*. Many restriction enzymes recognize the same site (isoschizomers) but do not necessarily cut at the same point. It is possible that you may try using an enzyme name which is not present in the restriction site file for your program. You might avoid this problem by having a listing or some form of catalogue of your restriction file, possibly with some form of cross reference between isoschizomers.

(ii)   *Incompletely specified bases*. Some restriction enzymes recognizes sites with some looseness, for example the enzyme *Hinc*II recognizes the sites GTPyPuAC (where Py is a pyrimidine and Pu a Purine). This situation is best handled by the use of the incompletely specified bases mentioned in Section 3.3 (shown in *Table 1*) provided that your program will recognize these codes for searching. In this case the *Hinc*II site would be expressed as GTYRAC.

(iii)   *Asymmetric sites*. Enzymes which recognize asymmetric sites in double-stranded DNA require two different recognition sites. For example, the enzyme *Mbo*II recognizes GAAGA on one strand and TCTTC on the other so that in restriction mapping both sites must be searched. Some sequence utilities handle this situation directly while others require that the enzyme is present twice in the restriction site file with each of its recognition sites.

(iv)   *Cutting versus recognition sites*. The restriction search will show the positions in a sequence at which the recognition sites occur — it does not necessarily indicate the points at which the enzyme will cut. Some programs will indicate true cutting sites but many do not so that DNA fragment sizes predicted may be somewhat inaccurate, especially if the cutting site is far from the recognition site. A good example of this is the enzyme *Mbo*II which cuts 12 bases after the start of one of its sites and 7 bases before the other. You should check whether your program indicates recognition sites or true cutting sites for restriction maps.

(v)   *Available versus known enzymes*. There is a need for flexibility in the contents

```
                        PROGRAM LDNA
                        ============
           A package for manipulating large DNA data files
               Version No. 1.4 run on   3-Feb-86
                 Mapping File: DK:CASA.DAT   for enzymes
                    From File SY:LABRST.DAT
                 FROM 1 TO 1349

      1          11          21          31          41
      |          |           |           |           |
      ATCTTCTTGA TCATCTCCCA GCTTCTCTCA CCCTACTCTT GGGTTCAAGA
       |-MBOII                |-ALUI    |-HPHI                    |-MBOI
           |-BCLI
           |-MBOI

     51          61          71          81          91
      |          |           |           |           |
      TCTTAGCAAC CATGAAACTT CTTATCCTCA CCTGCCTCGT GGCTGCTGCT
                                 |-MNLI     |-MNLI
                                 |-HPHI

    101         111         121         131         141
      |          |           |           |           |
      CTTGCTCTGC CTAGAGCTCA TCGTAGAAAT GCAGTCAGCA GTCAAACTCA
                 |-SACI
                 |-ALUI

    151         161         171         181         191
      |          |           |           |           |
      GCAAGAGAAT AGCAGCAGTG AGGAACAGGA AATTGTTAAA CAACCAAAGT
                            |-MNLI

    201         211         221         231         241
      |          |           |           |           |
      ATCTCAGTCT TAATGAGGAG TTCGTCAACA ACCTGAACAG ACAGAGAGAG
                 |-MNLI      |-HINCII                    |-ALUI

    251         261         271         281         291
      |          |           |           |           |
      CTTCTGACAG AACAGGATAA TGAAATCAAG ATAACTATGG ACTCATCAGC
                                                  |-HINFI |-ALUI

    301         311         321         331         341
      |          |           |           |           |
      TGAGGAACAA GCAACGGCAA GTGCTCAGGA AGATTCCTCC TCAAGCAGCT
       |-MNLI                            |-MBOII   |-MNLI   |-ALUI
                                         |-HINFI
                                            |-MNLI
```

Figure 10. A restriction map performed on the rat alpha casein cDNA sequence. The enzyme file used (LABRST.DAT) is a list of enzymes currently held in the laboratory stock. The position marked for each site is the start of the recognition site, not the cutting position.

of restriction site files in order to allow for the discovery of new enzymes or for changes in those commercially available. While the list of known restriction enzymes is large (in excess of 450 at the time of writing) and growing, only some 100 or so are commercially available at present. It is of rather academic interest to search for all known sites if you are unable to obtain some of the enzymes. In fact for some purposes you may wish to confine restriction maps to those enzymes which you currently hold in stock. The requirements of updating restriction site files or searching for a subset of enzymes can readily be met if your sequence utility or package provides the means to alter the contents of these files. If the file is in a form which can be directly edited by a text editor, yet still be read by the program then you can easily copy and alter the restriction file. Alternatively, the restriction site file may be in a special format which cannot be easily edited. In this case you will have to rely on a 'site update' function in the utility or package if it is provided.

7.1.2 *Full maps or single enzyme searches*

It is helpful to have a sequence utility which will perform automated searches for all of the enzymes in a given restriction site file, as well as possessing the facility to search for any specified enzyme. This is particularly useful when you want to know all of the possible fragmentation patterns of a DNA sequence with all of the restriction enzymes which you can obtain. *Figure 10* shows one form of output from just such a search by the LDNA program (7). While the complete listing shown (in part) here is useful for finding the exact position of a restriction site, it is less helpful when you wish to know the general distribution of sites for one or more enzymes. The summarized map shown in *Figure 11* conveys the overall distribution of sites in the form of a line printer graph, in this case showing the complete map shown partly in *Figure 10*.

As a first step in sequence analysis, line-printer graphs like *Figure 11* still remain a rapid and useful means of representing summarized information. With suitable hardware (e.g. graphics terminals or plotters) and suitable software you can go further and construct high quality restriction maps directly.

7.2 **Mapping other sequence features**

Any sequence feature which can be described to the computer can be mapped in the same way as restriction sites or sub-sequences. Unfortunately it is often difficult to produce succinct and accurate rules to describe nucleic acid sequence features, particularly those features which display an appreciable degree of variability. Features which fall into this category include prokaryotic and eukaryotic ribosomal binding sites, exon splice junctions and promoter sequences.

The methods employed to search for variable features are based on matrix methods (Chapter 10). Where the recognition algorithms use weighted frequency tables (14). Staden's ANALYSEQ (12) applies methods of this type to locate variable sequence features. The program scans through the sequence applying a weighted scoring matrix to successive windows at each point of the sequence and produces a graphical display of those positions which exceed a threshold value. Some features, (e.g. exon splice junctions) have further corollary tests which are included in establishing the most probable candidate positions in the sequence.

Best fit methods of this type are not totally reliable in that they can miss real features and can find false positives. This reflects the fact that our knowledge of how these

```
PROGRAM RMAP
============

Version 1.00

Makes restriction maps from LDNA summarised search output files

Written by Peter A. Stockwell, Otago University, NZ.

COPYRIGHT (c) 1985 by MRC (NZ) and P.A. Stockwell

Run on 3-Feb-1986 at 10:32

DK :CASA    Search listing file = try.lst
            .DAT  FOR ALL ENZ. from file SY:LABRST.DAT

Mapping sequence in file
Searched from   1 to 1349

         135    270    405    540    675    809    944   1079   1214   1349
         135    270    405    540    675    809    944   1079   1214   1349

 2 ACCI
16 ALUI
 1 BCLI
 1 HAEIII
 3 HHAI
 1 HINCII
 7 HINDIII
 6 HINFI
 4 HPHI
 5 MBOI
 4 MBOII
12 MNLI
 1 RSAI
 1 SACI

The following enzymes did not cut:

AVAI     PSTI
AVAII    SALI
BAMHI    SMAI
BGLII    TAQI
ECORI    XBAI
HPAI     XHOI
HPAII
KPNI
NCOI
```

Figure 11. The complete restriction map from *Figure 10* shown in a graphical form.

```
$ srchn
                            SRCHN
               D.J.Lipman & W.J.Wilbur
              Mathematical Research Branch
                       NIADDK, NIH
         Bldg. 31,Rm 4B-54, Tel.  (301) 496-4325

    This program compares a given nucleic acid
sequence with sequences in the Los Alamos Data Bank,
using the algorithm of Wilbur and Lipman.

Choose between a search against all MAMMALS    [1]
                        other VERTEBRATES       [2]
                            INVERTEBRATES       [3]
                                   PLANTS       [4]
                               ORGANELLES       [5]
       (not including mitochondrial genomes)
                                 BACTERIA       [6]
                          STRUCTURAL RNA        [7]
                                    VIRAL       [8]
                                    PHAGE       [9]
              (not including Lambda or T7)
                                SYNTHETIC      [10]

    another file in GenBank format               [11]

(1,2,3,4,5,6,7,8,9,10,11) [1] > 1
Output to disk file? (Y/N) [Y] > y
Output file > casa.lis
Query sequence
Sequence file > casa.dat
L=LDNA, S=Staden, Q=SEQ
Type of file? [S] > s
Reading sequence from casa.dat

Would you like to splice Query? (Y/N) [N] > n
Complement of Query? (Y/N) [N] > n
Type in K-tuple length (must be <7)
  (NOTE: Statistics only worked out for K=4) [4] > 4
Type in window size [20] > 20
Type in gap penalty [4] > 4
Do you wish dense or less dense local similarities? (D/L) [L]: 1
Searching database GENBANK:MAM.DAT
 . . . . . . . . . . . .
```

Figure 12. The start of a run with Lipman's SRCHN (4) program, in which the rat alpha casein sequence is compared with all mammalian entries in the Genbank database. The parameters K-tuple, window size, gap penalty and dense/less dense similarities all control aspects of the homology searching process. The user's responses are underscored.

features are recognized and used at the cellular level is incomplete. You may, however, still find value in such methods, and computer predictions may be confirmed or rejected by experimental observations.

8. SEQUENCE COMPARISON

One of the first questions which is asked about a new sequence is whether it resembles any other known sequence. A further point of interest will generally be whether the sequence contains internal repeats, palindromes, base-pairing regions and so forth. Chapter 13 contains a full description of sequence comparison so I shall give only a simple example of searching for similarities as you might expect to apply in the initial analysis of a newly-derived sequence.

8.1 **DNA/DNA similarities**

Wilbur and Lipman (5) have developed a series of programs which perform fast searches for similarities against sequence data bases. The method employed generates best fits by inserting gaps into the sequence and allowing mismatches as required. You are required to set the parameters which control the searching process at the start of each run, then the program will return a ranked list of database entries which are similar to the probe. *Figure 12* shows the start of a run with Lipman's SRCHN program in which the rat alpha casein cDNA sequence is compared with all of the mammalian sequences in Release 17 of the GENBANK database. *Figure 13* shows two examples from the program output showing how SRCHN inserts gaps to optimize the alignment. You can get an impression from this of how rapidly the matches deteriorate from an exact fit in this case through to a rather spurious matching. SRCHN also compares each individual score with the scores of all the matches to provide some indication of the significance of each fit, however the validity of such statistics is in some doubt (15,16).

The run performed here is only for one strand, in practice you would run the complementary sequence as well, particularly if you did not know which is the 'sense' and which the 'nonsense' strand of your sequence. There is one other important point about the Wilbur and Lipman programs: to achieve their computational speed they examine sequences as a set of k-tuples (groups of bases) and they may overlook certain types of similarity depending on the k-tuple and window parameters. In spite of this they provide a very useful first step in comparing your new sequence with databases of known sequences.

Other searching methods exist and may be provided in various sequence utilities. For example, the MOLGEN SEQ package (4) provides an implementation of a modified Needleman and Wunch algorithm (17) for homology comparison and searching. This provides gapped sequence alignments with estimates of evolutionary distance (or biological similarity) rather similar to the Wilbur and Lipman programs. This algorithm is not subject to the granularity imposed by k-tuple matching, but takes more processor time.

8.2 **Protein/protein similarities**

When you consider the biological significance of a DNA sequence, you could also look for any similarity between possible protein gene products of that sequence and known protein sequences. This process has already proved of value in establishing biological relationships for both nucleic acid (18) and protein (19) sequence information.

DNA sequence homology is established by direct matching of bases. However protein sequence homology is more complex if functional considerations are taken into account. It is possible to allow for substitution frequencies when calculating protein alignments, but this method requires large processor times when comparisons are made against large

(a)

```
LOCUS         RATCASA        1349 BP     MRNA              PRE-ENTRY 06/01/83
DEFINITION    RAT CASEIN-ALPHA MRNA.
ACCESSION     J00710
KEYWORDS      COMPLEMENTARY DNA;  CASEIN;  ALPHA-CASEIN;  SIGNAL PEPTIDE;
PHOSPHOPROTEIN.
X        10         20         30         40
ATCTTCTTGATCATCTCCCAGCTTCTCTCACCCTACTCTTGGGTTCAAGA
::::::::::::::::::::::::::::::::::::::::::::::::::::
ATCTTCTTGATCATCTCCCAGCTTCTCTCACCCTACTCTTGGGTTCAAGA
X        10         20         30         40


         60         70         80         90
TCTTAGCAACCATGAAACTTCTTATCCTCACCTGCCTCGTGGCTGCTGCT
::::::::::::::::::::::::::::::::::::::::::::::::::::
TCTTAGCAACCATGAAACTTCTTATCCTCACCTGCCTCGTGGCTGCTGCT
         60         70         80         90


        110        120        130        140
CTTGCTCTGCCTAGAGCTCATCGTAGAAATGCAGTCAGCAGTCAAACTCA
::::::::::::::::::::::::::::::::::::::::::::::::::::
CTTGCTCTGCCTAGAGCTCATCGTAGAAATGCAGTCAGCAGTCAAACTCA
        110        120        130        140
••••••••••••
```

(b)

```
LOCUS         RATPTRY1       804 BP                        PRE-ENTRY 03/01/83
DEFINITION    RAT PANCREATIC TRYPSINOGEN I MRNA.
ACCESSION     J00778
KEYWORDS      COMPLEMENTARY DNA;  TRYPSIN;  TRYPSINOGEN.
 50         60             70        80         90
GATCTTAGCAACCATGAA---ACTTCTTATCCTCACCTGCCTCGTGGCTG
         :::::::    ::::::  :::::  ::       ::::  :
  CCTTCTGCCACCATGAGTGCACTTCTGATCCTAGCCCTT---GTGGGAG
X        10         20         30         40


   100        110
CTGCTCTTGCTCTGCCTAGAG
:::::  :::::
CTGCTGTTGCTTTCCCTTTGG
   50         60
```

NUMBER OF MATCHED BASES=   35

Figure 13. Results from the SRCHN run shown in *Figure 12*. (**a**) The best fit (partly shown) is, naturally enough, the rat alpha casein sequence. (**b**) shows the fourth ranked fit (after rat gamma and beta caseins) to illustrate the gapping performed by SRCHN to achieve this fit. The remaining fits become even more spurious, with considerable gapping and isolated short regions of homology.

databases. In practice, useful rapid screening of proteins for similarities can be performed by considering exact amino acid matches.

Wilbur and Lipman use exact k-tuple matching in their SRCHGP program (5) to achieve rapid scanning through sequence databases. Lipman has extended this idea (20) in his FASTP program to use exact k-tuple matching to find a candidate set of best fits. A second step uses amino acid substitution frequencies to provide an optimized best alignment of these candidates thus providing the most likely best fit from a protein sequence database. Since FASTP is available on microcomputers as well as on mainframes you may find that it is a very useful early step in analysing your own DNA sequence information.

8.3 Repeats and palindromes

Rapid sequence comparison programs (like SRCHN) aim to find the single best fit between two different sequences. Other searching strategies differ from this (e.g. Needleman and Wunsch, 17) in that they can indicate all the similar regions between two sequences, or within the same sequence (simply by comparing the sequence with itself). Obviously it is a simple further step to compare a sequence with its complement, its reverse and its reverse complement, so that all types of internal repeats can be discovered.

A substantial range of programs will provide this type of search, including Staden's ANALYSEQ (12) and Brutlag's SEQ program (4). The identification of potential tRNA genes (12) is a specific application of a program which looks for base-pairing regions subject to certain constraints (see also Chapter 11).

The subjects of recognition of coding sequences and of nucleic acid secondary structure are covered in detail in Chapters 10 and 11 and the overall subject of sequence comparison in Chapter 13.

9. SEQUENCE CONVERSION

In Section 3 I have considered some of the sequence file formats in use. If you use a package or a unified suite of programs for your sequence analysis then you should find that all of your programs use the same file format and that conversion between one format and another is unnecessary. If you use a 'mix-and-match' set of programs, or if you import sequence files in different formats from other installations, you are faced with the problem of converting one file format to another. When you use sequence utilities in this type of environment, it is important to check that you have used the correct format for a given program, for example by checking that the program has read a sequence of the correct length. Trying to read the wrong type of file can produce results ranging from a program or system crash through to missing sequence regions depending on how the program has been written.

Approaches to file conversion vary. Some installations provide a utility specifically to translate one file format to another. In other cases, sequence editors may have the ability to read and write sequences in a variety of formats, so that conversion simply requires running the editor, without actually performing any edits. One solution to file formatting problems is to have sequence utilities which can themselves read a variety of formats. If you refer back to the run of Lipman's SRCHN program in *Figure 12* you will see an example of this type of input. Where the user is asked to specify the input file (CASA.DAT) the next prompt is for the file type and S (= Staden-type) is offered as a default. (This input routine has been added by this author to the local version of SRCHN as the original DEC-10/20 program only recognized files in MOLGEN/SEQ format.)

Whatever system you use for sequence analysis, it pays to be aware of different file formats and the possible side effects of these. If you are in the position of setting up a suite of programs you will certainly have to consider the different file types in use and look for programs which can handle various formats, or you should incorporate an effective file translation utility.

10. FORMATTING

Published DNA sequence information is generally printed as a numbered sequence with annotations alongside. Typical annotations include restriction sites, putative or known genes with the translation printed in parallel, comments about known or suspected pro-motor regions, TATA boxes, regions shown in double-stranded form, internal repeats, etc. Such listings can either be assembled by hand or through the use of special formatting programs, or even by a combination of both.

I should briefly mention that sequences of any appreciable length should be submitted to sequence databases or to journals in machine-readable form. This means on a flexible disk or a computer tape, both of which raise formatting problems of a different nature. In any event you must include details of your computer, the operating system, and any other relevant information to avoid giving nightmares to the recipients.

10.1 **Simple formatting**

Most sequence utilities or packages provide a sequence listing facility, and in better cases give you the option of writing a numbered sequence listing to a disk file. This file can act as the starting point for a hand-assembled annotated listing using any decent text editor to incorporate the additional information. This can be a little tedious when, for example, long translated sequences must be typed in, although the ingenious will undoubtedly find clever ways to automate this.

This simple approach to sequence formatting has some advantages. It only requires that you possess competence in using sequence utilities and a text editor, it does not require learning any special formatting command language. You can print your formatted listing on any printer and get exactly what you prepared with the text editor so that you do not need to have access to special graphics devices. If you use a high quality printer (whether it be a laser printer, a daisy-wheel, an ink-jet or a high-quality dot matrix printer) then you can produce publication quality listings directly.

Hand-annotated listings can be inefficient, particularly if you are obliged to insert or delete regions of the primary sequence. This will require that the whole listing be regenerated from scratch, wasting the previously invested time. In general, extensively annotated listings will only be produced late in the sequencing process, when the primary sequence is well established.

The method of hand-annotating sequence listings is also incapable of handling graphical information (e.g. diagrammatic restriction maps). If you lack computer tools for this type of work then you will have to use drawing pens in the traditional manner.

10.2 **Formatting programs**

If you are already using a computer to store and analyse your sequence data then it is tempting to automate the production of annotated listings for publication.

One approach to automated formatting is the DNADRAW program of Shapiro and Senapathy (21). DNADRAW takes an annotated sequence file into which you have written various commands and indicated, using a series of codes, which sections of the sequence are to be underlined or written in large letters. DNADRAW will write the sequence to a plotter in accordance with your instructions, producing a publication-quality figure without the need for hand drawing.

The problem with programs like DNADRAW is that they are performing a complex operation, and as a result, they have complex command systems. You may find that the effort required to learn and use such programs outweighs their value to you and that it is simpler to use the methods described above. The value of complex graphical programs is more likely to be realized when they are used by a local 'specialist' who, through practice, will become thoroughly conversant with the package and can provide a service for others.

11. ACKNOWLEDGEMENTS

I would like to thank Professor G.B. Petersen for suggestions and for proof-reading this chapter. I would also like to thank Dr. M.G. Smith for suggestions and discussions. I am employed on a research programme grant (to G.B. Petersen) from the Medical Research Council of New Zealand, to whom I am indebted.

12. REFERENCES

1. Korn,L.J. and Queen,C. (1984) *DNA*, **3**, 421.
2. Rawlings,C. (1986) *Software directing for molecular biology*, Macmillan, London.
3. Staden,R. (1977) *Nucleic Acids Res.*, **4**, 4037.
4. Brutlag,D.L., Clayton,J., Friedland,P. and Kedes,L.H. (1982) *Nucleic Acids Res.*, **10**, 279.
5. Wilbur,W.J. and Lipman,D.J. (1983) *Proc. Natl. Acad. Sci. USA*, **80**, 726.
6. Devereux,J., Haeberli,P. and Smithies,O. (1984) *Nucleic Acids Res.*, **12**, 387.
7. Stockwell,P.A. (1982) *Nucleic Acids Res.*, **10**, 115.
8. Cornish-Bowden,A. (1985) *Nucleic Acids Res.*, **13**, 3021.
9. Nomenclature Committee of the International Union of Biochemistry (1985) *Eur. J. Biochem.*, **150**, 4.
10. Staden,R. (1980) *Nucleic Acids Res.*, **8**, 3673.
11. Hobbs,A.A. and Rosen,J.M. (1982) *Nucleic Acids Res.*, **10**, 8079.
12. Staden,R. (1984) *Nucleic Acids Res.*, **12**, 521.
13. Staden,R. (1982) *Nucleic Acids Res.*, **10**, 141.
14. Minsky,M. and Papert,S. (1969) *Perceptrons*. The MIT Press, Cambridge Massachusetts and London, England.
15. Lipman,D.J., Wilbur,W.J., Smith,T.F. and Waterman,M.S. (1984) *Nucleic Acids Res.*, **12**, 215.
16. McLachlan,A.D. and Boswell,D.R. (1985) *J. Mol. Biol.*, **185**, 39.
17. Needleman,S. and Wunsch,C. (1970) *J. Mol. Biol.*, **48**, 443.
18. Gibson,T.J., Stockwell,P., Ginsburg,M. and Barrell,B.G. (1984) *Nucleic Acids Res.*, **12**, 5087.
19. Downward,J., Yarden,Y., Mayes,E., Scrace,G., Totty,N., Stockwell,P., Ullrich,A., Schlessinger,J. and Waterfield,M.D. (1984) *Nature*, **307**, 521.
20. Lipman,D.J. and Pearson,W.R. (1985) *Science*, **227**, 1435.
21. Shapiro,M.B. and Senapathy,P. (1986) *Nucleic Acids Res.*, **14**, 65.
22. Hill,D.F. and Petersen,G.B. (1982) *J. Virol.*, **44**, 32.

CHAPTER 3

Use of commercial software on IBM personal computers

PATRICIA HOYLE

1. INTRODUCTION

This chapter describes computer software packages for molecular biology written and designed specifically for self-contained and relatively inexpensive IBM PC or compatible microcomputers. Four commercially based software packages for molecular biology are described, with the hardware they require and the functions they offer.

To illustrate how this type of self-contained package works, an experimental project is described using one of these packages in an application to design a suitable gene for production of a vaccine (Section 4). The project is worked through in detail, demonstrating how usefully the software may be employed.

2. APPLICATIONS OF COMPUTING TO THE BIOLOGICAL SCIENCES

During the last five years, there has been a prolific expansion of computer applications in the biological sciences. In much the same time period, computer technology has developed rapidly to produce inexpensive microcomputers at a cost well within the limits of normal laboratory equipment budgets. This has given impetus to the development of a first generation of unified packages of programs specifically designed to be run on independent microcomputer workstations. Microcomputers are now powerful enough to perform many routine tasks, too time-intensive or tedious to carry out manually and which were once the unique province of remote main-frame computers accessed via a video display terminal and a modem.

There are now packages available that implement a wide range of data analyses covering practical and theoretical tasks in molecular biology. They are easy to use, since they have been designed specifically for the worker who has had little or no previous experience of a computer. Most commonly, the programs are executed via simple and easily understood selection routines written in straightforward English. With reliable customer support and maintenance, the use of these software packages for routine analyses in the molecular biology laboratory does not entail difficult introductory courses. Nor do they require serious diversion from the daily research activities in order to operate them routinely. Because they are self-contained, they obviate many of the innate problems incurred when using remote, time-sharing computing services, such as slow response times, inadequate support and accumulating expensive communication costs.

Considering the many other advantages of microcomputers, such as integration with other important tasks like word-processing and data management, together with the increasing amount of processing power available for proportionately decreasing costs, it is probable that the bulk of computing in molecular biology laboratories will be transferred to powerful personal microcomputers in the near future.

3. A COMPARISON OF COMMERCIAL SOFTWARE FOR IBM PERSONAL COMPUTERS

Four commercially available software packages for molecular biology are described and discussed in this chapter.

(i)　DNASIS DNA SEQUENCE INPUT AND ANALYSIS SYSTEM
Marketed by Hitachi Software Engineering Co. Ltd., 6 – 81, Onoemachi, Naka-ku, Yokohama 231, Japan; 950 Elm Avenue, San Bruno, CA 94066, USA and Genetic Research Instrumentation Ltd., Gene House, Dunmow Road, Felstead, Nr. Dunmow, Essex CM6 3LD, UK. Tel: (UK)-371-821082.

(ii)　DNASTAR COMPREHENSIVE MICROCOMPUTER SYSTEMS FOR MOLECULAR BIOLOGY (1)
Marketed by Dnastar Inc. at 1801 University Avenue, Madison, WI 53705, USA and Dnastar Ltd. at 8 Walpole Gardens, London W4 4HG, UK. Tel: (UK)1-994-0619.

(iii)　THE IBI/PUSTELL DNA AND PROTEIN SEQUENCE ANALYSIS SYSTEM (2)
Marketed by International Biotechnologies Inc., 275 Winchester Avenue, New Haven, CT 06535, USA
Distributed by IRL Press Ltd., PO Box 1, Eynsham, Oxford OX8 1JJ, UK. Tel: (UK)-865-882283 and IRL Press Inc., PO Box Q, Mclean, VA 22101, USA

(iv)　THE MICROGENIE SEQUENCE ANALYSIS PROGRAM (3)
Marketed by Beckman Instruments Inc., Palo Alto, CA 94304, USA and Beckman - RIIC Ltd., Progress Road, Sands Industrial Estate, High Wycombe, Buckinghamshire HP12 4SL, UK. Tel: (UK)-494-41181.

3.1 Hardware requirements

3.1.1 *Type and model of personal computer*

To run microcomputer packages for molecular biology, specific hardware requirements have to be met in terms of the type and model of microcomputer, size of computer memory and storage capacity. In addition, peripheral devices may be incorporated as features of the workstation, such as a digitizing data input device, telecommunications modem or the facility to connect the workstation to a local large computer. Each requires compatible entry ports on the computer.

All four packages described have been written for IBM microcomputers (personal computers), the IBM PC, IBM PC XT and IBM PC AT. The IBM PC XT has an integral hard disk and one floppy disk drive. The IBM PC AT can be equipped with one or two floppy disk drives (either low capacity, 360 Kbyte drives or high capacity, 1.2 Mbyte drives). A single hard disk drive of up to 120 Mbyte capacity can be added to the IBM AT and a second hard disk can optionally replace the second floppy drive.

The IBM PC model can be equipped with two floppy disk drives, but a range of hard disk drives can also be added in place of one of the floppy drives.

The DNASIS, MICROGENIE and IBI packages offer versions for use with a computer containing a 10 Mbyte hard disk. The full DNASTAR system requires a minimum storage capacity of 30 Mbytes because of its inclusion of the two complete and fully annotated databases (see below). Smaller independent DNASTAR packages and the MICROGENIE S Version package are available on floppy disks. The DNASIS and DNASTAR packages can run on any IBM personal computer and on IBM-compatible microcomputers (see Operating Systems below), such as the Compaq and Zenith personal computers. The MICROGENIE and IBI packages offer versions for different types of computer. The IBI package versions 81501 and 81502 run on IBM personal computers and other compatible machines which include the Compaq and the Eagle microcomputers and there is also a version for use with a DEC Rainbow computer.

Many IBM-compatible computers are cheaper and may possess technical superiority such as higher speed of performance. However, they may be less versatile in terms of their hardware components. The currently available range of alternative components such as cheap, high performance, large storage hard disks and high capacity floppy drives which can be installed into IBM machines, are not necessarily available for a compatible machine, so the choices may be narrower.

Stringent hardware requirements effectively mean that those are the configurations the software company will actively support. The concept of portability to other systems is attractive from the point of view of serving a wider market of users. However, it is unlikely that a software company would in practice offer active support on a continuing basis for a wide variety of different configurations and this inevitably reflects on the user. Advice and support for a particular hardware configuration should therefore be sought in each case.

3.1.2 *Memory requirements*

Memory (RAM) requirements vary. All three IBM personal computers and their compatibles can be equipped with 640 Kbytes of RAM for maximal use of the DOS operating system.

The DNASTAR and MICROGENIE packages require 640 Kbytes of RAM to be available. The IBI and DNASIS systems require 256 Kbytes of RAM. An upgraded DNASIS package shortly to be released will require 512 Kbytes and it may be anticipated that further additions of software to the IBI package will also increase the memory requirement up to 512 Kbytes.

3.1.3 *Operating systems*

All four systems are designed for the PC-DOS Operating System, versions 2.0 or higher and can also be used with the nearly identical MS-DOS Operating System used by many IBM-compatible systems. The IBI package is now the only package which offers a version for computers using the CP/M operating system. DOS cannot access more than 32 Mbytes of disk storage at a time so that hard disks larger than 32 Mbytes must be partitioned into units of 32 Mybtes or less, or utility software (now usually provided with larger hard disks) used to increase the addressable disk space.

3.1.4 *Use of colour*

The DNASIS, MICROGENIE and IBI packages use a standard colour screen. The DNASTAR package uses a monochrome screen, which except in the case of IBM's Enhanced Colour Monitor and Enhanced Colour Graphics Card, has a sharper definition. The use of colour in the IBI and MICROGENIE packages is effective for helping to direct the user's attention to particular aspects and for highlighting and prompting. Colour is less successfully used in the DNASIS package. Where the MICROGENIE and IBI packages use colour in blocks, DNASIS uses several different colours for individual words, lines and numbers and the effect is rather overwhelming. Graphic output such as dot matrix displays and two-dimensional physical structures depend upon the monitor for quality and tend to look fuzzy on a standard colour screen.

3.1.5 *Printers*

All the packages require a dot matrix printer. Ideally, the printer should have high speed normal data mode for the biological applications and near-letter quality mode for word-processing.

The DNASTAR systems support the APPLE laser printer and a variety of PC-compatible dot matrix printers. The MICROGENIE and IBI packages support any PC-compatible dot matrix printer. The DNASIS package uses IBM and EPSON dot matrix printers.

3.1.6 *Digitizer data input*

Three of the packages supply a sonic digitizing data input device (see Chapter 1). DNASIS supplies a Grafbar digitizer unit with optional stand-alone separate units, keyboard, light table and speech synthesizer. DNASTAR manufactures its own digitizing device, the SEQ.EASY Unit, which incorporates enhanced digitizing and speech synthesis electronics housed inside a light table with an integral keyboard. IBI supply a unit, the Gel Reader, consisting of a Grafbar digitizer unit and a light table. The MICRO-GENIE package does not supply its own digitizer hardware, but specifies a Grafbar sonic digitizer (Model GP-7). DNASIS and DNASTAR digitizing units both have sequence talk-back facilities. The DNASTAR SEQ.EASY unit has the option for using musical tones instead of voice synthesis. The IBI and MICROGENIE packages use an audible cue for prompting in the form of a computer-generated beep.

3.1.7 *Communications with other computers*

Modems can be incorporated into a workstation for linking up to local or remote large computers, either as a separate peripheral device or as an on-board component inside the computer. With the necessary software, the modem can be used for world-wide communications, with optional interchange between voice and data.

In practice, to operate inside a particular country, the modem has to possess the necessary specifications for communicating with at least one of the main national telecommunication standards, such as the CCITT V22 (UK) and Bell 212A (USA) standards. Furthermore, modems have in theory, to conform to national requirements (such as BABT in the UK).

DNASTAR is the only package which offers communciations software. Its software is written for the Hayes Smartmodem and uses a range of data speeds (up to 2400 baud).

3.1.8 *Copy protection*

The program source codes are not supplied by any of the companies marketing software and all the packages are protected in part by copyright and against unlawful duplication by protection systems.

The DNASTAR system uses a key disk protection system, consisting of a specially supplied floppy disk which must be used on entering the programs.

The DNASIS, MICROGENIE and IBI packages are all protected by a copy protection device called a dongle. The *dongle* is an *EPROM* (eraseable programmable read-only memory) chip sealed inside a case and plugged onto a serial port on the computer. If the software is to be transferred to another machine, the dongle has to be removed and reinstated in the other machine.

3.2 **Customer support**

3.2.1 *Installation and getting going*

A general description of the package and the programs it contains, will usually be available in brochure form. The next level of information is gained either by demonstration disks or by live demonstrations of how the package works. Demonstrations may be offered on-site or in a workshop. These can be expected to take a number of hours. Demonstration disks offered to the prospective user go some way towards giving a general idea of the design, but generally cannot supply enough information or evidence for a final decision to be made about the package. Workshops provide an informal halfway house between initial information about the package and a full on-site demonstration. Details should be sought concerning demonstrations and workshops from the companies producing the packages or their marketing agents.

Most companies that sell software packages will either offer to supply the hardware or will give advice on how to acquire and configure the necessary units. Installation of the software will either be performed on-site or straightforward installation procedures will be supplied. Getting the system 'up and running' will then usually be a joint exercise between the user following the prescriptions laid out in a manual and the company providing support, either on-site or at the end of a telephone. Generally, it is no more than a matter of hours or, at most, days before the user is relatively independent of help.

3.2.2. *Updates, support and maintenance*

Support to the user of a software package, a continuing commitment to maintenance of the package and the regular supply of updates and enhancements are variable features from package to package.

For support to the user, telephone contact is generally offered and the user should expect problems to be attended to within a few hours of notifying the company. Good support is easy to identify and future users should ask the opinion of others using the package for details of their experience of the support offered.

Updates and enhancements to existing software (as opposed to database updates, see below) and new applications, are regular features of any software package whose company is actively concerned with the development of the package. These implementations are sent out to users on transferrable media like floppy disks or cartridge tapes.

A laboratory about to invest in a package should find out as much as possible about the company's commitment to its users and the further development of its software. Without this assurance, it is less feasible to make the initial investment in favour of the self-contained microcomputer system rather than a well-supported central mainframe facility.

3.2.3 *Databases*

Two of the four packages described in this chapter offer nucleotide and protein sequence databases.

The DNASTAR package contains the complete GenBank (4) and PIR (5) databases with full annotation information. The MICROGENIE package includes sequence entries only from the GenBank and PIR databases in a compressed format on 12 floppy disks. The sequences can be accessed from each separate disk or transferred permanently onto a hard disk.

The IBI package does not supply any database but has plans to supply GenBank and PIR sequence entries in the future on 15 floppy disks. DNASIS has plans to supply full unabridged nucleotide and protein databases in the future and will incorporate appropriate hardware specifications for this.

3.2.4 *Storge of databases on microcomputers*

(i) *Databases distributed on floppy disk.* The GenBank database is doubling in size every two years. Complete revisions are issued quarterly and updates, monthly. The PIR database is also growing and new databases are being announced. This represents a large quantity of data which has to be updated frequently. For some vendors of molecular biology software, the problem is how to distribute and update this information on floppy disks. An uncompacted database would require over 40 high capacity (1.2 Mbyte) floppy disks. The distribution of data by phone via a modem is also not feasible, since the transmission of only 10 Mbytes of data at a baud rate of 1200 would take about 26 h.

The GenBank contracting organization, Bolt, Berenek & Newman Inc. (4) sells a floppy disk version of the database at a cost of $200 per issue. Data distributed on floppy disks is updated once every six months. To conserve space, these issues are abridged by removing fields of information from the data such as all taxonomic information, features tables, journal article titles and comment, but abridgement of the data reduces its effectiveness to a considerable extent. Data compaction techniques allow a more compact encoding of data without further loss of information. For example, the characters A, G, C, T can be compacted two or three to a byte instead of being represented as one character/byte. In DNASTAR's system, two levels of compression are implemented, (i) replacing one base per byte by two bases per byte, and (ii) replacing a hundred or so frequently repeated phrases with a single byte.

(ii) *Data distribution on tape cartridge.* An alternative medium for the distribution and storage of complete databases is the data tape cartridge technology, used currently in DNASTAR's TAPE-CACHE system. TAPE-CACHE works by making transient use of hard disk space with a software-driven copying facility from tape to disk. Only a small amount of hard disk space is needed, but its size allocation can be reset by the user at any time. Whenever data not currently on the hard disk is needed, TAPE-CACHE copies the needed pages from tape, replacing the historically least active ones.

For their TAPE-CACHE system, DNASTAR uses Tallgrass Technologies 4060 series tape drives with a capacity of 60 Mbytes. There are several reasons why competing technologies are not considered. The Tallgrass unit employs full cyclical redundancy coded error detecting/error correcting format known as PC/T. Data is coded on the tape in a redundant fashion that allows information to be recovered accurately even if the tape has suffered damage, wear or dropout. The high level of data security ensures that tapes written on one Tallgrass unit can be read by others. Less expensive tape units can often only restore data to the hard disk from which they were written and have difficulties with tapes written on other hardware units. In addition, the Tallgrass software allows single named file manipulation between tape and hard disk, not possible with tape streamers.

(iii) *Data security and back-up.* A Tallgrass tape unit is a good investment for the routine backup of the user's data. A hard disk 'crash' can wipe out data whose reacquisition costs can greatly exceed that of a tape unit. A single tape unit can also be shared within several laboratory groups by purchasing an interface card with each computer.

At GenBank's present rate of expansion, the 60 Mbyte capacity of the present tape cartridge provides for four years' expansion. This may be extended a year or two further by data compression techniques and by the development of higher density tapes or by the use of two tapes.

3.3 The software

Commercial packages differ widely in presentation, in design, in the range of features offered and in flexibility.

The broad categories of functions which can be generally expected to feature in a commercial package are as follows:

(i)   Data entry and editing
(ii)   DNA and protein analysis
(iii) Restriction enzyme site analysis and mapping
(iv)   Sequence comparison
(v)   Management and analysis of databases
(vi)   Management of sequencing projects

3.3.1 *Package design and handling*

All four packages can be started from the operating system level by typing a single command word and some familiarity with the operating system directory structure will be required. The user interacts with all programs using simple menus from which a selection is made in order to start each option.

DNASIS PROGRAMS

1. DNA Sequence Input/Update	2. DNA Sequence Connection
3. DNA Sequence Print	4. DNA Sequence Read Out
5. Open Reading Frame	6. Amino Acid Translation
7. Codon Usage Ratio	8. Restriction Enzyme Site Analysis
9. Stacking Site Search	10. Harr Plot Display
11. G/C Rich Region Ratio	12. Tandem Repeat Search
13. Codon Usage Ratio	14. Hairpin Loop Search
15. Palindrome Search	16. Homology Search
17. Restricition Enzyme Physical Map	18. DNA/AMino Acid Data Conversion
19. Hydrophilicity Analysis	20. Protein Homolgy Plot

Figure 1. DNASIS Package. Main menu of programs.

Output may generally be directed either to the printer or to the screen. Both of these options are not always available. The procedures for recording program output is dependent upon the function used. In these respects, the packages do not offer equal scope or flexibility and must be examined individually for performance.

(i) *The DNASIS package.* The current (1986) version of the DNASIS package originated in Japan and has been released in Europe and the USA within the last nine months. It is a comparatively small package and is constructed simply, is under development and is soon to be upgraded. At the time of writing, the maximum DNA sequence length which any individual function within the package can handle is 30 kb, although the upgraded version is expected to offer increased scope (60 kb). The programs included in the present version of the package consist of 20 different functions accessed from a main menu, shown in *Figure 1*. The program required is activated by selecting the appropriate number. Within each option, short submenus are given for selecting suboptions, or individual functions, within the program.

(ii) *The DNASTAR package.* The DNASTAR system is a flexible and comprehensive package entered via a main menu consisting of 10 major subject options, shown in *Figure 2*. Each subject option is highlighted with reverse video and the programs contained within the subject option displayed vertically. Selection of individual programs is made by horizontal and vertical cursor movement and by pressing the ENTER key. Within each program, submenus or question and reply routines enable the user to start each required function. There are also options for performing operating system level activities within the DNASTAR system (utilities such as getting directory information, carrying out DOS commands, reading in test files, setting pathways for downloading data files to remote directories or floppy disks, etc.) and for obtaining general Help and information (IUB and protein codes, etc.). Specific Help related to individual functions is obtained within the functions themselves. Accessing DNASTAR from a user's own subdirectory on the hard disk causes all data files to be down-loaded to that sub-

```
                           DNASTAR PROGRAMS

              EDITSEQ    input, edit, speak sequences
              READSEQ    input sequences from gels with digitizer
              CLONE      splice, catenate, reverse compliment
  EDIT        EDITCODE   create alternate genetic code
              EDITSITE   edit restriction site file
              SITEFILE   create enzyme subset file
              CHCOUNT    character distribution, file checksum
              EDIT       text editor

              MAPSEQ     display sequences with sites, translation, annotation
              QUIKLOOK   produce rapid DNA or protein sequence printout
              MODEL      create cylindrical 3D model of DNA sequence
              BASEDIS    plot nucleotide and dinucleotide frequencies
              TRANS      protein sequence from DNA, codon usage, mol.wt.
  DNA         FINDGENE   determines gene probability from codon usage pattern
              REVTRANS   DNA sequence from protein sequence
              FINDCODE   search for coding lke regions by Fickett method
              FINDPRO    find coding sequence for a known protein in a DNA
              ORF        search for open reading frames
              LOOPS      search for loop structures

              PROTEIN    plot hydrophilicity, charge and helix/turn potential
  PROTEIN     PROBE      design DNA probes from protein sequences
              TITRATE    calculate charge vs pH for proteins

              RETABLE    display restriction enzyme recognition sequences
              SITELIST   scan sequence, list restriction sites
              SITELOOK   display restriction may graphically or in lists
  R-SITES     GELWIZE    predict restriction gel electrophoresis patterns
              DIGIGEL    measure restriction fragments vs standard with digitizer
              RMAP       create restriction map from restriction fragment digests
              SIZEGEL    measure restriction fragments vs standard from keyboard

              COMPARE    compare two DNA sequences by sliding window method
              AACOMP     compare two protein sequences by sliding window method
  COMPARISONS ALIGN      align two DNA sequences by Wilbur & Lipman method
              AALIGN     align two protein sequences by Wilbur & Lipman method
              SEQCMP     align two or many DNA sequences by Martinez/Needleman Wunsch method
              GAP        align DNA or protein sequences interactively, display homologies graphically

              SEQMAN     merge gel readings into a sequence project
  SEQUENCING  SEQMANED   modify a sequence project
              STRATEGY   design optimum Maxam Gilbert sequencing experiments

              GENEMAN    search GenBank and PIR databases
  DATABASE    INDEX      index a database
              NUCSCAN    search GenBank database with a query DNA sequence
              PROSCAN    search PIR database with a query protein sequence

  PHONE       CALL       modem control program
```

Figure 2. DNASTAR Package. Main menu of programs.

directory. There is no upper limit to sequence length in DNASTAR for operations performed on single sequences, such as scanning for restriction sites.

(iii) *The MICROGENIE package.* The MICROGENIE package consists of two main programs called ENTRY and ANALYSIS each of which is entered directly and separately from the operating system level by typing either the word ENTER or the word ANALYZE.

The ENTRY program contains all the data management functions. In the hard disk version of MICROGENIE, a *user password* is requested which then usefully directs all further operations to the user's own *subdirectory*. Within each of the two sections, a menu of options is displayed. Within the ANALYZE section, there are a number of functions for the analysis of DNA and protein sequences. Within the option COMPARE TWO SEQUENCES, a further menu allows a choice of four methods of comparison. The complete set of functions is shown in *Figure 3*.

The maximum sequence length handled by any MICROGENIE function is 60 kb.

(iii) *The IBI package.* The IBI package consists of a main menu containing the editing, DNA sequence analysis, protein analysis and sequence comparison functions contain-

MICROGENIE PROGRAMS

ENTRY SECTION

Record a new Sequence	Organize Search Files	Transform a sequence
List the Directory	Erase a Sequence	Proofread a Sequence
Display a Sequence	Copy a Sequence	Merge two Sequences
Alter a Sequence	Backup a Sequence	Join several Sequences

ANALYSIS SECTION

Analyze a Sequence	Describe Program Procedures
Compare two Sequences	List Directory of Sequences
Examine Program Output	Make Search of Data Bank
Print Program Output	Set Program Parameters

List of Program Procedures in ANALYZE

1. List and number the sequence in various formats
2. Determine nucleotide or amino acid frequencies
3. Determine codon frequencies
4. Translate a DNA sequence in reading frame or reverse translate a protein
5. Translate a DNA sequence in all reading frames
6. Locate sites in a sequence, find fragment lengths and display in tables
7. Locate sites in a sequence and display graphically
8. Locate sites in a sequence and display along sequence
9. Find repeated regions in a sequence
10. Find dyad symmetries in a DNA sequence
11. Find sequence regions rich in specific nucleotides or amino acids
12. Predict protein physical secondary structure (Garnier algorithm)
13. Predict hydrophobic and hydrophilic regions (Hopp/Woods algorithm)

COMPARE SECTION   (subset of Compare two Sequences)

Homology comparison	Alignment of two sequences
Matrix comparison	Protein alignment by similarity

Figure 3. MICROGENIE Package. Main menu of programs and list of procedures.

IBI   PROGRAMS

1	2
Enter/Edit Sequences & Directories	Translate Sequence/Codon Bias Table
Enter/Edit Restriction Enzyme File	Restriction Sites/Map/Fragments
Translate a Sequence & Store Peptide	Search for Homology with Subsequence
Create Codon Bias File	Calculate Base Composition

3	4
Plot Amino Acid/Hydropathy	Forward Homology Matrix
AA compose/Fragments/pI of Peptide	Reverse Homology Matrix
Locate Protein Coding Regions	Automatic Aligned Sequence Segments

Figure 4. IBI Package. Main menu of programs.

ed within four blocks, as shown in *Figure 4*. Within each block, each option is selected by typing the appropriate number of letters and a series of submenus set up the procedure required by the user. The complex subsubmenu options, however, can be tedious to use and mistakes made in starting a required function send the user back to the main menu level, or worse, out of the software package and into the operating system. A useful feature is the availability of separate routines in each block for visualizing a file or redirecting the output (to a file or to the printer).

The maximum length of sequence that can be handled is 60 kb.

3.3.2 *Data management and analysis*

(i) *The DNASIS package.* The data management functions in the DNASIS package include a comprehensive, easy to use sequence editor with a menu of options. DNASIS allows data entry using the digitizer or from the keyboard and also allows the use of speech synthesizer for reading out (talking back) a DNA sequence. DNASIS's present analysis functions, as shown in *Figure 1*, include open reading frame searches in all three protein translation phases, translation of both DNA strands with the option to use 1- or 3-letter protein codes and calculation of codon usage ratios. The current release does not permit the use of alternative genetic codes. There is also an option for displaying G/C rich regions graphically and separate functions for predicting DNA secondary structure, including scanning a sequence for tandem repeats, hairpin loops, palindromes and regions of unusual nucleotide and dinucleotide distribution.

Analyses of protein sequences in DNASIS allow protein hydrophilicity to be determined although there is no option for secondary structure prediction. The present version of DNASIS is further restricted by the limitation that parts of the sequence may not be selected for individual treatment and only the entire data file may in general be analysed.

(ii) *The DNASTAR package.* DNASTAR allows DNA sequence data entry by digitizer, keyboard or modem using the programs READSEQ, EDITSEQ and CALL (see *Figure 2*) and a full screen editor (EDITSEQ) is available for both DNA and protein sequences. Sequence files may be viewed onscreen or printed by the program QUIKLOOK and within each program, there are also options for viewing files or directories. The DNA analysis programs are grouped within the DNA section, protein analysis programs within the PROTEIN section and output may be directed either to the printer or the screen. The option to send output to a file is always possible and files are automatically given a 3-letter extension (as in DOS) defining the contents [e.g. an enzyme subset file (.ENZ), a restriction fragment marker file (.MRK) or a DNA sequence (.SEQ)]. The user supplies an 8-letter file name.

(iii) *The MICROGENIE package.* MICROGENIE's data acquisition and management functions are accessed from the ENTER section of the package. Nucleotide sequences can be entered into the computer using a digitizer or from the keyboard. The editing and management of sequence files is achieved through various functions in the ENTRY section, many of which are well designed and attractively presented. There are useful functions in the ENTRY section, such as PROOFREAD A SEQUENCE to verify data entered via the digitizer, BACKUP SEQUENCES for transfer to a floppy disk in addition to general utilities like erasing and copying files.

The ANALYSIS section contains the DNA and protein analysis functions whose options are invoked from a list specifying the available operations which can be performed on a sequence. Each of the options (Program Procedures) is controlled by a set of user-specified parameters. There is a separate option for creating a complete set of the 24 'Program Parameters' although the default set is provided with the package. The 'Parameter Set' may be used by any of the functions in the ANALYSIS section to avoid having to set individual parameter values each time. This is useful when the same analysis must be repeated frequently and forms a basis for running MICROGENIE in an automatic unattended mode.

The main problems with using 'Parameter Sets' is that they must be pre-specified before a sequence is analysed and they affect a diverse range of MICROGENIE's behaviour, from file formats to the detailed control of the sequence searching algorithm. This diversity can be a source of confusion for new users. The use of 'Parameter Sets' is therefore valuable when a repetitive task can be defined in detail beforehand and a custom 'Parameter Set' constructed, but awkward if the user needs to experiment with the program prior to a more extensive analysis.

MICROGENIE's approach to obtaining program output is cumbersome. To view the output from procedures such as a comparison between two sequences a separate function has to be invoked beforehand (EXAMINE A PROGRAM OUTPUT). The recorded data goes to a universal file called OUTFILE which must be renamed by the user at the operating system level (by an DOS command) in order to prevent it being overwritten by the next recorded file. Other limitations in MICROGENIE are that ENTER options other than Copy and Merge can only be executed after copying files to the user's own directory and there is also a somewhat arbitrary approach to use of the ENTER key. In all editing functions, pressing the ENTER key records changes, so accidental use of the key will overwrite the previous file. In some places, the ENTER key has to be pressed to execute a command, in others typing the command alone will do and pressing the ENTER key will exit the user from the procedure.

(iv) *The IBI package.* IBI functions for data management, selected from Block 1 (see *Figure 4*), include all the required utilities for creating and editing DNA and protein sequences. There is a full screen editor that can also translate a DNA sequence to protein and create a codon usage table. Direct entry of DNA sequence data is accepted from the IBI Gel Reader or from the keyboard. Sequences are stored in files using a format compatible with the GenBank data library and it is therefore easy to incorporate sequences extracted from GenBank into the user's own collection. A minor limitation in the IBI sequence management system is that before a newly entered sequence may be analysed it must be explicitly catalogued in the data management system using a special program function. This introduces an additional step into what is normally a two stage process of data entry and analysis. There is a facility to use alternative genetic codes. It also has functions for determining base composition and mono-, di- and trinucleotide frequencies, strand asymmetry and the pyrimidine/purine ratios. In the protein analysis option, there are functions to perform the reverse translation of a protein sequence, plot hydropathy and antigenic domains, calculate amino acid composition, determine the molecular weight, estimate the protein isoelectric point and map the position of trypsin and cyanogen bromide cleavage sites.

Difficulties were encountered using the IBI software because it is sensitive to the format of typed commands and data. The format of a file name or a restriction enzyme name must be exactly correct and lower case letters cannot be interchanged with upper case letters as the input will be rejected. There is no means of backing up to previous commands and a mistaken input will often take the user back to main block menu. An incorrectly entered command may only be re-entered by returning to the top level block menu. To use a function a second time, for instance to repeat an incorrectly specified analysis, the entire question and answer dialogue must be repeated.

3.3.3 *Restriction enzyme site analysis*

All four packages have comprehensive facilities for scanning sequences for restriction sites. They also include databases of restriction enzyme recognition sites from which customized subsets can be made by the user. The enzyme databases may generally be edited, altered and updated by the user.

The DNASTAR and DNASIS packages allow the direct measurement of the mobilities of DNA fragments recorded on autoradiographs from restriction enzyme digests using a digitizer. The size of the fragments are calculated from standard curves fitted to measurements of mobility of fragments of known length.

Only the DNASIS and DNASTAR packages have programs for the calculation and assembly of a physical map directly from the measurement of restriction enzyme digest data. DNASTAR has recently patented its own algorithm performing this task (R-MAP).

(i) *The DNASIS package.* DNASIS has functions for listing restriction enzyme site data for up to 100 enzymes and for presenting site data graphically along the sequence or in reduced form. The graphic display of sites along the sequence gives enzyme names stacked below the sequence at the position of the cutsite. The usefulness of the restriction site analysis is limited by not allowing selected regions of the sequence to be investigated individually.

(ii) *The DNASTAR package.* DNASTAR has comprehensive facilities for the analysis of restriction sites, which include programs that search for, in addition to definite sites found, 'possible' sites in sequences containing IUB ambiguous base codes (see Section 4), and others for editing enzyme site data, constructing hypothetical mutation sites and for customizing enzyme subset files to use in sequence anlaysis. There are several different functions for the analysis of sites in a sequence. MAPSEQ gives a full display of sites and protein translations in any phase and there are additional options in the R-Site Section for listing and graphically displaying in various ways those sites that were found to be present or absent from the sequence. The program GELWIZE calculates the predicted size and order of fragments in a digest. The program R-MAP assembles a physical map directly from restriction enzyme digest data.

(iii) *The MICROGENIE package.* MICROGENIE performs all searching of sequences for restriction sites and other sequence patterns with pre-constructed 'Search Files' which may be controlled using pre-set parameters. This can be convenient and time-saving if a number of routines are set up in advance for re-scanning the same sequence with different subsets of enzymes or for calculating the results of different digests, but can as stated previously, be a tedious requirement for single searches. There is, at the time of writing, no facility for the measurement of gel fragments by digitizer and gel data has to be entered from the keyboard for further analysis.

(iv) *The IBI package.* The IBI package offers functions for listing restriction enzyme sites in linear and circular sequences using sets of enzymes using either a default set (those which the program will use if no changes are installed by the user) or a customized set of enzymes. A maximum of 20 enzymes is allowed for any searching operation. There is also a function for predicting digestion fragment sizes from up to six enzymes and another to search sequences containing a limited number of IUB base ambiguity codes.

3.3.4 *Sequence comparisons*

All four of the packages reviewed offer more than one method of comparing sequences for structural similarity. The differences between the comparison techniques generally reflect their suitability for either rapid but insensitive searches through a collection of sequences or a sensitive but exhaustive analysis of just two sequences (see Chapter 13). It is left up to the user to select the most appropriate function for his task.

(i) *The DNASIS package.* The DNASIS package has two methods of sequence comparison accessed directly from the main menu (HOMOLOGY SEARCH and PROTEIN HOMOLOGY PLOT). They enable a dot matrix display of compared DNA or protein sequences to be viewed on the screen and can display sequence alignments, with gaps inserted automatically to maximize the alignment.

(ii) *The DNASTAR package.* DNASTAR has separate comparison programs for DNA and protein sequences. In the COMPARISONS section, the ALIGN programs for both DNA and protein sequences offer linear displays of the sequences being compared together with the user selected parameters governing the stringency of the comparison. The protein version of this function performs an assessment of the similarity relationship between non-identical amino acids. The translation from nucleotide to amino acid is scored on the basis of the probability of acceptable mutation and is called a PAM identity (6,7). In the output, identical amino acids and two levels of PAM-related identities are shown in an alignment of the two proteins being compared. There is also a facility to assess similarity on the basis of genetic relatedness, scoring similarity according to the number of point mutations involved in the amino acid replacement. The automatic alignment functions are fast but will reject candidates having too little homology to qualify for output.

The COMPARE programs for both DNA and protein sequences allow the comparison of two sequences with possibly poor overall similarity but with significant and perhaps widely dispersed homologous segments. The comparison is performed upon the pair with one of the sequences offset with respect to the other. Each subsequent comparison is made when this offset has been shifted by an amount defined by a window of aligned characters. The window being compared at any one time is determined by parameters such as size and match percent and the user determines how many times this operation shall be performed. The output consists of a set of quality ranked comparison results. Sequences with poor overall homology can be compared in this way when an automatic comparison optimizing one best result would either create an artificial alignment or reject the comparison. The program GAP is an extension of the operation performed in the COMPARE function and allows the similarity relationships between two sequences to be explored further. The program is interactive and allows the user to scrutinize the structure of specific alignments. There are several different output features, one of which is a dot matrix in which alignments are signified by individual bases or amino acids rather than dots and because of the structure of this matrix, there are no limitations to the length of sequences compared in this way. Finally, there is a function for comparing groups of DNA sequences with one another (SEQCOMP). For these functions, there is an upper limit of a total of 20 kb of sequence material compared.

In the Database section, there is a further set of options for comparing a query sequence with all the sequences in the databases or their subsets, which are described below. The longest sequence compared with a database of sequences is 12 kb.

(iii) *The MICROGENIE package.* MICROGENIE's comparison functions are accessed from the ANALYSIS section under the option to COMPARE TWO SEQUENCES. The HOMOLOGY COMPARISON can then be performed upon either DNA or protein sequences. The parameters which govern the type and stringency of the comparison are set by the user beforehand and matches which qualify are then presented as a list. This comparison function is similar to one of the 'Program Procedures' that allows a search for repeats in a single sequence and it can also be used for searching for a region in a DNA sequence coding for a known protein by comparing the DNA sequence with the protein sequence.

The MATRIX COMPARISON can be used for comparing DNA or protein sequences and the output is shown by a dot matrix display. The maximum total length of two sequences that can be compared by this method is 60 kb.

The final function offered is searching for an optimal alignment between two DNA or protein sequences (ALIGNMENT OF TWO SEQUENCES and PROTEIN ALIGNMENT BY SIMILARITY). Specific recommendations are made about the lengths of sequences that may be compared by this method. The maximum amino acid sequence length is 2000. The protein sequence comparison also assesses their similarity using a single PAM-related matrix (7). The alignment function is also controlled by values in the 'Parameter Set' and is a fast method of confirming homology between two sequences known to possess a certain degree of identity. Candidates are rejected if results of the comparison fall below the qualifying level of homology set by the user.

(iv) *The IBI package.* In this package, the FORWARD HOMOLOGY MATRIX allows comparisons to be made between DNA sequences (maximum length 30 kb) or protein sequences. The program will find an optimum alignment and display its results in a dot matrix form. The user may filter out the less significant matches in the matrix and an additional facility, the REVERSED HOMOLOGY MATRIX (for DNA sequences only) allows the user to search automatically for similarity between reversed sequences (palindromes) or between one sequence and its complement (hairpins). It is also possible to produce an alignment of two sequences based on regions selected using the dot matrix analysis.

The IBI package offers in its manual a helpful introduction to the subject of sequence comparison by computer.

3.3.5 *Sequence database management and analysis*

The reviewed releases of the DNASIS and IBI packages have no provision for supporting any of the nucleotide or protein sequence databases, although both plan to do so in the future. DNASIS will shortly provide abridged versions on floppy disks and eventually plans to incorporate complete nucleotide and protein databases on an optical disk. This presumably anticipates the resolution of current limitations in optical disk technology, which at present is not appropriate for the storage of material which has to be frequently updated. IBI plans to include the GenBank and PIR protein databases in a forthcoming version of the package. It also plans in the longer term, to include database management software and the facility to search the databases for similar sequences.

(i) *The DNASTAR package.* The DNASTAR package includes complete versions of both the GenBank and PIR databases and both are accessed using the program GENEMAN (8). The program opens with a menu with three sublevels; the first allows the user to choose which database to search; the second, what type of search to make and the third; what to do with the data. The search may be initiated over an entire database or any subset, or a subset constructed by the user. Search specifications can be designed to look for keywords in the annotation fields or to look for specified patterns of bases or amino acids, allowing built-in variations such as ambiguous base codes, degrees of permitted mismatch or sequence construction variations, such as specified inserts in the pattern. This option is particularly useful for finding database entries with regions of biological significance, or signal sequences. Both strands of sequence entries are scanned. Having identified candidate sequences in the database, any of the database entry annotation fields may be selectively viewed, printed or extracted as a database subset for subsequent analysis.

Other programs included in the database package are NUCSCAN and PROSCAN for the comparing every sequence in a database or subset with a query nucleotide or protein sequence. The upper limit of sequence length allowed by the program NUSCAN is 12 kb. The PROSCAN program uses the same methods as the ALIGN program to assess protein similarity.

(ii) *The MICROGENIE package.* The MICROGENIE package is distributed with the GenBank and PIR sequence entries which can be accessed using the LIST and DISPLAY functions in the ENTRY section. Query sequences can be compared with the database sequence entries using the MAKE SEARCH function in the ANALYZE section.

The database is organized on disk using several DOS *directories*. The GenBank sequence directories are grouped according to organism of origin (Mammal, Virus etc.) and the PIR sequence directories are organized alphabetically by protein common name. The sequence files within all the directories are organized alphabetically according to the common name of the sequence. There is no duplication of entry, so for example, a sequence stored in the Mammal group will not be found repeated in the Primate group.

Initiating a database search using MAKE SEARCH is very simple. The output on screen consists of the name of the entry found as a candidate match. The function uses only a single match condition for each run however (75% identical bases in the case of nucleic acids and 40% in the case of amino acids). The file containing query sequence must be modified in order to change the nature of the search. There is an upper limit of 2.2 kb for the length of the query sequence in the MAKE SEARCH function.

A useful feature in MICROGENIE is the ability to initiate multiple analyses using 'Program Procedures' and 'Parameter Sets'. Thus several searches may be started either using one query sequence and several subsets of the database, or using several query sequences and one database subset. These may be left to run unattended overnight or through the weekend if required.

3.3.6 *Sequencing project management programs*

The DNASIS, DNASTAR and MICROGENIE packages include automatic sequencing project management programs. The IBI package intends to include one in a future release. The sequencing project management systems are easy to use and can process data almost as quickly as similar systems running on more powerful computers. For example, the data from 12 sequencing gels each containing 200−300 nucleotides can be assembled automatically into one contiguous sequence on a personal computer in a matter of minutes. Since it takes approximately one minute to process the data from 28 similar gels in an equivalent way on a VAX 11/780 (9) the microcomputer based systems can be considered a viable alternative to programs available on more powerful computers (Chapter 8).

The time taken to enter the primary data into the computer is considerably shortened by using a digitizer. To enter data from sequencing gel autoradiogram of a dozen clones using the digitizer and assembling them with a sequence project manager program would take between two and three hours.

(i) *The DNASIS package.* DNASIS's DNA SEQUENCE CONNECTION program automatically assembles contiguous sequence segments from gel files. Some degree of manual intervention by the user is permitted for editing and deleting connected files in the project.

The size of project is limited to the maximum size of file accommodated by the package, currently 30 kb. The sum of all contiguous sequence files assembled is limited to this sequence length.

(ii) *The DNASTAR package.* DNASTAR's program SEQMAN is a fully automated sequencing project assembly program, supporting both Sanger (10) and Maxam−Gilbert (11) sequencing methods. The program includes the facility to exclude a portion of a sequence from being entered into the project database (for example a portion of the vector). It is also possible to seed the project at any point with a previously generated consensus sequence. The status of the project and the data in it can be examined using a number of different options within the program and output may be directed to the screen, printer or files.

SEQMAN will handle files containing up to 1000 bases at a time and will assemble a maximum of 100 contigs in a single project.

Two other programs in DNASTAR's sequencing section are a dedicated sequencing project editor (SEQMANED) and the program STRATEGY, for determining the optimum sequencing strategy when using the Maxam and Gilbert method.

(iii) *The MICROGENIE package.* MICROGENIE's program MERGE is a fully automated sequencing project management program which supports both the Sanger (10) and the Maxam−Gilbert (11) methods. The manual addition or removal of a sequence from the project independently of the automated assembly process requires the user to exit from the program MERGE and to select other functions which allow these operations (JOINING or TRANSFORMING). Recorded results have to be viewed by invoking a separate function, DISPLAY.

MERGE allows files of up to 400 bases to be assembled although seed sequences of any length can be used. The total length of all sequences in the project cannot exceed 60 kb. MICROGENIE's function allows an 'excluded' sequence to be entered into the project and then gels containing it will be excluded from the assembly process. However, this also counts towards the total sequence length handled by the project and therefore the exclusion of a large vector sequence in this way is not advisable.

4. GENE DESIGN WITH THE DNASTAR SYSTEM

In order to illustrate the power and flexibility of microcomputer software for the management and analysis of DNA and protein sequence data, the use of the DNASTAR package to solve a non-trivial laboratory problem in synthetic gene construction is presented. The gene to be designed is intended to yield a hypothetical vaccine for the human respiratory syncytial virus although the methods could equally well apply to the design of many biologically active proteins.

The advent of rapid and efficient methods for oligonucleotide synthesis has made it possible to produce proteins by expression of synthetic genes. One advantage of direct gene synthesis over reliance on a natural gene is lower project costs, but more importantly, gene synthesis affords flexibility. A synthetic gene can be specially designed so that it is easy to manipulate and to improve the yield or characteristics of the protein product. This is normally difficult, or impossible, to achieve by modification of a natural gene.

Since a given protein sequence can be coded for by a large number of alternative gene sequences (a 100 amino acid protein, for example, could be coded for by about 10^{43} different DNA sequences), the choice from among these possibilities with the aid of a computer allows the investigator to design genes incorporating useful features beyond the minimal requirement that the gene code for the desired product.

The starting point for this design is the amino acid sequence of the capsid protein of the human respiratory syncytial virus (12) and the DNA sequence of a suitable expression vector to express the protein in *Escherichia coli*. The design objective is a reasonably short DNA sequence coding for the most probably antigenic sites of the nucleocapsid protein that can be synthesized and inserted into the vector. The product expressed in *E. coli*, could be purified and in principle, used as the active ingredient of a vaccine to confer immunity against the virus itself. The many practical problems are ignored in the interest of demonstrating the computational principles.

4.1 The design objectives

(i)   We wish to identify and target for expression the regions of the protein most likely to confer immunity, i.e. antigenic sites. These are most likely to be on

```
CODE    >VHNZ        467 amino acids              complete       2      13
TITLE   Nucleocapsid protein - Human respiratory syncytial virus
REFER   Elango, N., and Venkatesan, S.; Nucl. Acids Res. 11, 5941-5951, 1983
        (Sequence transtated from the genomic RNA sequence)

SEQ     M A L S K V K L N D T L N K D Q L L S S S K Y T I Q R S T G D S I D T P
        N Y D V Q K H I N K L C G M L L I T E D A N H K F T G L I G M L Y A M S
        R L G R E D T I K I L R D A G Y H V K K A N G V D V T T H R Q D I N G K E
        M K F E V L T L A S L T T E I Q I N I E I E S R K K S Y K K M L K E M G E
        V A P E Y R H D S P D C G M I I L C I A A L V I T K L A A G D R S G L T
        A V I R R A N N V L K N E M K R Y K G L L P K D I A N S F Y E V F E K H
        P H F I D V F V H F G I A Q S S T R G G S R V E G I F A G L F M N A Y G
        A G Q V M L R W G V L A K S V K N I M L G H A S V Q A E M E Q V V E V Y
        E Y A Q K L G G E A G F Y H I L N N P K A S L L T Q F P H F S S V V
        L G N A A G L G I M G E Y R G T P R N Q D L Y D A A K A Y A E Q L K E N
        G V I N Y S V L D L T A E E L T L K T T K D P K P Q T T K S K E V P T
        T K P T E E P T I N T T K T N I T T L L T S N T T G N P E L T S Q M E
        T F H S T S S E G N P S P S Q V S T T S E Y P S S P P N T P R Q
```

Figure 5. PIR protein sequence entry used in designing the vaccine (GENEMAN output).

65

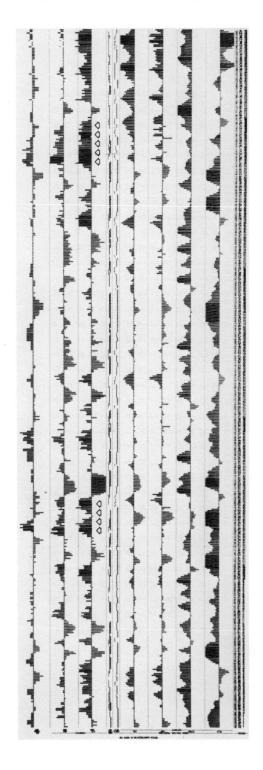

Figure 6. Prediction of the antigenic sites using the PROTEIN program (compressed output).

```
                Editing screen


         ANTSITE.PRO      POS:72              ^E Exit    ^S Search   ^T Talk
                                              ^C Codes   ^B Block    ^G Gel
COMMENTS:
Candidate RSV antigen sites protein sequence

SEQUENCE:
        v          v          v          v       50v
NIEIESRKSYKKMLKEMGEVAPEYRHDSPDCGMEELTLKTTKKDPKPQTT

        v          v          v          v      100v
KSKEVPTTKPTEEPTINTTKTN



                Onscreen Menus


   EXIT/SAVE                  SEARCH                    TALK

Q QUIT                     S SPECIFY SEARCH          P PROOFREAD
X EXIT                     F SEARCH FORWARD          Q QUIETMODE
S SAVE                     B SEARCH BACKWARD         B BEEP MODE
N SAVE AND RENAME          P GOTO POSITION           T TALK MODE
D DONE                     R RELOCATE BEGINNING
C CLEAR                                              Esc to CANCEL
                           Esc to CANCEL             F1 for HELP
Esc to CANCEL              F1 for HELP
F1 for HELP
                                                     GEL

                           BLOCK
   HELP                                            A ASSIGN LANES
                           B SET BLOCK BEGIN       L DEFINE LANE
I IUB DNA CODES            E SET BLOCK END         V VIEW LANE
P PROTEIN CODES            U UNDO BLOCK            F FIND YOUR PLACE
C CURSOR CONTROL           C COPY BLOCK
N NO HELP                  M MOVE BLOCK           G GEL MODE         ON
                           D DELETE BLOCK         T TEXT MODE        OFF
Esc to CANCEL              P PRINT BLOCK
F1 for HELP                W WRITE BLOCK          D DIRECT MODE      ON
                           R READ FILE INTO BLOCK C COMPLEMENT MODE  OFF

                           Esc to CANCEL          N NORMAL ADD       ON
                           F1 for HELP            B BACKWARD ADD     OFF

                                                  ^Home SET KEYPAD
                                                        ORIGIN

                                                  Esc to CANCEL
                                                  F1 for HELP
```

Figure 7. Construction of the working protein sequence using the screen editor EDITSEQ (editing screen and command screen shown).

the surface of the protein and therefore, areas of high hydrophilicity with nearby areas having potential for secondary structure involving β-turn (13,14) are chosen.

(ii)   The synthesized gene is tailored for expression in *E. coli*, even though the virus evolved in the human host. This is done by selection of the codon usage pattern favoured by *E. coli*.

(iii) The synthetic gene is designed for insertion into the expression vector by providing appropriate *Eco*RI and *Hin*dIII restriction sites at either end of the gene and avoiding them internally.

(iv)   To facilitate the introduction of specific mutations into the antigenic site at a later date, regularly spaced unique restriction sites are designed and incorporated into the gene. This allows subsequent excision of small regions (cassettes) for replacement with alternative tailored sequences.

4.2 **The steps of the gene design**

The working protein is accessed from the PIR protein database using the program GENEMAN. A search for the words *Human, Nucleocapsid* and *Syncytial* in the title information field (t) of the database with the GENEMAN command

human/t

AND nucleocapsid/t

AND syncytial/t

allows the protein sequence to be withdrawn (*Figure 5*).

As indicated in the comments associated with the PIR database entry, this protein sequence was determined from an RNA sequence. Although not absolutely necessary, since there is a GenBank entry for the gene, it is worth extracting it since the sequence will provide an interesting comparison with the final designed gene.

To identify candidate antigenic sites in the capsid protein, the PROTEIN program is used to identify hydrophilic stretches of amino acids near to regions that have a potential for β-turns. The graphic output of the program, presented at a much reduced scale in *Figure 6*, shows the amino acid sequence, probable secondary structures indicated as α-helix, β-sheet, β-turn and random coil potentials according to Garnier and Robson (15 – 17), potential secondary structure according to Chou and Fasman (18 – 20) (see also Chapter 12), the indexes of hydrophilicity calculated according to Kyte and Doolittle (21) and Hopp and Woods (13), and the portions of charged residues. Using a single graph that combines all this information allows the user to easily perceive strong correlations between the various potential structures. Scanning this output shows several regions with the desired characteristics. For this vaccine design we have chosen two of these, amino acids 126 to 158 and amino acids 373 to 411, which are indicated with arrows on the figure. These are the two candidate regions with highest hydrophilicity and probably offer the best possibility of serving as antigenically active regions. The working protein sequence file combining only the two antigenic sites is constructed with the aid of the sequence editor, EDITSEQ (*Figure 7*).

4.3 **The expression vector**

The expression vector chosen is the *E. coli* plasmid cloning vector, pUC19 (22) whose sequence is accessed from the GenBank database and automatically filed with the GENEMAN program.

For historical reasons, the numbering system conventionally used for this vector is unusual and rather non-intuitive. The expression module is numbered 'backwards', the numbers decreasing with the distance from the promoter. The cloning sites in the polylinker region are numbered 396 – 447 (see *Figure 8*). The CLONE program is used to create a new vector sequence called P19, derived from the original pUC19 but numbered so that it begins with the promoter and increasing in the usual direction. The CLONE command to accomplish this is

P19 = PUC19 (553 < 554)

The command indicates that the new sequence file is to begin at position 553 in the original vector and to proceed leftwards, taking the reverse complement of the sequence, continuing to position 554. The result is a new sequence file, exactly the same length

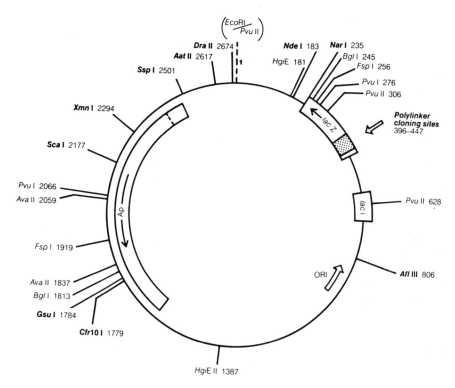

Figure 8. *E. coli* plasmid vector pUC19 (20).

as the original vector sequence but which starts at the beginning of the *lac* promoter, reading in the conventional direction.

The MAPSEQ program is then used to print out the components of the expression module, beginning with the first base of the new working sequence, P19, shown in *Figure 9*. The translation shows that expression in this vector implies the presence of an extra peptide of sequence MTMITP which must inevitably be attached to the beginning of any expressed protein.

4.4 Sites to design into the gene

The next step is to determine which restriction sites are absent from the chosen vector that could serve as unique sites in a gene coding for the vaccine protein. Since the cloning operation will replace the vector's polylinker segment with the target sequence, the CLONE program was used again to produce a new sequence which we call 'p19nolnk'. To excise the polylinker segment between base 100 and base 159 the CLONE command

$$p19nolnk = P19 (1, 100) + (159, rend)$$

is used which catenates the segments from bases 1 to 100 and, bases 159 to the end of the sequence. The new sequence file, P19nolnk, is then scanned with the program SITELIST to obtain a list of all Class II restriction sites not present in the sequence. Since

```
                 lac promoter                    lac operator          Ribosome  Translation
           minus 35                 minus 10               m-RNA start  binding   start
           <---->                   <---->   <-------------------v------>  <---->    v    lac Z gene

CACCCCAGGCTTTACACTTTATGCTTCCGGCTCGTATGTTGTGTGGAATTGTGAGCGGATAACAATTTCACACAGGAAACAGCTATGACCATGATTACGC
----.----+----.----+----.----+----.----+----.----+----.----+----.----+----.----+----.----+----.----+  100
GTGGGGTCCGAAATGTGAAATACGAAGGCCGAGCATACAACACACCTTAACACTCGCCTATTGTTAAAGTGTGTCCTTTGTCGATACTGGTACTAATGCG

                                                                           m  t  m  i  t  p


           POLY LINKER REGION

     H       S       P SAH   X     B     AXS A    K     SBE
     I       P       S ACI   B     A     VMM S    P     SAC
     N       H       T LCN   A     M     AAA P    N     TNR
     3       1       1 112   1     1     111 1    1     121
CAAGCTTGCTAGCCTGCAGGTCGACTCTAGAGGATCCCCGGGTACCGAGCTCGAATTCACTGGCCGTCGTTTTACAACGTCGTGACTGGGAAAACCCTGG
----.----+----.----+----.----+----.----+----.----+----.----+----.----+----.----+----.----+----.----+  200
GTTCGAACGATCGGACGTCCAGCTGAGATCTCCTAGGGGCCCATGGCTCGAGCTTAAGTGACCGGCAGCAAAATGTTGCAGCACTGACCCTTTTGGGACC

   m  l  h  m  c  r  m  t  l  m  d  p  r  v  p  m  m  n  m  l  m  v  v  l  q  r  r  d  w  m  n  p  m
```

Figure 9. Promoter and polylinker regions of the expression vector pUC19 (MAPSEQ output).

ACC1__GTvMK^AC	DRA3__CAC^XXXvGTG	RSR2__CGvGWCCG^
AFL2__CvTTAA^G	EC47__AGCGCT	SAL1__GvTCGA^C
APA1__G^GGCCvC	ECR1__GvAATT^C	SFI1__GGCCX^XXXvXGGCC
ASP1__GvGTAC^C	ECRV__GAT:ATC	SMA1__CCC:GGG
ASU2__TTvCG^AA	ESP1__GCvTXA^GC	SNA1__GTATAC
AVA1__CvYCGR^G	HIN2__GTY:RAC	SNAB__TAC:GTA
AVA3__ATGCAT	HIN3__AvAGCT^T	SPE1__AvCTAG^T
AVR2__CvCTAG^G	HPA1__GTT:AAC	SPH1__G^CATGvC
BAL1__TGG:CCA	KPN1__G^GTACvC	SST1__G^AGCTvC
BAM1__GvGATC^C	MLU1__AvCGCG^T	SST2__CC^GCvGG
BAN2__G^RGCYvC	MST2__CCvTXA^GG	STU1__AGG:CCT
BCL1__TvGATC^A	NAE1__GCC:GGC	STY1__CvCWWG^G
BGL2__AvGATC^T	NCO1__CvCATG^G	TTH1__GACXvX^XGTC
BSM1__GAATG^CXv	NHE1__GvCTAG^C	XBA1__TvCTAG^A
BSS2__GvCGCG^C	NOT1__GCvGGCC^GC	XHO1__CvTCGA^G
BSTE__GvGTXAC^C	NRU1__TCG:CGA	XMA1__CvCCGG^G
BSTX__CCAX^XXXXvXTGG	NSI1__A^TGCAvT	XMA3__CvGGCC^G
CLA1__ATvCG^AT	PST1__C^TGCAvG	

Figure 10. Class II restriction enzymes which do not cut in the vector portion of pUC19 (SITELIST output).

the aim is to use restriction sites for precise genetic manipulations, the feature of SITELIST that creates enzyme subsets is invoked to avoid sites of Class I cutters (see *Figure 10*). The program SITEFILE is then used to create an enyme file containing only those enzymes. In this case, it was inappropriate to select a subset of enzymes automatically, so the manual method was used to stage the specific desired enzymes as shown in *Figure 11*.

4.5 Designing a DNA sequence that codes for the protein

The next step is the creation of DNA sequences by reverse-translation that code for the target protein. This can be done with either of the two programs REVTRANS or EDITCODE. The degeneracy of the genetic code implies that many different DNA sequences can be envisaged, all of which code for the same protein. This ambiguity can be represented in the DNA sequence using the IUB code of 16 degeneracy symbols (23).

Using the IUB code to reverse translate a protein with the standard genetic code is shown in *Figure 12*. For each amino acid, all the unambiguous codons are shown together with the ambiguous triplet. The DNA sequence obtained by reverse translation,

```
                    Enzymes  from  'class2.enz'

AAT2    AVR2-*   CFR1    HAE1    MAE1     NLA3     SCA1     THA1
ACC1-*  BAL1-*   CLA1-*  HAE2    MAE2     NLA4     SCR1     TTH1-*
AFL2-*  BAM1-*   DDE1    HAE3    MAE3     NOT1-*   SFAN     TTH2
AFL3    BAN1     DPN1    HGA1    MBO1     NRU1-*   SFI1-*   XBA1-*
AHA2    BAN2-*   DRA1    HGE2    MBO2     NSI1-*   SMA1-*   XHO1-*
AHA3    BBE1     DRA2    HGIA    MLU1-*   NSP2     SNA1-*   XHO2
ALU1    BBV1     DRA3-*  HHA1    MNL1     NSPB     SNAB-*   XMA1-*
APA1-*  BCL1-*   EC47-*  HIN2-*  MST1     NSPC     SPE1-*   XMA3-*
APY1    BGL1     ECR1-*  HIN3-*  MST2-*   PSS1     SPH1-*   XMN1
ASP1    BGL2-*   ECR2    HINP    NAE1-*   PST1-*   SSP1
ASU1    BIN1     ECRV-*  HNF1    NAR1     PVU1     SST1-*
ASU2-*  BSM1-*   ESP1-*  HPA1-*  NCI1     PVU2     SST2-*
AVA1-*  BSS2-*   FNUH    HPA2    NCO1-*   RSA1     STU1-*
AVA2    BSTE-*   FOK1    HPH1    NDE1     RSR2-*   STY1-*
AVA3-*  BSTX-*   GDI2    KPN1-*  NHE1-*   SAL1-*   TAQ1
```

Selected enzyme is displayed in reverse, those tagged with -* are saved.

```
-------------------------------------------------------------------
  Type arrow keys to move or:

  ^E  up     ^S left   ^A left end   ^R top      <return>    to leave
  ^X  down   ^D right  ^F right end  ^C bottom               L leave
                                                 T tag/untag
                                                 K kill/delete tagged enz.
                                                 U use/save tagged enz.
-------------------------------------------------------------------
```

Figure 11. Selection of the designed enzyme set, using the program SITEFILE.

AA's		Codons	IUB code
A	(Ala)	GCA,GCG,GCC,GCT	GCN
C	(Cys)	TGC,TGT	TGY
D	(Asp)	GAC,GAT	GAY
E	(Glu)	GAA,GAG	GAR
F	(Phe)	TTC,TTT	TTY
G	(Gly)	GGA,GGG,GGC,GGT	GGN
H	(His)	CAC,CAT	CAY
I	(Ilu)	ATA,ATC,ATT	ATH
K	(Lys)	AAA,AAG	AAR
L	(Leu)	CTA,CTG,CTC,CTT,TTA,TTG	YTN
M	(Met)	ATG	ATG
N	(Asn)	AAC,AAT	AAY
P	(Pro)	CCA,CCG,CCC,CCT	CCN
Q	(Gln)	CAA,CAG	CAR
R	(Arg)	AGA,AGG,CGA,CGG,CGC,CGT	MGN
S	(Ser)	AGC,AGT,TCA,TCG,TCC,TCT	WSN
T	(Thr)	ACA,ACG,ACC,ACT	ACN
V	(Val)	GTA,GTG,GTC,GTT	GTN
W	(Trp)	TGG	TGG
Y	(Tyr)	TAC,TAT	TAY

Figure 12. Derivation of the reverse translation codons for the standard genetic code, using EDITCODE programs.

i.e. by substitution of the corresponding IUB triplet for each amino acid, is guaranteed to represent all possible sequences that could code for the protein. In addition, the sequences may also be represented that do not code for the correct protein. This problem is discussed below.

The program EDITCODE allows any genetic code assignments to be used in the reverse translation. This feature is of interest not only for use of unusual genetic codes such as those found in mitochondria, but also because it allows the level of ambiguity in the gene design to be varied by changing the choice of codons used in the reverse translation step. In the current example, three levels of ambiguity were considered:

(i) the fully degenerate standard genetic code (*Figure 12*)
(ii) an intermediate set of codons that occur with reasonable frequency in a large sample of genes highly expressed in *E. coli* (see below)
(iii) the completely non-degenerate sequence obtained by selecting the best *E. coli* codon for each amino acid.

The derivation of these three reverse translation protocols, using the EDITCODE program is shown in *Figure 13*. The selections prepared with the program TRANS, shown in *Figure 14*, are based on the codon usage table derived from a statistical analysis of a large sample of genes strongly expressed in *E. coli* (24).

4.6 Designing the restriction sites into the sequence

The design of a DNA sequence using alternative levels of degeneracy may now begin.

The key information required at this stage is which of the restriction sites found to be absent from the vector can be incorporated into the protein coding segment and where they cut. For this purpose, the program MAPSEQ is used to scan the sequence for

GCA Ala(A)170	CAG Gln(Q)225	UUG Leu(L) 15*	UAA Ter(.)
GCC Ala(A) 57*	Total (Q)258	Total (L)555	UAG Ter(.)
GCG Ala(A)156	GAA Glu(E)384	AAA Lys(K)310	UGA Ter(.)
GCU Ala(A)204	GAG Glu(E)106	AAG Lys(K)105	ACA Thr(T) 14*
Total (A)587	Total (E)490	Total (K)415	ACC Thr(T)198
AGA Arg(R) 0*	GGA Gly(G) 2*	AUG Met(M)164	ACG Thr(T) 25*
AGG Arg(R) 0*	GGC Gly(G)234	Total (M)164	ACU Thr(T)126
CGA Arg(R) 1*	GGG Gly(G) 10*	UUC Phe(F)158	Total (T)363
CGC Arg(R) 91	GGU Gly(G)360	UUU Phe(F) 46	UGG Trp(W) 51
CGG Arg(R) 0*	Total (G)606	Total (F)204	Total (W) 51
CGU Arg(R)282	CAC His(H) 79	CCA Pro(P) 34	UAC Tyr(Y)152
Total (R)374	CAU His(H) 15	CCC Pro(P) 0*	UAU Tyr(Y) 48
AAC Asn(N)261	Total (H) 94	CCG Pro(P)188	Total (Y)200
AAU Asn(N) 17*	AUA Ilu(I) 0*	CCU Pro(P) 18*	GUA Val(V)143
Total (N)278	AUC Ilu(I)339	Total (P)240	GUC Val(V) 37*
GAC Asp(D)286	AUU Ilu(I) 67	AGC Ser(S) 62	GUG Val(V) 90
GAU Asp(D)144	Total (I)406	AGU Ser(S) 8*	GUU Val(V)284
Total (D)430	CUA Leu(L) 1*	UCA Ser(S) 7*	Total (V)554
UGC Cys(C) 22	CUC Leu(L) 35*	UCC Ser(S)126	
UGU Cys(C) 21	CUG Leu(L)469	UCG Ser(S) 14*	
Total (C) 43	CUU Leu(L) 23*	UCU Ser(S)116	
CAA Gln(Q) 33	UUA Leu(L) 12*	Total (S)333	

Data from a total of 6645 amino acids in highly expressed E. Coli Genes
'*' denotes less than 10 % usage for the amino acid

Figure 13. Codon usage of a large set of genes highly expressed in *E. coli* (22) using the EDITCODE program.

	Partially Degenerate Code (better than 10 % usage in Coli)		Non Degenerate Code (only the best in E. Coli)	
AA's	Codons	IUB code	AA's	Codons=IUB codes
A (Ala)	GCA,GCG,GCT	GCD	A (Ala)	GCT GCT
C (Cys)	TGC,TGT	TGY	C (Cys)	TGT TGT
D (Asp)	GAC,GAT	GAY	D (Asp)	GAT GAT
E (Glu)	GAA,GAG	GAR	E (Glu)	GAG GAG
F (Phe)	TTC,TTT	TTY	F (Phe)	TTT TTT
G (Gly)	GGC,GGT	GGY	G (Gly)	GGT GGT
H (His)	CAC,CAT	CAY	H (His)	CAT CAT
I (Ilu)	ATC,ATT	ATY	I (Ilu)	ATT ATT
K (Lys)	AAA,AAG	AAR	K (Lys)	AAG AAG
L (Leu)	CTG	CTG	L (Leu)	CTG CTG
M (Met)	ATG	ATG	M (Met)	ATG ATG
N (Asn)	AAC	AAC	N (Asn)	AAC AAC
P (Pro)	CCA,CCG	CCR	P (Pro)	CCG CCG
Q (Gln)	CAA,CAG	CAR	Q (Gln)	CAG CAG
R (Arg)	CGC,CGT	CGY	R (Arg)	CGT CGT
S (Ser)	AGC,TCA,TCC,TCT	WSH	S (Ser)	TCT TCT
T (Thr)	ACA,ACC,ACT	ACH	T (Thr)	ACT ACT
V (Val)	GTA,GTG,GTC,GTT	GTN	V (Val)	GTT GTT
W (Trp)	TGG	TGG	W (Trp)	TGG TGG
Y (Tyr)	TAC,TAT	TAY	Y (Tyr)	TAT TAT

Figure 14. Reverse translation codes derived from codon usage of genes expressed in *E. coli*, using the TRANS program.

restriction sites that are potentially present in a degenerate sequence (the MAYBE option) and to simultaneously translate the sequence, highlighting any ambiguity in the protein produced with question marks.

MAPSEQ map outputs for five different versions of the gene is presented in *Figure 15*. The first is the natural gene taken from GenBank (Section 4.2) which has typical mammalian codon usage and only a single restriction site, that of *Xba*I. The natural gene would probably be poorly translated in *E. coli* and is certainly not amenable to cassette mutagenesis. The second map is constructed using the best codon for each amino acid according to the codon selection determined from a group of highly expressed genes of *E. coli*. This product is likely to be expressed well, but still has only a single restriction site, *Bam*I. The next two panels show respectively, a moderately degenerate DNA sequence in which the best 10% of the preferred *E. coli* codons have been selected and a fully degenerate sequence allowing all possibilities within the standard genetic code. As expected, the number of potential restriction sites, in addition to the translational ambiguities, increases with the degree of degeneracy allowed.

Inspection of *Figure 15* allows the investigator to identify which recognition sites can readily be incorporated into the sequence. From the possibilities offered, those which would give the best distribution for cassette manipulation are selected. The final gene is then derived by introducing the minimal number of changes to the sequence, based on optimal *E. coli* codon usage, consistent with the introduction of the required sites. In this way, the final codon preference will be as close as possible to that of the highly expressed genes of *E. coli*, a condition favourable to high levels of expression in the bacterial host.

4.7 Checking candidate sites for consistency with the protein

The details of the selection process used in our example are presented in *Figure 16*, which shows the 21 possible restriction sites to be considered in the order they appear in the proposed gene. Since they fall into tightly clustered groups, it is clear that at most, eight well spaced and potentially useful sites could be included in the designed gene.

The eight possible regions and their sites must be examined individually for conflict with one another (since only one of each kind of site is required) and for compatibility with the correct protein sequence.

It is necessary to examine the designed sequences for compatibility with the correct protein, since translational ambiguities occur at serine, leucine and arginine codons which can not be reversibly back-translated. For example, leucine can be coded for by six codons: CTA, CTC, CTG, CTT, TTA and TTG which are represented by YTX in the IUB degeneracy code. The ambiguous triplet YTX could forward translate to either leucine or phenylalanine (TTY). Arginine has six codons back-translating to MGN, which can be translated as either arginine or serine. The six codons of serine back-translate to the triplet WSN, which can be translated to serine, cystine, arginine, threonine, or a terminator. The partially degenerate code restores reversibility to all but serine. In the final gene design, all translational ambiguities must be resolved in favour of the correct protein structure.

In *Figure 16*, it can be seen that eight of the 21 sites are excluded because they imply

DESIGN OF A GENE FOR THE EXPERIMENTAL ANTIGENIC SITE

Restriction Maps Using Enzymes That Don't Cut the Proposed Expression Vector

MAP 1: Natural Gene Sequence of Two Putative Antigenic Sites of the Nucleocapsid Protein of Human Respiratory Syncytial Virus

```
                  X
                  B
                  A
                  1
                                                                                     v 100                                                                                               v 200
AACATTGAGATAGAATCTAGAAAATCCTACAAAAAAATGCTAAAAGAAATGGGAGAGGTAGCTCCAGAATACAGGCATGACTCTCCTGATTGTGGGATGGAAGAACTAACCCTCAAGACAACCAAAAAAGATCCCAAACCTCAAACCACTAAATCAAAGGAAGTACCCACCACCAAGCCCACAGAAGAGCCAACCATCAACACCACCAAAACAAAC
----.----+----.----+----.----+----.----+----.----+----.----+----.----+----.----+----.----+----.----+----.----+----.----+----.----+----.----+----.----+----.----+----.----+----.----+----.----+----.-
TTGTAACTCTATCTTAGATCTTTTAGGATGTTTTTTTACGATTTTCTTTACCCTCTCCATCGAGGTCTTATGTCCGTACTGAGAGGACTAACACCCTACCTTCTTGATTGGGAGTTCTGTTGGTTTTTTCTAGGGTTTGGAGTTTGGTGATTTAGTTTCCTTCATGGGTGGTGGTTCGGGTGTCTTCTCGGTTGGTAGTTGTGGTGGTTTTGTTTG

n  i  e  i  e  s  r  k  s  y  k  k  m  l  k  e  m  g  e  v  a  p  e  y  r  h  d  s  p  d  c  g  m  e  e  l  t  l  k  t  t  k  k  d  p  k  p  q  t  t  k  s  k  e  v  p  t  t  k  p  t  e  e  p  t  i  n  t  t  k  t  n
^                                                                                              ^  ^                                                                                                  ^
126 (-------------------- First Putative Antigenic Segment ------------------------------) 158  373 (---------------------------------- Second Putative Antigenic Segment ----------------------------------) 411
```

MAP 2: Non Degenerate Reverse Translation Using the Best E. Coli Codons

```
                                                                                                                                       B
                                                                                                                                       A
                                                                                                                                       M
                                                                                                                                       1
                                                                                     v 100                                                                                               v 200
AACATTGAGATTGAGTCTCGTAAGTCTTATAAGAAGATGCTGAAGGAGATGGGTGAGGTTGCTCCGGAGTATCGTCATGATTCTCCGGATTGTGGTATGGAGGAGCTGACTCTGAAGACTACTAAGAAGGATCCGAAGCCGCAGACTACTAAGTCTAAGGAGGTTCCGACTACTAAGCCGACTGAGGAGCCGACTATTAACACTACTAAGACTAAC
----.----+----.----+----.----+----.----+----.----+----.----+----.----+----.----+----.----+----.----+----.----+----.----+----.----+----.----+----.----+----.----+----.----+----.----+----.----+----.-
TTGTAACTCTAACTCAGAGCATTCAGAATATTCTTCTACGACTTCCTCTACCCACTCCAACGAGGCCTCATAGCAGTACTAAGAGGCCTAACACCATACCTCCTCGACTGAGACTTCTGATGATTCTTCCTAGGCTTCGGCGTCTGATGATTCAGATTCCTCCAAGGCTGATGATTCGGCTGACTCCTCGGCTGATAATTGTGATGATTCTGATTG

n  i  e  i  e  s  r  k  s  y  k  k  m  l  k  e  m  g  e  v  a  p  e  y  r  h  d  s  p  d  c  g  m  e  e  l  t  l  k  t  t  k  k  d  p  k  p  q  t  t  k  s  k  e  v  p  t  t  k  p  t  e  e  p  t  i  n  t  t  k  t  n
```

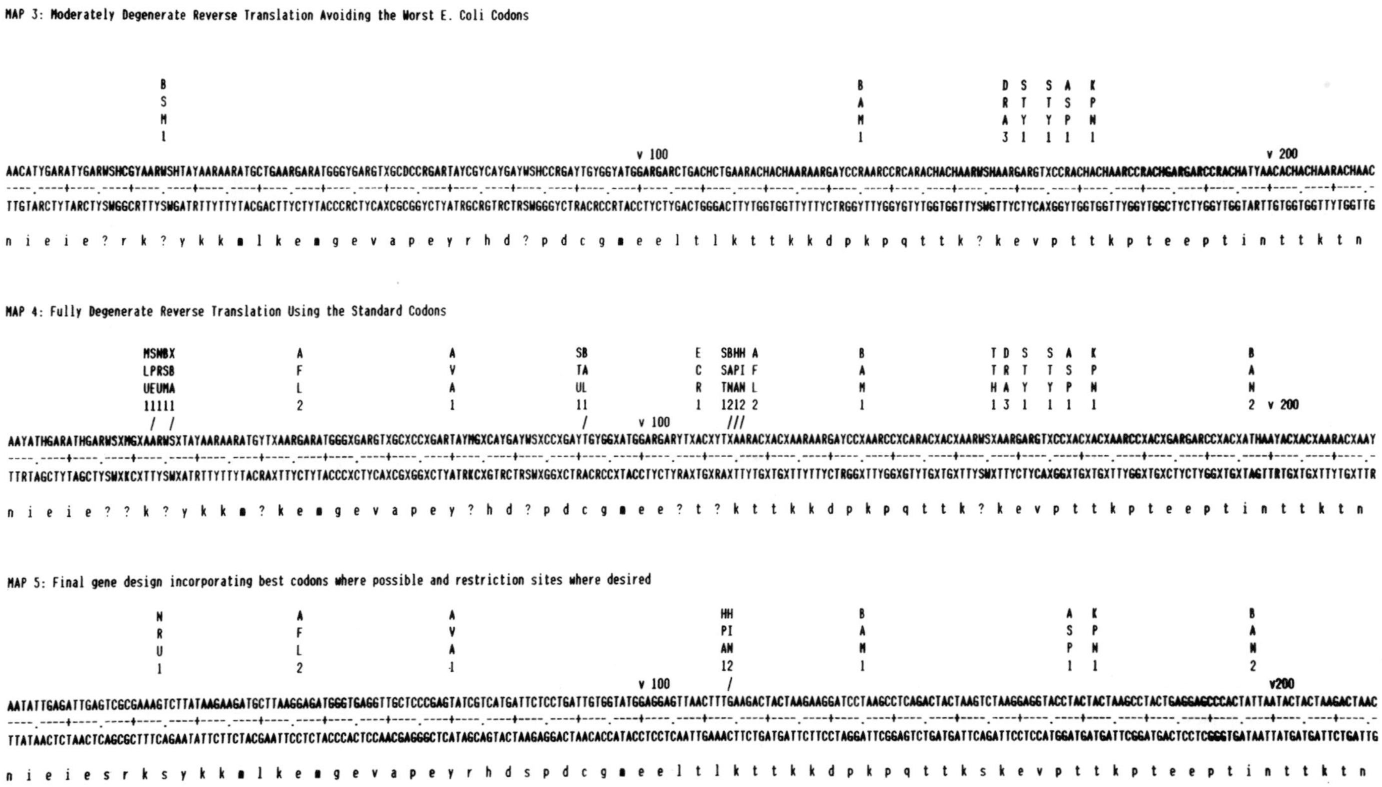

MAP 3: Moderately Degenerate Reverse Translation Avoiding the Worst E. Coli Codons

MAP 4: Fully Degenerate Reverse Translation Using the Standard Codons

MAP 5: Final gene design incorporating best codons where possible and restriction sites where desired

Figure 15. Maps 1—5 (generated by the program MAPSEQ) showing the stages of design of the genes for the experimental antigenic site.

DETAILED EXAMINATION OF POTENTIAL RESTRICTION SITES FOR COMPATIBILITY WITH THE PROTEIN

Restriction Enzyme	Recognition Sequence		Cutsite Position	Translation Status
XBA1	_____TvC T A G^A		16	ok (but not with NRU1)
SPE1	_____AvC T A G^T		16	s-->t not ok
MLU1	_____AvC G C G^T		16	r-->s not ok
NRU1	_____T C G:C G A		18	ok (but not with XBA1)
BSM1_ G A A T G^C Xv			19	s-->c not ok
G A R W S X M G X				
• s r				
AFL2	_____CvT T A A^G		40	ok
Y T X A A R				
l k				
AVA1	_____CvY C G R^G		64	ok
C C X G A R				
p e				
BAL1	_____T G G:C C A		84	s-->w not ok
STU1	_____A G G:C C T		84	s-->r not ok
W S X C C X				
s p				
ECR1	_____GvA A T T^C		103	l-->f not ok
HIN2	_____G T Y:R A C		107	ok if Y is a T (not with SST1)
HPA1	_____G T T:A A C		107	ok (not with SST1)
SST1	_____G^A G C TvC		107	ok (not with HIN2/HPA1)
AFL2	_____CvT T A A^G		112	ok (but dont use because of another AFL2 site above)
G A R Y T X A C X Y T X A A R				
e l t l k				
BAM1	_____GvG A T C^C		129	ok
R G A Y C C X				
d p				
TTH1	____G A C XvX^X G T C		150	s-->v not ok
DRA3	____C A C^X X XvG T G		152	s-->v not ok
STY1	_____C:C W W G G		155	k-->?? not ok
ASP1	_____GvG T A C^C		162	ok
KPN1	_____G^G T A CvC		166	ok
X A C X A A R W S X A A R G A R G T X C C				
t k s k e v p				
BAN2	_____G^R G C YvC		191	ok if R is A and Y is C
G A R C C X				
e p				

CONCLUSION: The gene can be constructed with

NRU1 at 18 or XBA1 at 16 / AFL2 at 40 / AVA1 at 64 / HIN2 at 107 or SST1 at 107
HPA1 at 107 or SST1 at 107 / BAM1 at 129 / ASP1 at 162 / KPN1 at 166 / BAN2 at 191

Figure 16. Examination of potential restriction sites for compatibility with the protein.

an incorrect protein sequence. In one of the groups, neither of the two candidates is possible, therefore only seven well spaced sites can be accommodated. At the remaining positions, at least one site is possible and in some cases more than one choice is available. After elimination of one of the two possible *Afl*II sites, the final design choice is between *Nru*I and *Xba*I in the first group and between *Hind*II/*Hpa*I and *Sst*I in the fourth group. In reality, therefore there are four distinct designs acceptable for this

INSERTION OF THE DESIGNED GENE FOR ANTIGENIC SITES INTO THE EXPRESSION VECTOR

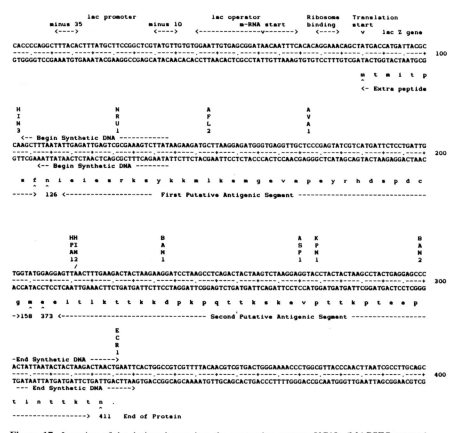

Figure 17. Insertion of the designed gene into the expression vector pUC19. (MAPSEQ output.)

gene and these are summarized as follows:

> *Nru*I at 18 or *Xba*I at 16/*Afl*II at 40/*Ava*I at 64
> *Hind*II at 107 or *Sst*I at 107/*Hpa*I at 107 or *Sst*I at 107
> *Bam*I at 129/*Asp*I at 162/*Kpn*I at 166/*Ban*II at 191

To finish the design, we have chosen the *Nru*I and the *Hind*II/*Hpa*I possibilities and constructed the final gene presented in map 5 of *Figure 15*. The desired pattern of well distributed cassette boundaries is present and the codon usage is almost entirely composed of the best *E. coli* choices.

4.8 Insertion of the gene into the vector

The final step in this hypothetical project is the insertion of the gene into the expression vector as illustrated in *Figure 17* using the MAPSEQ program output. In order to replace the polylinker component of the vector with the designed gene, a few residues are added to each end of the strands designed in order to provide the four bases overhangs a   Quiklook printout of VIRUS.PRO (1, 286)

```
  1   NIEIRSRKSYKKMLKRMGEVAPRYRHDSPDCGMIILCIAALVITKLAAGDRSGLTAVIRRANNVLKNEMKRYKGLLPKDIANSFYEVFEKHPHFIDVFVH  100
101   FGIAQSSTRGGSRVRGIFAGLFMNAYGAGQVMLRMGVLAKSVKNIMILGHASVQAEMBQVVEVBYAQKLGGEAGFYHILNNPKASLLSLTQFPHFSSVVL  200
201   GNAAGLGIMGEYRGTPRNQDLYDAAKAYAEQLKENGVINYSVLDLTAEELTLKTTKKDPKPQTTKSKEVPTTKPTEEPTINTTKTN  286
```

b   Search of RTVIRUS.SEQ (1, 858) for the 100 best regions of fewer than 25 residues to be synthesized as a mixture with fewer than 1024 distinct species none having a melting temperature less than 50 deg C.

RANK	SIZE	T-M	DEGENERACY	RESIDUES			SEQUENCE (5' to 3')
1	20	56	64	457	===>	476	CARGCXGARATGGARCARGT
2	19	54	64	457	===>	475	CARGCXGARATGGARCARG
3	19	52	64	458	===>	476	ARGCXGARATGGARCARGT
4	20	50	64	34	===>	53	AARATGYTXAARGARATGGG
5	17	50	64	46	===>	62	GARATGGGXGARGTXGC
6	18	50	64	457	===>	474	CARGCXGARATGGARCAR
7	18	50	64	458	===>	475	ARGCXGARATGGARCARG
8	18	50	64	459	===>	476	RGCXGARATGGARCARGT
9	23	56	128	31	===>	53	AARAARATGYTXAARGARATGGG
10	20	56	128	43	===>	62	AARGARATGGGXGARGTXGC
11	20	56	128	454	===>	473	GTXCARGCXGARATGGARCA
12	20	56	128	460	===>	479	GCXGARATGGARCARGTXGT
13	22	54	128	32	===>	53	ARAARATGYTXAARGARATGGG
14	19	54	128	44	===>	62	ARGARATGGGXGARGTXGC
15	20	54	128	364	===>	383	TTYATGAAYGCXTAYGGXGC

any key for more---esc for menu--- ^R to restart list--- ^P to turn on the printer

Figure 18. a. The working protein sequence (QUICKLOOK output). b. Output of a screen for best probes using a reverse translation of the working protein sequence (PROBE program).

needed to clone the synthetic gene into the *Hin*dIII and *Eco*RI sites of the vector. A translation terminator, TGA, is added at the end of the gene and one extra residue is also required at the beginning to ensure the proper phase of translation after attachment of the gene.

4.9 **Further investigations**

The program LOOPS can be used to examine the experimental sequence of the vector (bases 10 to 84 in *Figure 9*) for possible secondary RNA structure which might be unfavourable because of potential steric hindrance around the promotor and translational start site regions.

An alternative approach to this project would be to design a probe from the original protein sequence and to screen DNA sequences published in GenBank for potential similarity. To do this, a DNA sequence obtained by back-translating VIRUS.PRO, the file made from the two antigenic regions of the original protein sequence, can be used with the program PROBE to give a selection of the best probes (ranked according to the melting point, degeneracy of the probe and length of the oligonucleotide string). *Figure 18* shows the output of a PROBE scan of the reverse translated VIRUS.PRO protein sequence. GENEMAN can then be used to screen GenBank with the selected probe (in this case, a 20mer) with the command,

CARGCXGARATGGARCARGT %90/q

which initiates a search of the database sequence entries (q) for similarities with the sequence pattern allowing two possible mismatches out of 20 matched bases. A search using 100% match level to the probe sequence retrieves the native DNA sequence entry for the virus. An 80% search (four out of five matches) retrieves 35 entries with similarity to the probe. These entries could then be filed and directly entered as a project into the SEQMAN program to examine their similarities in more detail.

Alternatively, the PROSCAN program could be used to screen the PIR database or the users own database for similarity with the viral protein sequence. Similarity relationships existing between the DNA or the protein sequences could then be further investigated using the programs COMPARE and GAP.

4.10 **Conclusion**

We have used some of the more advanced features of the DNASTAR package to design a gene for a segment of a viral protein that may have a high probability of being an immunogen for the respiratory syncytial virus.

It is clearly attractive to approach projects such as this using a gene sequence generated hypothetically with a computer compared to the difficulty and cost of extracting the immunogenic segments from the natural gene without the convenience of appropriate restriction sites. The complete synthesis of the designed gene requires two strands of 224 residues which must each be divided into pieces of under 100 residues to accommodate the current technological limits of DNA synthesizers. At a cost of about £10 per base, this would still be economical considering the alternative high costs of experimental time and labour involved in the cloning of a natural gene.

5. ACKNOWLEDGEMENTS

I thank Fred Blattner, Department of Genetics, University of Wisconsin, USA and Carol Richardson, DNASTAR Inc., Madison, USA for their collaboration in this project.

6. REFERENCES

1. Blattner,F.R. and Schroeder,J.L. (1984) *Nucleic Acids Res.*, **12**, 615.
2. Queen,C. and Korn,L.J. (1984) *Nucleic Acids Res.*, **12**, 581.
3. Pustell,J. and Kafatos,F.C. (1982) *Nucleic Acids Res.*, **10**, 51.
4. GenBank, The Los Alamos National Laboratory, Bolt Beranek and Newman Inc., 10 Moulton Street, Cambridge, MA 02238, USA.
5. Protein Identification Resource, National Biomedical Research Foundation, Georgetown University Medical Center, 3900 Reservoir Road, N.W., Washington, DC 20007, USA.
6. Lipman,D.J. and Pearson,W.R. (1985) *Science*, **227**, 1435.
7. Dayhoff,M. (1978) *Atlas of Protein Sequence and Structure*. Natural Biomedical Research Foundation, Silver Spring, MD, Vol. 5, supplement 3.
8. Doggett,P.E. and Blattner,F.R. (1986) *Nucleic Acids Res.*, **14**, 611.
9. Staden,R. (1982) *Nucleic Acids Res.*, **10**, 4731.
10. Sanger,F., Nicklen,S. and Coulson,A.R. (1977) *Proc. Natl. Acad. Sci. USA*, **74**, 5463.
11. Maxam,A.M. and Gilbert,W. (1977) *Proc. Natl. Acad. Sci. USA*, **74**, 560.
12. Elango,N. and Venkatesan,S. (1983) *Nucleic Acids Res.*, **11**, 5941.
13. Hopp,T.P. and Woods,K.R. (1983) *Mol. Immunol.*, **20**, 483.
14. Rose,G.D. and Roy,S. (1980) *Proc. Natl. Acad. Sci. USA*, **77**, 4643.
15. Garnier,J. (1978) *J. Mol. Biol.*, **120**, 97.
16. Robson,B. (1974) *Biochem. J.*, **141**, 853.
17. Nishikawa,K. (1983) *Biochim. Biophys. Acta*, **748**, 285.
18. Chou,P.Y. and Fassman,G.D.(1978) *Adv. Enzymol.*, **47**, 45.
19. Chou,P.Y. and Fassman,G.D. (1974) *Biochemistry*, **13**, 222.
20. Chou,P.Y. and Fassman,G.D. (1977) *J. Mol. Biol.*, **115**, 135.
21. Kyte,J. and Doolittle,R.R. (1982) *J. Mol. Biol.*, **157**, 105.
22. Yanisch-Perrou,C., Vieira,J. and Messing,J. (1985) *Gene*, **33**, 103.
23. Nomenclature for incompletely specified bases in nucleic acid sequences (1985) *Eur. J. Biochem.*, **150**, 1.
24. Hawley,D.K. and McClure,W.R. (1983) *Nucleic Acids Res.*, **11**, 2237.

CHAPTER 4

Molecular sequence databases

M.J.BISHOP, M.GINSBURG, C.J.RAWLINGS and R.WAKEFORD

1. INTRODUCTION

Nucleic acid and protein sequences contain a wealth of information of interest to molecular biologists, not surprisingly, as the genome forms the blue-print of the cell. The advent of molecular sequence databases held on computers provides an opportunity for the computer analysis and comparison of all the available sequences. Apart from their obvious role in recording genetic sequences, the sequence databases serve two main functions: to facilitate comparisons of newly determined sequences and to act as a source of data for the generation and testing of hypotheses about molecular sequence organization, function and evolution.

1.1 Growth of sequence databases

Collections of protein sequences have been accumulating since the mid-1960s, the best known publication being the Dayhoff Atlas (1). The ability to rapidly sequence DNA developed in the mid-1970s and has led to an explosive growth in the sequence data available. This has resulted not only in a growth in the amount of nucleic acid sequence information, but also of protein sequence information determined via the DNA sequence. Some putative proteins discovered in this way have yet to be identified and characterized.

Linear at first, the rate of determination of new sequences is entering a phase of exponential growth (*Figure 1*). With serious consideration being given to the determination of the human genome sequence (2) (3×10^9 nucleotides), we can expect a revolution in experimental methods, inlcuding hardware and software, for sequence acquisition (see Chapters 7 and 9). Advances are also being made in the automation of the entire process of DNA sequencing (3). In the future, storage of, and retrieval from, the sequence databases will require innovative hardware and software. Parallel computer processing and optical rather than magnetic storage media are likely to be used. Above all, there must be a re-adjustment in mental attitudes to sequence data. It is unlikely that hard copy publication in scientific journals will continue to be the primary source. The volume of the data and the error rate of transcription will prohibit this. Rather, sequence data will be both collected and transmitted electronically to the central repositories. This is possible today, but the majority of research workers are reluctant to spare the time to do it, or are unaware of the mechanisms (see Section 1.3 and Chapter 5).

1.2 Co-ordination of sequence data

Molecule sequence databases are usually supplied as 'flat' files, i.e. they consist of sequential information tagged with field labels with little or no hierarchical structure.

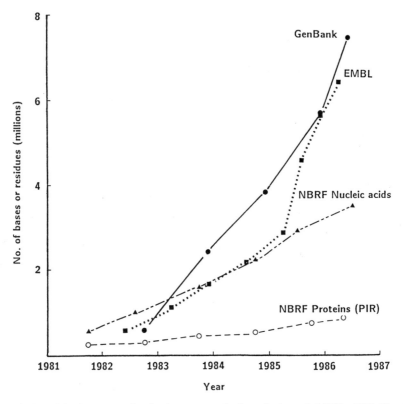

Figure 1. Growth in the content of molecular sequence databases in the period 1982 – 1986. The data for this figure were compiled with the help of Graham Cameron (EMBL), June Bossinger (GenBank) and Katherine Sidman (NBRF).

The intention is that users can further process the databases to a form suitable for locally available database management systems.

In addition to the molecular sequence itself the sequence databases contain textual information (such as bibliographic details) and sequence annotation (describing sequence function or conflicts or uncertainties in the sequence itself as determined by different authors). The activities at the data centre fall into two main classes, data acquisition and database distribution. First, there is the data acquisition step. This requires the publication of the sequence by authors, survey of the literature and extraction of the published data into the format used by the data bank. Extraction of the information is the most time consuming step and results in the compilation of the databases being highly labour intensive. An educational campaign is required to encourage authors to publish their sequence data in a recognized format. Too often the details of organism, strain, biological features of the sequence, etc. are buried deep in the text. It would be to the benefit of all if common protocols were followed. Data bank staff have to assess the quality and reliability of the data, format the data as required by the data bank, and finally enter the new material into both sequence files and indexes. The second task is the distribution of the data bank, currently on magnetic media. Both ½ inch magnetic tapes and 5¼ inch floppy disks are used.

Co-ordination of the different data bank projects is a major difficulty at present (4). It is made more difficult if the information contained within a database cannot be transformed by a computer program into one of the other database formats. This inability to transform the data arises from the lack of standardization of conventions and descriptions. The CODATA Task Group have gone a long way towards solving this problem for the protein sequence databases by establishing a common format for sequence data exchange (4). For nucleic acid sequences, the problem is less acute as the two major data banks (EMBL and GenBank, see Section 3) work in collaboration. Though they use somewhat different formats, they are working towards a common definition and content of the database fields.

Not only do the nucleic acid and protein sequence databases need to be co-ordinated, but there should be pointers to other biologically important data collections such as genetic maps, and databases of cultures, strains and taxonomies of organisms. Integration of all these sorts of information should be a major goal for the future. Even now, about half the known protein sequences are contained within the nucleic acid sequence databases and protein should be derivable from DNA by running a computer program (5).

1.3 Submission of sequence data to the banks

It is to the advantage of everyone if new sequences can be incorporated into the databases as swiftly as possible. To this end, we offer some guidelines on submission of data to the banks.

A laboratory equipped for DNA or protein sequencing contains a considerable amount of costly apparatus. For a few thousand dollars, a microcomputer and a digitizer input device (Chapter 8) will make sequence reading faster and more accurate once the necessary experience has been gained. Sequences may be entered twice to help detect reading errors. It may soon be routine that sequences are read directly into a computer without human intervention (Chapter 9).

The next step (described in detail in Chapter 8) is to assemble the sequences of individual clones into a consensus DNA sequence. It is essential to ensure that the sequence does not represent one of the cloning vectors that has been used to manipulate the DNA. This caution has not always been observed, and some data bank entries do contain vector sequences. It is also necessary to check that the sequence is not meaningless because it represents the joining of two ends of a linear piece of DNA (Chapter 8). Then, the cloned pieces of DNA are assembled, when any errors in sequencing or omissions in coverage of the DNA to be sequenced become apparent. If directed, rather than random 'shotgun' techniques are used, it is still desirable to sequence in both directions to keep a check on errors.

The next step is to annotate the finished sequence. It is sensible to do this in one of the current database formats, as they have been designed for the purpose. A computer program is the most systematic way to do the annotation, as it ensures consistency. The data centres do not, at present, provide such a program to users, but it would be

Table 1. Addresses of the major data banks.

EMBL Nucleotide Sequence Data Library

Address:
European Molecular Biology Laboratory, Postfach 10 22 09, D-6900 Heidelberg, FRG.
Telephone: 6621 387 258
EARN: DATASUBS@DHDEMBL (for data submission), DATALIB@DHDEMBL (for other matters).

GenBank Genetic Sequence Data Bank

Address for contributing to GenBank:
Los Alamos National Laboratory, Los Alamos, NM 87545, USA.
Telephone: 505 667 7510
BITNET: WBG@LANL.ARPA

Address for ordering a copy of GenBank:
Bolt, Beranek and Newman Laboratories Incorporated, 10 Moulton Street, Cambridge, MA 02238, USA.
Telephone: 617 492 2742

NBRF-PIR Protein Sequence Database

Address:
Protein Identification Resource, National Biomedical Research Foundation, Georgetown University Medical Center, 3900 Reservoir Road, N.W. Washington, DC 20007, USA.
BITNET: NBRF@GUVM

---

helpful if they would do so. [The Cambridge online system (Chapter 5) provides a program called NEWSEQ to prepare nucleic acid sequences for submission to EMBL.] It is a common practice in the literature to annotate sequences in such a way that a visually pleasing display is produced (see Chapter 2). Such forms are unsuitable as primary input to the databases, though they can of course be produced from the database information.

Sequences may be sent to the data centres by electronic mail, on magnetic tape, on floppy disk, or even on paper. Inquiries about contributing data should be made to the appropriate data centre, as this may be subject to change. The present details are presented in *Table 1*. Electronic mail is the most convenient route for small amounts of data. For larger amounts it may be possible to use some kind of file transfer protocol. Magnetic tapes should be standard unlabelled tapes written in card image format (fixed block records of length 80, padded with spaces if necessary) in the ASCII code. To ensure that no difficulties are encountered reading the tape, it is best to use modest block sizes (800, 1600 or 3200 bytes) and to record at a density of 1600 bytes per inch. Floppy disks are not very suitable because of the alarming number of formats. They should always be accompanied by a hard copy of the contents in case they cannot be read by machine. IBM PC format is likely to be readable at all centres. In the case of lack of any of these facilities, your contribution will still be welcome in hard copy.

1.4 **The storage and retrieval problem**

We have already discussed the present size and the rate of growth of the molecular sequence databases in Section 1.1. The large size of the databases places constraints on their manipulation in small computers. The present rate of increase should be taken into account when planning to purchase computer equipment to hold the databases. What can sensibly be done depends on the nature of the task (Section 6). To make use of bibliographic information it is, at present, possible to use the major nucleic acid sequence databases in book form (6) or in a compressed form on floppy disks in 360 Kbyte IBM PC format. Retrieval and analysis of sequence databases on the IBM PC is possible as described in Chapter 3. More powerful computers are in routine use for the purpose and access to some of these is described in Chapter 5.

1.5 Supplementary information

As well as distributing the sequence files with their annotation, the data centres distribute a user manual and a number of indexes. The user manual explains the conventions used in the database and describes each of the line types employed. There may also be notes specific to the current release. The format of the databases is not yet stabilized and changes are to be expected in the future, particularly with regard to the mode of description of functional features of the sequences. The indexes may be derived from the sequence databases with an appropriate program. However, it is useful to have them immediately available in machine readable form. They are usually distributed in hard copy form also, which is very useful to those with small computers or if computer access is limited. Indexes usually supplied are by species, by keyword, by author of publication, by journal of publication, by identifier in the database, by accession number in the database, and by single line short descriptions of entries.

1.6 Sequence features

Feature tables provide a precise mechanism for the annotation of sequence data. Another approach is to use punctuation within the sequence itself as found in the PIR database. Although the concept of a feature table is simple enough, trying to represent the more complex features of nucleotide sequences is not easy. The CODATA group recommends a feature table modelled on the EMBL or GenBank table. However, these latter centres are considering the development of a form of feature table which may be significantly different from the present.

2. SEQUENCE SYMBOLS

2.1 Nucleic acid sequence symbols

The recommended symbols for nucleic acid sequences, including uncertainty codes, are shown in *Table 2*. Either upper or lower case letters are permitted. It has been found that human readers tend to make less mistakes if lower case letters are used for sequences, and therefore the use of lower case letters is to be encouraged. For example, c and g are more distinctive than C and G. The uncertainty codes may be represented in computer programs by the four digit binary numbers shown in *Table 2*. Two symbols match if a bit-wise AND operation on their binary representations produces a non-zero result. For further discussion of sequence symbols see Chapter 2.

There is no standard symbol to indicate padding when sequences are aligned. The symbols '-', '*' or '.' are often used.

The Staden codes for uncertainties (Chapter 8) relate to problems of interpretation

Table 2. Codes for nucleotides, including uncertainties.

Code	Nucleotides	Binary ACGT
A	Adenine	1000
C	Cytosine	0100
G	Guanine	0010
T	Thymine	0001
U	Uracil	0001
R	Purine (A or G)	1010
Y	Pyrimidine (C or T/U)	0101
M	A or C	1100
W	A or T/U	1001
S	C or G	0110
K	G or T/U	0011
D	A, G, or T/U	1011
H	A, C, or T/U	1101
V	A, C, or G	1110
B	C, G, or T/U	0111
N	A, C, G, or T/U	1111
	No base (padding)	0000

These symbols are currently under consideration by a commission of the IUPAC-IUB and are expected to be recommended as the standard. In addition, the representation of these uncertainties as four bit binary numbers is given.

of DNA sequencing gels and so serve a different purpose from the standard code.

Codes for the modified bases found in tRNA molecules are shown in *Table 3*. There is a two-letter code and a longer code. The feature table entries in the EMBL database use the two-letter code.

2.2 Protein sequence symbols

The standard symbols for the amino acidcs of genetic sequences are shown in *Table 4*. The one-letter code is used in the PIR database because it is more compact. Human readers find the mnemonic three-letter code more helpful if they do not deal regularly with protein sequences. There is no standard symbol to represent a termination codon (when proteins are displayed in translation alongside DNA). The symbols '*' or '.' are often used.

In the PIR database, punctuation symbols are used in addition to amino acid symbols. The punctuation symbols are described in *Table 5*. The information they convey could instead be coded in a feature table.

3. NUCLEIC ACID SEQUENCE DATABASES

General descriptions of nucleic acid sequence databases have been given by Kneale and Bishop (10) and by Mount (11). The major databases are the EMBL data library and the GenBank genetic sequence data bank. The NBRF also supply a nucleic acid sequence database (12). The DNA database of Japan has been established at the National Institute of Genetics (for information write to Hisao Uchida, Teikyo University, 2-11-1, Kaga, Itabashi-ku, Tokyo 173, Japan).

Table 3. Codes for modified bases found in tRNA molecules taken from Sprinzl and Gauss (7).

Codes		Modified bases
A1	M1A	1-methyladenosine
A2	M2A	2-methyladenosine
A4	I6A	N^6-isopentenyladenosine
A5	MS2I6A	2-methylthio-N^6-isopentenyladenosine
A6	M6A	N^6-methyladenosine
A7	T6A	N-[(9-beta-D-ribofuranosylpurine-6-yl)carbamoyl]threonine
A8	MT6A	N-[(9-beta-D-ribofuranosylpurine-6-yl)N-methylcarbamoyl]threonine
A9	MS2T6A	N-[(9-beta-D-ribofuranosyl-2-methylthiopurin-6-yl)carbamoyl]threonine
G1	M1G	1-methylguanosine
G2	M2G	2-methylguanosine
G3	GM	2'-O-methylguanosine
G4	M22G	2,2-dimethylguanosine
G7	M7G	7-methylguanosine
I1	M1I	1-methylinosine
Q	Q	queuosine
Q1	MAN Q	β-D-mannosylqueuosine
Q2	GAL Q	β-D-galactosylqueuosine
Y1	YW	wybutosine
Y2	O2YW	wybutoxosine
C2	S2C	2-thiocytidine
C3	CM	2'-O-methylcytidine
C4	AC4C	4-acetylcytidine
C5	M5C	5-methylcytidine
C6	M3C	3-methylcytidine
T1	T1	5-methyluridine
T2	S2T	5-methyl-2-thiouridine
T3	TM	2'-O-methyl-5-methyluridine
F	F	pseudouridine
F1	M1F	1-methylpseudouridine
D	D	dihydrouridine
X	X	2-(3-amino-3-carboxypropyl)uridine, (ACP3)U
U1	MAM5U	5-methylaminomethyluridine
U2	S2U	2-thiouridine
U3	UM	2'-O-methyluridine
U4	S4U	4-thiouridine
U7	MCM5U	5-methoxycarbonylmethyluridine
U8	MAM5S2U	5-methylaminomethyl-2-thiouridine
U9	MCM5S2U	5-methoxycarbonylmethyl-2-thiouridine
V1	O5U	uridine-5-oxyacetic acid, (V)
V2	MO5U	5-methoxyuridine
V3	MV	uridine-5-oxoacetic acid methylester
V4	CMNM5U	5-carboxymethylaminomethyluridine
V5	CMNM5S2U	5-carboxymethylaminomethyl-2-thiouridine

Both the two-letter code and the longer code for each base is shown. The two-letter code is used in feature table entries in the EMBL database.

Molecular sequence databases

Table 4. The one-letter and three-letter codes for amino acids.

Codes		Amino acid
A	Ala	Alanine
B	Asx	Aspartic acid or asparagine
C	Cys	Cysteine
D	Asp	Aspartic acid
E	Glu	Glutamic acid
F	Phe	Phenylalanine
G	Gly	Glycine
H	His	Histidine
I	Ile	Isoleucine
K	Lys	Lysine
L	Leu	Leucine
M	Met	Methionine
N	Asn	Asparagine
P	Pro	Proline
Q	Gln	Glutamine
R	Arg	Arginine
S	Ser	Serine
T	Thr	Threonine
V	Val	Valine
W	Trp	Tryptophan
X	X	any amino acid
Y	Tyr	Tyrosine
Z	Glx	Glutamine or glutamic acid

These codes conform to the rules adopted by the Commission on Biochemical Nomenclature of the IUPAC-IUB (8,9).

Table 5. Annotation of protein sequences by punctuation as used in the PIR database.

	Two adjacent amino acids with no punctuation between are connected, as determined experimentally. Spaces are not significant.
( )	Encloses a region, the composition but not the complete sequence of which has been determined experimentally, or encloses a single residue that has been tentatively identified.
=	Indicates)(, the juxtaposition of two regions of indeterminate sequence, while preserving proper spacing between residues (padding character).
/	Indicates that the adjacent amino acids are from different peptides, not necessarily connected. When the amino end of a protein has not been determined, / precedes the first residue. When the carboxyl end has not been determined, / follows the last residue. When) / , / (, or) / (are needed, only / is used.
.	Outside parentheses, indicates the ends of sequenced fragments. The relative order of these fragments was not determined experimentally but is inferred from comparison with similar sequences or other indirect evidence.
.	Within parentheses, indicates that the amino acid to its left has been inferred with at least 90% confidence by similarity with known sequences.
,	Indicates that the amino acid to its left could not be positioned with confidence by comparison with the other sequences. If the structure of related proteins is not known, the position of the amino acids within parentheses is arbitrary.

Table 6. Files on the distribution tape of the EMBL data library.

File	Contents	Name
1	Table of contents.	contents.txt
2	The user manual.	usrman.txt
3	Release Notes.	relnote.txt
4	Species index.	specind.txt
5	Keyword index.	keyind.txt
6	Author index.	autind.txt
7	A list of short descriptions of the entries.	oneln.txt
8	A list of the entry names in the release. Entries new in the current release and those that have been modified are marked.	entries.txt
9	An accession number index.	acno.txt
10	An index to the literature cited.	litlist.txt
11	The entries of the data library.	datfile.dat

3.1 The EMBL data library

The EMBL data library was established in October 1980. The three primary goals were as follows: to make freely available a reliable and comprehensive collection of the published nucleic acid sequence data, to encourage standardization and free exchange of data in the international molecular biology community, and to serve as a European focus for efforts relating to computing and information services in molecular biology (13). Release 1 of the database appeared in April 1982. A list of the files on the distribution tape is given in *Table 6*. The sequence data is supplied either in a single large file, or each entry may be supplied as a separate file. An example of an entry from the EMBL data library is given in *Figure 2*.

3.2 The GenBank genetic sequence data bank

The GenBank genetic sequence data bank was created in 1982 and grew out of the earlier Los Alamos sequence library. GenBank was set up by the National Institute of General Medical Sciences of the National Institutes of Health (NIH). US co-sponsors include the National Cancer Institute, the National Institute of Allergy and Infectious Diseases, the National Institute of Arthritis, Diabetes and Kidney Diseases, and the Division of Research Resources of the NIH, as well as the National Science Foundation, the Department of Energy, and the Department of Defense. The Theoretical Biology and Biophysics Group of Los Alamos National Laboratory gathers, annotates and organizes the data. The data bank is distributed by BBN (see *Table 1*). The project sponsors appoint database curators who assist the team at Los Alamos in collection, organization and accuracy of the information. There is also a government appointed advisory panel to steer the project. The data bank has been described in detail in a number of publications (14−16).

A list of the files on the distribution tape is given in *Table 7*. The sequence data are broken down into separate files according to the organism from which they were derived. An example of an entry from the GenBank data bank is given in *Figure 3*.

A reduced version of GenBank on 5¼ inch floppy disks in IBM PC format is also available. The sequence data, selected annotation and the index files are in compressed form (17), and programs are provided to decode the data. Some software packages

```
ID   HSIL02            56 41    24 JUN 86 14.38.08
**   HSIL02       standard; RNA; 801 BP.
XX
AC   V00564;
XX
DT   19-AUG-1985  (ref 2 added)
DT   11-APR-1983
XX
DE   Human mRNA encoding interleukin-2 (IL-2)
DE   a lymphozyte regulatory molecule.
XX
KW   interleukin; signal peptide.
XX
OS   Homo sapiens (man, homme, Mensch)
OC   Eukaryota; Metazoa; Chordata; Vertebrata; Tetrapoda; Mammalia;
OC   Eutheria; Primates.
XX
RN   [1]   (bases 1-801; enum. 1 to 801)
RA   Taniguchi T., Matsui H., Fujita T., Takaoka C., Kashima N.,
RA   Yoshimoto R., Hamuro J.;
RT   "Structure and expression of a cloned cDNA for human
RT   interleukin-2";
RL   Nature 302:305-310(1983).
XX
RN   [2]
RA   Devos R., Plaetinck G., Cheroutre H., Simons G., Degrave W.,
RA   Tavernier J., Remaut E., Fiers W.;
RT   "Molecular cloning of human interleukin 2 cDNA and its expression
RT   in E. coli";
RL   Nucl. Acids Res. 11:4307-4323(1983).
XX
FH   Key          From     To        Description
FH
FT   MSG           1       801       messenger RNA of interleukin-2
FT   CDS          48       506       coding sequence of interleukin-2
FT                                   (48 is 1st base in codon)
FT                                   (506 is 3rd base in codon)
FT   CDS          48       107       coding sequence of signal peptide
FT                                   (48 is 1st base in codon)
FT                                   (107 is 3rd base in codon)
XX
SQ   Sequence  801 BP;   282 A;   147 C;   114 G;   258 U.
     AUCACUCUCU UUAAUCACUA CUCACAGUAA CCUCAACUCC UGCCACAAUG UACAGGAUGC
     AACUCCUGUC UUGCAUUGCA CUAAGUCUUG CACUUGUCAC AAACAGUGCA CCUACUUCAA
     GUUCUACAAA GAAAACACAG CUACAACUGG AGCAUUUACU GCUGGAUUUA CAGAUGAUUU
     UGAAUGGAAU UAAUAAUUAC AAGAAUCCCA AACUCACCAG GAUGCUCACA UUUAAGUUUU
     ACAUGCCCAA GAAGGCCACA GAACUGAAAC AUCUUCAGUG UCUAGAAGAA GAACUCAAAC
     CUCUGGAGGA AGUGCUAAAU UUAGCUCAAA GCAAAAACUU UCACUUAAGA CCCAGGGACU
     UAAUCAGCAA UAUCAACGUA AUAGUUCUGG AACUAAAGGG AUCUGAAACA ACAUUCAUGU
     GUGAAUAUGC UGAUGAGACA GCAACCAUUG UAGAAUUUCU GAACAGAUGG AUUACCUUUU
     GUCAAAGCAU CAUCUCAACA CUAACUUGAU AAUUUAAGUGC UUCCCACUUA AAACAUAUCA
     GGCCUUCUAU UUAUUUAAAU AUUUAAAUUU UAUAUUUAUU GUUGAAUGUA UGGUUUGCUA
     CCUAUUGUAA CUAUUUAUUUCU UAAUCUUAAA ACUAUAAAUA UGGAUCUUUU AUGAUUCUUU
     UUGUAAGCCC UAGGGGCUCU AAAAUGGUUU CACUUAUUUA UCCCAAAAUA UUUAUUAUUA
     UGUUGAAUGU UAAAUAUAGU AUCUAUGUAG AUUGGUUAGU AAAACUAUUU AAUAAAUUUG
     AUAAAUAUAA AAAAAAAAAA C
//
```

Figure 2. An entry from the EMBL data library.

for the IBM PC support this format. GenBank Release 44 of August 1986 required
26 disks and this form of distribution may not remain feasible for long. It could remain
useful for subsets (for example, virus sequences only).

Table 7. Files on the distribution tapes of the GenBank genetic sequence data bank.

1.   A file containing the user notes.
2.   A short directory of the data bank.
3.   A long directory of the data bank.
4.   A brief listing of those entries in the release that are new or that have undergone substantial revisions in the course of the two most recent releases.
5.   An index of the entries in the database according to their accession numbers.
6.   An index of the entries in the database according to the keyword phrases assigned to them.
7.   An index of the entries in the database according to the authors of the citations listed in each entry.
8.   An index of the entries in the database according to the journal citations listed in each entry.
9.   Primate sequence entries.
10.   Rodent sequence entries.
11.   Other mammalian sequence entries.
12.   Other vertebrate sequence entries.
13.   Invertebrate sequence entries.
14.   Plant sequence entries (including fungi and algae).
15.   Sequence entries for eukaryotic organelles.
16.   Bacterial sequence entries.
17.   Structural RNA sequence entries.
18.   Viral sequence entries.
19.   Phage sequence entries.
20.   Synthetic and chimeric sequence entries.
21.   Unannotated sequence entries.
22.   Complete Epstein−Barr virus sequence entry.

3.3 The CODATA recommendations

As a result of a meeting in Paris in December 1984, the CODATA Task Group on Coordination of Protein Sequence Data Banks recommended a common format for sequence data exchange. This is closely modelled on the EMBL and GenBank database formats and so will be described along with them. It would be desirable if a common format were accepted for all sequence databases, but this should not be so rigid as to forbid the incorporation of new ideas.

In addition to line identifiers, subidentifiers introduced by a ' # ' (number-sign) character are used. Multiple items under a single identifier are separated by ' \ ' (backslash). Spaces separate items and are not otherwise significant except that at the beginning of a line they indicate continuation. (This is like GenBank, and unlike EMBL where all text lines have a two letter identifier and all sequence lines start with 5 spaces.) An example of an entry in the proposed CODATA format is shown in *Figure 4*.

3.4 Line structure

A database file may contain many concatenated entries, so the beginning and end of each entry has to be clearly distinguished. The beginning is distinguished by an entry identifier and the end by '//'. Each record (i.e. line) contains characters from the ASCII set and may be up to 80 characters long. The equivalences between database line identifiers are given in *Table 8*. An account of the line types follows.

```
LOCUS       HUMA1AT4      292 BP    DNA              UPDATED   03/12/84
DEFINITION  HUMAN ALPHA 1-ANTITRYPSIN GENE: 3' TERMINUS.
ACCESSION   J00067
KEYWORDS    ALPHA-1-ANTITRYPSIN; ANTITRYPSIN; PROTEASE INHIBITOR.
SEGMENT     4 OF 4
SOURCE      HUMAN CDNA TO LIVER MRNA [1],[3]; GENOMIC DNA [2].
  ORGANISM  HOMO SAPIENS
            EUKARYOTA; METAZOA; CHORDATA; VERTEBRATA; TETRAPODA; MAMMALIA;
            EUTHERIA; PRIMATES.
REFERENCE   1  (BASES 1 TO 274)
  AUTHORS   KURACHI,K., CHANDRA,T., FRIEZNER DEGEN,S.J., WHITE,T.T.,
            MARCHIORO,T.L., WOO,S.L.C. AND DAVIE,E.W.
  TITLE     CLONING AND SEQUENCE OF CDNA CODING FOR ALPHA 1-ANTITRYPSIN
  JOURNAL   PROC NAT ACAD SCI USA 78, 6826-6830 (1981)
REFERENCE   2  (BASES 113 TO 292)
  AUTHORS   LEICHT,M., LONG,G.L., CHANDRA,T., KURACHI,K., KIDD,V.J.,
            MACE,M.JR., DAVIE,E.W. AND WOO,S.L.C.
  TITLE     SEQUENCE HOMOLOGY AND STRUCTURAL COMPARISON BETWEEN THE CHROMOSOMAL
            HUMAN ALPHA 1-ANTITRYPSIN AND CHICKEN OVALBUMIN GENES
  JOURNAL   NATURE 297, 655-659 (1982)
REFERENCE   3  (BASES 113 TO 287)
  AUTHORS   ROGERS,J., KALSHEKER,N., WALLIS,S., SPEER,A., COUTELLE,CH.,
            WOODS,D. AND HUMPHRIES,S.E.
  TITLE     THE ISOLATION OF A CLONE FOR HUMAN ALPHA 1- ANTITRYPSIN AND THE
            DETECTION OF ALPHA 1- ANTITRYPSIN IN MRNA FROM LIVER AND LEUKOCYTES
  JOURNAL   BIOCHEM BIOPHYS RES COMMUN 116, 375-382 (1983)
COMMENT     ALPHA 1-ANTITRYPSIN IS AN IMPORTANT PROTEASE INHIBITOR PRESENT IN
            MAMMALIAN BLOOD.  CORRESPONDING REGIONS OF HUMAN AND BABOON ALPHA
            1-ANTITRYPSIN MRNAS AND THEIR AMINO ACID SEQUENCES ARE GREATER THAN
            96% HOMOLOGOUS.  SEE <MNKA1AT> AND OTHER <HUMA1AT> LOCI.  [3]
            CALCULATES THAT HUMAN LEUKOCYTES PRODUCE 0.15% AS MUCH ALPHA 1-AT
            MRNA AS HUMAN LIVER.
                THE MUTATION NOTED AT BASE 154 CHANGES A GLU CODON.  TO AN ASP
            CODON.  THE RESULTING CHANGE IS NEUTRAL.  THERE IS ALSO A SLIM
            POSSIBILITY THAT THE BASE CHANGE IS MERELY DUE TO A CONFLICT.
FEATURES       FROM  TO/SPAN    DESCRIPTION
     PEPT   <    1     211      ALPHA 1-ANTITRYPSIN (EXON 4, PARTIAL)
     SITES
       PEPT/PEPT   1      0     ALPHA 1-AT MATURE PEPT UNSEQUENCED/SEQUENCED
       REFNUMBR    2      3     NUMBERED CODON 326 IN [1]
       REFNUMBR  113      1     NUMBERED 1231 IN [2]; ZERO NOT USED
       REFNUMBR  113      3     NUMBERED CODON 363 IN [3]
       MUT       154      1     A IN [1],[2]; C IN [3]
       PEPT<-    211      1     ALPHA 1-AT MATURE PEPT (EXON 4) END
       CONFLICT  258      1     C IN [1],[3]; T IN [2]
       MRNA<-    292      1     ALPHA 1-AT MRNA END (POLY-A SITE)
BASE COUNT       72 A    94 C     59 G      67 T
ORIGIN      ABOUT 3 KB AFTER <HUMA1AT3>.
        1 ACCCCTGAAG CTCTCCAAGG CCGTGCATAA GGCTGTGCTG ACCATCGACG AGAAAGGGAC
       61 TGAAGCTGCT GGGGCCATGT TTTTAGAGGC CATACCCATG TCTATCCCCC CCGAGGTCAA
      121 GTTCAACAAA CCCTTTGTCT TCTTAATGAT TGAACAAAAT ACCAAGTCTC CCCTCTTCAT
      181 GGGAAAAGTG GTGAATCCCA CCCAAAAATA ACTGCCTCTC GCTCCTCAAC CCCTCCCCTC
      241 CATCCCTGGC CCCCTCCCTG GATGACATTA AGAAGGGGTT GAGCTGGTCC CT
//
```

Figure 3. An entry from the GenBank genetic sequence data bank. In this example, only upper case letters are used. The data bank is also supplied in a form which uses both upper and lower case letters.

Table 8. Line structure of nucleic acid sequence databases. Where they exist, the EMBL, GenBank and CODATA equivalents are shown.

EMBL	GenBank	CODATA
ID	LOCUS	ENTRY
		# Length
		# Checksum
AC	ACCESSION	ACCESSION
DE	DEFINITION	NAME
		ALTERNATE-NAME
		INCLUDES
		GENE-NAME
		MAP-POSITION
	SEGMENT	
DT		DATE
OS	SOURCE	SPECIES
	ORGANISM	
		# Strain
		# Plasmid
		# Clone
		# Tissue
		# Life-cycle
		HOST
OC		TAXONOMY
		SUPERFAMILY
	REFERENCE	REFERENCE
RN		# Number
RA	AUTHORS	# Authors
RL	JOURNAL	# Citation
RT	TITLE	# Title
CC	COMMENT	COMMENT
KW	KEYWORDS	KEYWORDS
FH	FEATURE	FEATURE
	SITES	
FT		
		INTRONS
		START-CODON
	BASE COUNT	SUMMARY
		# Molecular-weight
		# Length
		# Checksum
XX		
SQ		SEQUENCE
	ORIGIN	
//	//	//

3.4.1 *The entry line*

This is always the first line of an entry. The line identifier is 'ID' (EMBL), 'LOCUS' (GenBank) or 'ENTRY' (CODATA). The second item is a unique entry identification code. If each entry is a separate computer file, then this code usefully serves as a file name. Unfortunately, some computer systems do not allow file names longer than eight characters, while CODATA recommend 10 characters as the maximum code length.

```
ENTRY          OKBOG        Protein  #Length 670 #Checksum 5530
NAME           cGMP-dependent protein kinase (EC 2.7.1.37) - Bovine
DATE           17-May-1985  #Sequence 17-May-1985  #Text 27-Nov-1985
SPECIES        Bos taurus #Common-name ox
REFERENCE      Sequences of residues      1-17, 89-374, and 407-670
   #Authors    Takio K., Wade R.D., Smith S.B., Krebs E.B., Walsh, K.A.
                  Titani, K.
   #Journal    Biochemistry (1984) 23: 4207-4218
REFERENCE      Sequence of residues 13-104
   #Authors    Takio, K., Smith, S.B., Walsh, K.A., Krebs, E.G., Titani, K.
   #Journal    J. Biol. Chem. (1983) 258:5531-5536
REFERENCE      Sequence of residues 373-409
   #Authors    Hashimoto, E., Takio, K., Krebs, E.G.
   #Journal    J. Biol. Chem. (1982) 257: 727-733
COMMENT        The protein, isolated from lung, is a dimer of identical chains.
SUPERFAMILY
   #Name       cAMP-dependent protein kinase regulatory chain
                  #Residues 102-340
   #Name       kinase-related transforming protein
                  #Residues 475-599
KEYWORDS       acetylation\ phosphoprotein\ cGMP\
                  serine-specific protein kinase
FEATURE
   1                          #Modified-site acetylated amino end\
   42                         #Disulfide-bonds interchain\
   58                         #Binding-site phosphate\
   1-101                      #Domain dimerization <DIM>\
   102-219                    #Domain cGMP-binding 1 <GB1>\
   320-340                    #Domain cGMP-binding 2 <GB2>\
   341-474                    #Domain ATP-binding <APB>\
   475-599                    #Domain catalytic <CAT>
COMMENT        These boundaries are approximate.
SUMMARY        #Molecular-weight 76287 #Length 670 #Checksum 5530
SEQUENCE
                    5        10        15        20        25        30
     1 S E L E E D F A K I L M L K E E R I K E L E K R L S E K E E
    31 E I Q E L K R K L H K C Q S V L P V P S T H I G P R T T R A
    61 Q G I S A E P Q T Y R S F H D L R Q A F R K F T K S E R S K
    91 D L I K E A I L D N D F M K N L E L S Q I Q E I V D C M Y P
   121 V E Y G K D S C I I K E G D V G S L V Y V M E D G K V E V T
   151 K E G V K L C T M G P G K V F G E L A I L Y N C T R T A T V
   181 K T L V N V K L W A I D R Q C F Q T I M M R T G L I K H T E
   211 Y M E F L V S V P T F Q S L P E E I L S K L A D V L E E T H
   241 Y E N G E Y I I R Q G A R G D T F F I I S K G K V N V T R E
   271 D S P N E D P V F L R T L G K G D W F G E K A L Q G E D V R
   301 T A N V I A A E A V T C L V I D R D S F K H L I G G L D D V
   331 S N K A Y E D A E A K A K Y E A E A A F F A N L K L S D F N
   361 I I D T L G V G G F G R V E L V Q L K S E E S K T F A M K I
   391 L K K R H I V D T R Q Q E H I R S E K Q I M Q G A H S D F I
   421 V R L Y R T F K D S K Y L Y M L M E A C L G G E L W T I L R
   451 D R G S F E D S T T R F Y T A C V V E A F A Y L H S K G I I
   481 Y R D L K P E N L I L D H R G Y A K L V D F G F A K K I G F
   511 G K K T W T F C G T P E Y V A P E I I L N K G H D I S A D Y
   541 W S L G I L M Y E L L T G S P P F S G P D P M K T Y N I I L
   571 R G I D M I E F P K K I A K N A A N L I K K L C R D N P S E
   601 R L G N L K N G V K D I Q K H K W F E G F N W E G L R K G T
   631 L T P P I I P S V A S P T D T S N F D S F P E D N D E P P P
   661 D D N S G W D I D F
//
```

Figure 4. An example of the proposed CODATA common format representation.

EMBL give the data class (a measure of reliability: standard, unreviewed or preliminary), whether DNA or RNA, and the length of the sequence. GenBank give sequence length, molecule type, and date of last update of the entry. CODATA recommend giving molecule type, and provide subidentifiers for length and a sequence check-sum calculated by the method used in the UWGCG package of Devereux *et al.* (18).

3.4.2 *The accession line*

The accession line provides a stable way of identifying entries from release to release. This allows the tracking of data when entries are merged or split. The first listed accession number, not the entry code, should be used when citing information from the data banks. The line identifiers are 'AC' (EMBL) and 'ACCESSION' (GenBank and CODATA). This line is used to produce the accession number index.

3.4.3 *The description line*

This is a short description of the contents of the database entry. These lines are useful for listing the contents of the database and are output by retrieval software (Section 6). They are therefore concise and give the source organism and the name of the gene or molecule. These lines are used to produce the entry description or short directory index. The line identifier is 'DE' (EMBL), 'DEFINITION' (GenBank) and 'NAME' (CODATA). CODATA suggest an 'ALTERNATE-NAME' line for cases where a standard nomenclature has not been established (e.g. cytochrome f, cytochrome c553), 'INCLUDES' when a molecule consists of distinct subunits and 'GENE-NAME' and 'MAP-POSITION' to refer to genetic map databases. The GenBank line 'SEGMENT' gives information on the order in which an entry appears in a series of discontinuous sequences from the same molecule.

3.4.4 *The date line*

The date line gives the date of addition or update of a database entry. The EMBL identifier is 'DT' and the CODATA recommendation 'DATE'. GenBank have latest update information on the 'LOCUS' line.

3.4.5 *The source organism line*

This line specifies which organism was the source of the sequenced material. The line identifiers are 'OS' (EMBL), 'SOURCE' (GenBank) with subidentifier 'ORGANISM', and 'SPECIES' (CODATA) together with 'HOST' where appropriate. Subidentifiers 'Strain', 'Plasmid', 'Clone', 'Tissue', and 'Life-cycle' are suggested by CODATA.

3.4.6 *The organism classification line*

This line gives the classification of the source organism listed as the nodes of a taxonomic tree from root to tip. This means that the broadest category is quoted first. The line identifier is 'OC' (EMBL) or 'TAXONOMY' (CODATA). In GenBank the information is under 'ORGANISM'. The organism classification line is used to produce a taxonomic index.

3.4.7 *The literature citation lines*

These lines contain the literature citations for published material in the databases. GenBank and CODATA introduce this with 'REFERENCE' and continue with subidentifiers

Change indicators

UNSURE   authors report uncertain assignment of base(s). This key is also used when the sequence has not been unambiguously readable during data entry, and verification has not yet been possible.

CONFLICT   different papers report differing sequences.

VARIANT   authors report that sequence variants exist.

MUTANT   paper reports existence of a mutation at this point.

ALLELE   paper reports existence of mutations not necessarily expressed in phenotype.

GAP   indicates the extent of a sequence gap whose length is approximately known.

REVISION   extent of bases that have been corrected by later publications or author communications.

RPT   direct repeat.

Regions

MSG   extent of spliced messenger RNA.

TRANSCR   extent of primary non spliced transcript.

IVS   extent of intervening sequence (spliced out of mature messenger).

CDS   extent of coding sequence from the first base of the first codon (or the first base after an IVS) to the last base of the last translated codon (or the last base before an IVS). Note that programs written to perform sequence translation must consider intervening sequences as well to avoid frame shift errors.

TRNA   extent of bases which are transcribed and included in the mature tRNA.

RRNA   extent of bases which are transcribed and included in the mature rRNA.

CAP   first base of a messenger RNA where a cap is added after transcription. The chemical nature of the cap is given in the description.

Signals

PRM   extent of reported promoter (RNA polymerase binding site, Pribnow box, etc.)

RBS   extent of reported ribosome binding site (e.g. Shine/Dalgarno region).

OPR   extent of transcription regulation signal.

ATTEN   extent of attenuation signal.

ORGRPL   reported origin of replication (either first base replicated or extent of regulatory signal).

POLYA   base at which polyadenylation occurs (in eukaryotes).

MODBASE   modified base reported at this site. The description field gives the code for the modification (see Table 3).

Other features

TPOSON   extent of reported transposon.

INSSQ   extent of reported insertion sequence.

INVREP   extent of reported inverted repeat.

SOMREC   base (in the listed strand) to the left of a reported somatic recombination event.

PROVRL   extent (in reported sequence) of proviral DNA.

CELL   extent (in reported sequence) of cellular DNA if there is a compound listing of proviral and host sequences.

SITE   extent of any other signal or interesting region that has been reported.

Figure 5. Definition of the feature table keys used in the EMBL data library entries. Other key names may be added to this list if required in the future.

```
Field 1.  Key for the kind of feature that is coded.
          pept. Protein coding region ( initiation and
                termination codons are included.)
          sigp. Signal peptides (initiation codon is not
                included.
          matp. Coding regions that correspond to mature
                protein products ( initiation and
                termination codons are not included )
          trna. Transfer RNA coding regions.
          rrna. Ribosomal RNA coding regions.
Field 2.  Number of the first base in the coding region.
Field 3.  Number of the last base in the coding region.
No reference is made to the numbering in the original
citation. When a feature is known to extend beyond the
sequenced region the endpoint specification is preceded by
< ( for features continuing beyond the 5' end ) or by > ( for
features continuing beyond the 3' end ). Unknown endpoints
are designated by ?.
Field 4.  Description of the gene product, enzyme
classification number, or cross-references to other databases.
```

Figure 6. Feature table description for the GenBank genetic sequence data bank. This relates to coding regions for protein, tRNA or rRNA. For a description of the information under the 'SITES' subidentifier see *Figure 7*.

'AUTHORS', 'JOURNAL' and 'TITLE'. CODATA recommend a separate subidentifier 'Number' to introduce each citation. The EMBL lines are 'RN' (reference number), 'RA' (author), 'RT' (reference title) and 'RL' (reference location) for the citation.

3.4.8 *The comment lines*

These are textual comments about the entry and may convey any information which is considered to be of value. The line identifiers are 'CC' (EMBL) or 'COMMENT' (GenBank and CODATA). EMBL uses an 'XX' line for spacing, to improve readability.

3.4.9 *The keywords line*

This line provides the information to produce a keyword cross-reference index. The line identifier is 'KW' (EMBL) or 'KEYWORDS' (GenBank and CODATA).

3.4.10 *Feature tables*

These provide a means of annotating sequence data. They are introduced by 'FH' (EMBL) or 'FEATURES' (GenBank and CODATA).

EMBL feature tables have a 'Key', 'From' and 'To' fields and a 'Description' (see *Figure 2*). The keys are described in *Figure 5*. The 'From' and 'To' endpoint specifications designate (inclusively) the endpoints of the feature named in the key field. In general, these fields simply contain base numbers indicating positions in the sequence as listed. These positions are always specified assuming a numbering of the listed sequence from 1 to n, no reference being made to the numbering used in the original reference(s). If the 'From' specification is a larger number than the 'To' specification, the feature is usually on the strand complementary to that listed. (The only exceptions occur in circularly closed sequences, where features sometimes cross the enumeration origin.) A feature which refers to the complementary strand is always indicated by the characters '(C)' following the 'To' field. If the 'From' and 'To' specifications are equal, the feature indicated consists of the single base at that position. When a feature is known

```
Field 1.  Site locations within the sequence.
Field 2.  Keywords describing the type of feature coded.
          allele     allelic variation of base
          anticdn    anticodon of tRNA
          attack     site of DNA damage caused by non-
                     metabolic chemical
          binding    protein binding site
          cell       cellular DNA ( versus a foreign insert )
          conflict   different authors disagree
          cutds      site of double-stranded scission
          cutss      site of single-stranded scission
          d-loop     displacement loop in mitochondria
          glossary   a line to explain abbreviations used in
                     the rest of the entry
          idna       intervening DNA sequence, removed
                     during development
          irna       identifier RNA
          ivs        intervening sequence or intron
          ltr        a group of bases repeated at the ends
                     of an inserted element
          methyl     site of nucleotide methylation
          modified   site of naturally occurring modified base
          mrna       messenger RNA
          mult       coding regions overlap in more than one
                     reading frame
          mut        mutation ( insertion, deletion, or point
                     mutation )
          orgrpl     origin of replication
          pept       peptide coding sequence
          prov       sequence of provirus ( versus cellular )
          recomb     site of recombination
          refnumbr   numbering scheme of publication
          orgpl      replication initiation
          rpt        repeat ( perhaps approximate ) of a
                     group of bases
          revision   sequence has been revised by the same
                     laboratory
          rna        RNA primary transcript
          rrna       mature ribosomal RNA sequence
          signal     signal sequence ( eg promotor, operator,
                     attenuator )
          site       wild card site key
          trns       transposable element
          trna       mature transfer RNA sequence
          unsure     author(s) unsure of base
          urna       small nuclear RNA ( mature )
          variation  base difference between homologous
                     sequences
          virion     sequence corresponding to RNA genome of
                     virus
Field 3.  Span of bases starting from the site location given
in field 1. If the value is 0 then the position given is at a
transition point between two different regions which are
indicated by two keywords in Field 2 separated by a slash.
The position given is the first base in the region indicated
by the key to the right of the slash.
```

Figure 7. Locations of sites as given in feature tables of the GenBank genetic sequence data bank, including keys describing the sites.

to extend beyond the end(s) of the sequenced region, the endpoint specification will be preceded by < (for features which continue 5′ to the left end) or > (for features which continue 3′ to the right end). Unknown endpoints are denoted by '?'. When the relevant reading frame is ambiguous 'base in codon' information is given as shown in *Figure 2*. A more satisfactory way of representing this information is being sought.

GenBank feature tables are similar (see *Figure 3*). They have a section which documents the regions coding for proteins, tRNA or rRNA (see *Figure 6*). There is a sub-identifier 'SITES' for information on specific locations within the sequence (see *Figure 7*).

3.4.11 *Sequence composition*

GenBank have a line called 'BASE COUNT' to summarize the composition of the sequence. CODATA recommend the identifier 'SUMMARY' with subidentifiers 'Molecular-weight', 'Length' and 'Checksum' as well as base counts. EMBL put the composition on the 'SQ' line.

3.4.12 *The last text line*

The last text line is followed immediately by the sequence. The EMBL line identifier is 'SQ' and it has a summary of sequence composition. GenBank end the text with an 'ORIGIN' line which specifies the location of the sequence in the genome. CODATA have no way of marking the last text line.

3.4.13 *The sequence data*

Symbols used have been described in Section 2. Sequence lines are introduced with one or more spaces and may contain spaces for layout. GenBank number the sequence positions on each line to aid human readers. CODATA suggest the sequence be introduced with the line identifier 'SEQUENCE' which if it stands on a line of its own becomes a 'last text line'.

3.4.14 *The termination line*

The end of each entry is marked by a '//' on a line by itself.

3.5 **Difficulties**

It is difficult to keep track of the large literature of molecular biology in a diversity of journals and some sequences published several years ago have not yet found their way into the databases which are distributed. If authors notice any deficiencies, would they please inform the data bank centres. The problems caused by the large volume of new data has been discussed in Section 1.

Duplication of data is also a problem. Sequences tend to be published as small regions of biological interest, rather than systematically. Not all the sequence data in a collection are unique. For example, in the EMBL data library many of the smaller adenovirus entries have yet to be merged with the genomic entries. Adequate software for identification of identical sequences in the databases and their merging together is still under development. This problem exists not only at the level of the sequences, but affects the annotation of entries. Very long sequence entries may have to be split into sections and the best way to do this has not been established. Merely splitting the sequence into equal pieces will disrupt biological features.

More co-ordination between data banks will aid coherence. The appropriate choice of keywords is essential to ensure adequate information retrieval. The taxonomic classification of organisms presents a problem which is being tackled on a much wider plane. Strain, clone and culture data banks also exist (see Section 5). Correlation between genetic maps and sequences is very important. Further collaboration between centres with these interests needs to be encouraged.

4. PROTEIN SEQUENCE DATABASES

Protein sequence databases have been available (in book form) for 20 years (1). The major databases are the NBRF-PIR database and the PSD-Kyoto database. The NEWAT collection is a useful subset for rapid retrieval. It played a key role in the discovery of the link between platelet-derived growth factor and a viral oncogene (19,20). S. Sakakibara, Protein Research Foundation, Osaka, Japan maintains a protein sequence collection. There are numerous specialized collections such as the KABAT database of sequences of proteins of immunological interest published by the US National Institutes of Health, and the International Haemoglobin Information Center directed by R.N. Wrightstone of the Medical College of Georgia (4). PGtrans is a derivative of GenBank. PseqIP is claimed to be a non-redundant collection derived from PIR, PSD, NEWAT and PGtrans. These databases use a variety of formats which will merely be of historical interest if the CODATA recommendations are accepted.

4.1 The NBRF-PIR database

The Protein Identification Resource (PIR) originated from the work of Margaret Dayhoff and was established in 1984 at the National Biological Research Foundation (NBRF) under the sponsorship of the Division of Research Resources of the National Institutes of Health. The PIR database has been described in a number of publications (21−23). The address is given in *Table 1* and the files on the distribution tape are listed in *Table 9*.

All proteins in the database are organized into a hierarchy of superfamilies, families, subfamilies, entries and subentries, based primarily on sequence similarity (22). Each protein entry contains a unique identifier, a title, the amino acid sequence and at least

Table 9. Files on the distribution tape of the PIR database.

1. Listing of superfamily number and entry title
2. Listing of the table of contents
3. Sequence entries
4. Author index
5. Feature table index
6. Superfamily name index
7. Species name index
8. Keyword index
9. Listing of new entries in update
10. Listing of entries revised in update
11. Release notes
12. Software for displaying sequence entries
13. Listing of the table of contents for sequences in preparation
14. Provisional sequence entries

one bibliographic citation from which the sequence was derived. An example of an entry from the database in CODATA format is shown in *Figure 4*. The entry may contain additional information, such as alternative names of the protein ('ALTERNATE-NAME'), gene name ('GENE-NAME'), map position ('MAP-POSITION'), segment number (viruses), chromosome number, location of introns, secondary structural information, references for X-ray crystallography, key words and various comments. The following feature subidentifiers are recommended by CODATA: 'Active-site', 'Binding-site', 'Cleavage-site', 'Disulphide-bonds', 'Domain', 'Duplication', 'Inhibitory-site', 'Modified-site', 'Peptide', 'Protein', 'Region'. These are currently in use in the PIR database. 'Protein' and 'Peptide' refer to the residues that form the mature protein or peptide. 'Domain' refers to residues that make up discrete functional domains. Amino acid symbols have been discussed in Section 2.2, as well as the punctuation symbols which are used to annotate sequences.

4.2 The PSD-Kyoto database

This database is about the size of the PIR database (see *Figure 1*), and is in a format not dissimilar to the CODATA recommendation. For further information write to T.Ooi, Chemical Research Institute, Kyoto University, Uji, Japan.

4.3 The NEWAT database

This database was originally compiled to supplement the Atlas of Protein Sequence and Structure (1) before the PIR database became available. It was also compiled as a sub-set which would be useful for examining protein relationships (24). The sequences were collected principally from work published between 1979 and 1984. The database contains the sequences, a set of literature citations for the sequences and an accessory set of programs written in the C language for examining and manipulating the data. The database is divided into six sections representing: enzymes from all sources, non-vertebrate eukaryotic sequences excluding enzymes, prokaryotic sequences excluding enzymes, vertebrate sequences, animal virus sequences and ribosomal sequences. For further information write to R.F.Doolittle, Chemistry Department, University of California at San Diego, La Jolla, CA 92093, USA.

4.4 The PGtrans and PseqIP collections

PGtrans is a computer-generated collection of protein sequences obtained by translation of GenBank (25,26). There are difficulties in machine translation (27) which may now have been cured (28). Rather than distributing the translated proteins, Fickett (5) distributes two programs written in the C language for the translation of GenBank proteins. To obtain these programs write to James W.Fickett, Los Alamos National Laboratory, Los Alamos, NM 87554, USA.

PseqIP (29) represents an attempt to make a non-redundant protein sequence collection by merging PIR, PSD, NEWAT and PGtrans. For further information about PGtrans and PseqIP write to Jean Michel Claverie, Computer Science Unit, Institut Pasteur, 28 rue du Dr.Roux, 75724 Paris Cedex 15, France.

5. OTHER DATABASES RELEVANT TO MOLECULAR BIOLOGY

While this chapter is concerned with molecular sequence databases, we have seen that these inevitably contain more information than the sequence itself. In this section we briefly outline some of the other sources of related information. Note that while some of these services are well established, others are still under development. There is a directory of commercially available online databases (30), and a directory of molecular biology software (31).

5.1 Macromolecular structure databases

The Brookhaven Protein Data Bank (32) is an international repository for the results of macromolecular structural studies. The Data Bank includes the results of structural studies of proteins, tRNAs, polynucleotides and polysaccharides. The data are compiled at the Brookhaven National Laboratory (which acts under contract to the US Department of Energy) and the project is funded by the US National Science Foundation and the US National Institutes of Health. A quarterly newsletter gives details of the data which are distributed on magnetic tape or on microfiche.

Two classes of information are collected, stored and distributed. These are atomic coordinates and structure factor-phase data. There are also literature citations for macromolecular structures for which coordinates are not yet available. Computer programs in FORTRAN which operate on the atomic coordinate entries are also distributed.

Data should be sent to Brookhaven on magnetic tape. Distribution of the data bank is from Brookhaven, except that there are national centres in Australia and Japan. For further information write to Protein Data Bank, Chemistry Department, Brookhaven National Laboratory, Upton, NY 11973, USA.

5.2 Enzyme databases

ENZIDEX is a data bank of the characteristics and availability of commercial enzymes with worldwide coverage. One index gives enzyme names, international number, synonyms, trade names, supplier, biological source, substrate and reaction product. A second gives details of activity, unit definition, assay, physical optima, specificity, activators and inhibitors. For further information write to Biocatalysts Ltd., 430 Victoria Road, South Ruislip, Middlesex HA4 0HF, UK.

Users of the BIONET facility (see Chapter 5) have access to a restriction enzyme database which is provided and updated by Dr R.J.Roberts (33). As well as a comprehensive listing of all known restriction enzymes (now numbering over 400), databases for use specifically within the Intelligenetics suite of programs have also been generated. The user has at his disposal the fully comprehensive listing of enzymes, a list of commercially available enzymes, or one of about 10 files listing enzymes specifically provided by individual manufacturers. The databases contain simple lists of the restriction enzymes, where the format is the name of the enzyme, the name of a prototype enzyme (where applicable) and the recognition sequence. The cut site is indicated where known by a ˆ symbol. The databases are intended for use within the SEQ or PEP programs available on the BIONET facility. No information is provided on the experimental conditions for use of the enzymes.

5.3 **Genetic map databases**

For those researchers involved in human genetics, the Howard Hughes Medical Institute Gene Mapping Library provides five databases of interest (see Chapter 5 for details of access). These databases are a collection of literature references containing citations and abstracted information, a map database with information pertaining to mapped genes, information on available probes, a database of information on restriction fragment length polymorphisms (RFLPs), and a database listing persons to contact for obtaining various probes. Some of these databases are also cross-referenced.

The LIT database is a rapidly growing collection of information references, currently standing at more than 5000. The database is organized on citation information including authors, key words and journal. A limited amount of abstracted information is also available such as gene symbols, Enzyme Commission numbers (34) and McKusick numbers (35). (McKusick numbers indicate the dominant, recessive or X-linked nature of a variant.) These may be used to retrieve information. Examining the database at the command level shows that some 20 indexes are available. For example, information on organelle, syndrome or mapping method may be used to set up search parameters.

The MAP database contains over 1500 entries, each representing one gene or one locus. The entries are cross-referenced to the LIT database by a 5-digit number. Each entry contains information as found in the table of the Nomenclature and Chromosome Committee of the Human Gene Mapping (HGM) Workshop (36). This includes the agreed HGM Workshop symbol, status of assignment, chromosomal location and name of the gene or locus. The information also includes Enzyme Commission numbers and McKusick numbers.

The PROBE database contains some 2000 entries each of which represents one probe or clone. Whilst not purporting to be a fully comprehensive collection of probes used for gene mapping, the database aims to provide a contact for obtaining each probe and is cross-referenced to the LIT database. The information includes such details as the enzymes used to generate the probe, whether the probe is cDNA or genomic, and the HGM Workshop symbol for the gene or locus that is identified by the probe, whether the probe has been sequenced, which vectors were used, the GenBank entry name and a contact name.

The RFLP database entries are in three parts. The first gives the localization and name of the system, the second gives the definition of the polymorphism and the third gives the allele frequency information for each population sampled and allele identified. Each site (defined as a probe−enzyme combination) that is polymorphic is grouped into those polymorphisms which function as a single locus in linkage studies and this constitutes a database entry. There are cross-references to the LIT and PROBE databases.

The CONTACT database provides the names, addresses and telephone numbers of over 1300 individuals from whom a probe or additional information may be obtained.

In book form, Human Gene Mapping (36) lists 1390 genes and markers, 808 of which were mapped with the aid of recombinant DNA probes.

There is a repository of human DNA probes and cloned genes at the American Type Culture Collection (ATCC). Holdings and background data will be supplied from an online computer data bank (37). For further details write to Harold D.Hatt, American Type Culture Collection, 12301 Parklawn Drive, Rockville, MD 20852, USA.

S.Karger AG of Basel, Switzerland, have a 'Genetic Information Retrieval System'.

5.4 Hybridoma databases

The Hybridoma and Monoclonal Antibody Data Bank is a joint project of CODATA and the International Union of Immunological Societies (38). The data bank was established in 1983 and data on hybridoma characteristics may be obtained from the nodes established in the US, France and Japan. For further information write to Lois Blaine, American Type Culture Collection, 12301 Parklawn Drive, Rockville, MD 20852, USA. The ATCC also publishes a catalogue of cell lines and hybridomas (39).

5.5 Cloning vector databases

VectorBank is a database containing information about commonly used cloning vectors as well as restriction maps. The database is intended for use within the CLONER program provided as part of the BIONET facility (see Chapter 5).

The database consists of a directory whose individual files relate to specific vector maps. Each file contains a comments section providing information on such features as vector size, the location of sites of genetic or biological interest, as well as literature references. The files do not contain complete DNA sequences, instead they provide the locations of restriction enzyme recognition sites. Multiple maps are generally available for each vector. Thus the file PBR322_6CUT.MAP will contain lists of the locations of all the sites found for enzymes with a recognition sequence 6 bp long within pBR322. Similarly, the file PBR322_UNQ.MAP will contain a list of the sites for those enzymes which cut uniquely within pBR322.

To use the information contained within the database the required map is loaded into the CLONER program. Information about the insert and the enzyme recognition sites is contains may either be generated within the SEQ program or alternatively loaded from the keyboard having given the NEW command. By giving the INSERT command together with the restriction sites to be used the insertion of fragment into vector is accomplished. The construction is then available for further analysis or manipulation.

5.6 Culture collection databases (including strains and cell lines)

The Microbial Strain Data Network (MSDN) will be a worldwide network of databases. It is being implemented as a result of recommendations of a task group set up by CODATA and the World Federation of Culture Collections. The MSDN will be a locator service for clone (strain, cell line) repositories. For further information write to Dr Michah I.Krichevsky, Microbial Systematics Section, National Institutes of Health, Bethesda, MD 20014, USA.

The Microbial Information Network Europe (MINE) will provide an integrated catalogue of holdings of culture collections in Europe. MINE is funded by the CEC Biotechnology Action Programme. Entries for each strain will be in the form of an agreed minimum dataset for each main group of organisms such as fungi and yeasts, bacteria, algae, protozoa and animal cell lines. For information write to Rhonda Platt, MINE Project Administrator, Commonwealth Agricultural Bureau, International Mycological Institute, Culture Collections and Industrial Services, Ferry Lane, Kew, Surrey TW9 3AF, UK.

The Microbial Culture Information Service (MiCIS) collects strain data from UK national collections (40) and will contain more detailed information than MINE. For information write to Mrs G.V.Alliston, Biotechnology Group, Laboratory of the Government Chemist, Cornwall House, Waterloo Road, London SE1 8XY, UK.

The Microbiological Resource Databank (MIRDAB) is a data bank of cells and their holders set up by Elsevier. It includes animal cells, plant cells and animal viruses. For information contact P.Bruin, Elsevier Science Publishers, PO Box 211, 1000 AE Amsterdam, The Netherlands.

5.6 Databases for taxonomy and identification

Zoological Record Online provides taxonomic information about animals. For information write to BIOSIS User Services, 2100 Arch Street, Philadelphia, PA 19103, USA. BIOSIS has also launched a prototype information system for biologists called the Taxonomic Reference File (RF). Initially, the project will concentrate on bacteria (a group of about 10 000 known organisms). Write to Jean M.Walat at the above address or to Michael Dadd, BIOSIS UK Ltd., 54 Micklegate, York YO1 1LF, UK.

The Computer Assisted Yeast Identification System (COMPASS) provides an online identification by comparing the properties of an unknown yeast with the properties of known taxa. For further information write to Dr B.Kirsop, AFRC Institute of Food Research, Norwich NR4 7UA, UK.

5.7 Molecular biology software databases

The GenBank Software Clearing House is a database available through online access to the GenBank system (see Chapter 5) and contains information pertaining to software packages for sequence analysis. The information for each entry comprises 15 fields of basic information. Also available is a more expanded form of entry which contains a comprehensive description of the software package and its functional capabilities, as well as information on how to obtain the package, any likely charges and a bibliography. The limited fields comprise such information as the name and authors of the package, the operating systems, computers supported and language used to write the package are also indicated. The remainder of the entry fields are concerned with how the documentation is presented, what databases, if any, are provided and the mode of software distribution. When using the database, any of the 15 fields may be used to locate the entries of interest after which the expanded description may be used to obtain more detailed information. The database is uncritical and presents information as supplied by the authors of the packages.

There is a proposal to set up a European Bank of Computer Programs in Biotechnology. The task force is chaired by K.Ch.A.M.Luyben, TH Delft, The Netherlands.

6. USING MOLECULAR SEQUENCE DATABASES

The unusual nature of molecular sequence data (e.g. in terms of the variability in the size of data entries) and the specialist operations applied to the data in order to access and analyse it have generally prevented the widespread use of general purpose database management systems for the storage of molecular sequences. Instead, specialist programs have been developed providing the data retrieval functions required for molecular sequence data. Such systems are often closely linked to suites of sequence data manipulation and analysis programs.

Whilst no general agreement on all the features essential for sequence library management and retrieval has emerged, a number of basic operations are available on most systems and these, together with more sophisticated data retrieval functions are discussed in the following sections, together with descriptions of their potential uses.

6.1 **Organization of data**

The organization of the sequence databases on your local computer system will greatly influence the ways in which it may be accessed and the types of programs likely to be available to manage and retrieve data. The nature and the variety of uses to which sequence data is put means that there is no 'best way' to organize the sequence data collections. Therefore, the organization on your system will probably be a compromise that reflects the most common requirements of the users, the need for compatibility with existing analysis software and the constraints on those who have the responsibility for keeping the data collection up to date.

The following general comments are intended to briefly outline the three basic data organizations which may be encountered and indicate how the organization might influence the ways in which the data can be used.

6.1.1 *One entry — one file*

The simplest organization of the sequence databases is to store each entry in a separate disk file. Files will generally be named using the identification code of each entry: the 'ID' (EMBL) or 'LOCUS' (GenBank). The management of the database is then achieved using the file organization methods provided by the computer operating system. For example the hierarchical directory structure of MS-DOS or the UNIX operating systems may be used to group entries into directories according to some partial taxonomic classification. Alternatively, most operating systems (although this does not necessarily apply to application programs) allow groups of files (and directories) to be described using so-called wild-cards in the file name. Wild-card characters under UNIX and MS-DOS are '*' which matches any character or string of characters and '?' which matches any single character in a file name. Since the identification codes in EMBL and GenBank are derived in a fairly systematic manner, combining the source organism and a mnemonic indicating the biological function of the sequence, wild-cards may be used to define required subsets of the data. For example, sequences derived from humans could be described by 'HUM*' in a GenBank collection and 'HS*' in an EMBL collection. Similarly, '*ACT*' would select the actin entries in the GenBank database. Some operating systems also allow groups of files to be defined by listing the file names in an 'indirect file'.

6.1.2 *Many entries — one or more files*

Perhaps the most common organization for the molecular sequence databases is to store many sequence entries in one file. This is the way that the EMBL, GenBank and the PIR databases are distributed and is the organization adopted for the databases on BIONET (Chapter 5). The BIONET files do not correspond to the original database distribution files but are a reorganization with a small number of entries per file, each file representing an organism or species. Where the number of entries for a species is very high, the entries are split between a number of files based on other aspects of the entry (e.g. biological role of the sequence). The advantage of regrouping the sequence entries into a large number of files each with relatively few entries is that tools provided by the operating system can be exploited for inspection of the entries and features like wild-cards and indirect files can be used to specify large groups of entries.

6.1.3 *Database management systems*

The most sophisticated method of organizing the sequence databases is to use a database management system. If based on techniques of relational databases then the organization would be one entry to many files. Many of these systems closely couple access to the databases with sequence analysis software (12,41−44) and have developed specialist database management systems, whereas in other systems (e.g. 45,46) a general purpose relational database system has been used. See Chapter 3 for descriptions of the methods for database management and retrieval used by commercial packages available for the IBM PC.

The advantages of using database management systems are that the organization of the data within the computer system need not concern the user, and all aspects of the entry are immediately accessible for creating subsets of the database. The overheads the user must pay for this flexibility are the time taken to learn how to use a possibly complex program and having to use the database system every time that data is required from the database.

6.2 **Retrieval of entries**

The four most important methods for identifying sequences within a database that are required for retrieval or subsequent analysis involve using the unique sequence identifier, identifying those sequences whose annotation lines contain certain key words or whose sequence contains a certain pattern of bases and fourthly those entries with a sequence similar to a query sequence.

An important class of retrieval functions provided by some sequence database management systems uses the sequence features tables (Section 3.4.10) to enable sections of a sequence to be retrieved. These functions may be used, for example, to retrieve introns or exons from a group of sequences or to select a region around a particular site (e.g. a splice junction). Such functions are vital to the selection of data to build consensus sequence patterns in order to search for control signals (Chapter 10) or for statistical analysis of sequence data.

6.2.1 *Retrieval by sequence identifier*

The simplest type of retrieval is that used when the sequence identifier is known. In the case of those systems which do not use a sequence database manager, the extraction of subsequences based on the feature table would be facilitated by the system text editor or a sequence editor. Most sequence data editors permit simple extraction of subsequences into a file. The retrieval of a complete sequence using the sequence identifier in a sequence database management system is normally a very simple command.

6.2.2 *Retrieval by key words*

Retrievals of sequence entries by key words, either in the key word lines (Section 3.4.9) or else based on the occurrence of character strings in other lines of the sequence annotations are most efficiently processed by sequence databases management systems. A typical query for selecting adenovirus sequences from the EMBL database is that used in the GENAS system (46):

FIND KW = ADENOVIRUS

Note that for this type of retrieval the line type ('KW') must be specified. The use of the PSQ program (12) to retrieve data from the PIR database is illustrated in Figure 10 of Chapter 5. The reason that database management systems are so efficient at processing this type of query is because their organization of data exploits the fact that the user will specify where to look for the appropriate information. The database may further increase performance by keeping indexes of entries for line types specified by the designer of the database.

In a system where there is no database management system available for sequence data and yet retrieval by key words is required, there are various alternative methods for achieving the same result in a slightly less efficient, more labour intensive way. The first of these is to consult one of the indexes compiled by the database distributors. Indexes are currently available for authors and key words in the EMBL, GenBank and PIR databases, for citations in EMBL and GenBank, for taxonomic classification in EMBL and PIR and for structural and functional features in PIR. The indexes may be searched manually to find the unique sequence identifiers of the relevant sequence entries or, in preference, the data files containing the indexes may be searched using a system text editor or an efficient text searching program.

Fast, efficient text searching utilities exist on most computer systems as general purpose utilities. They are variously called FIND (MS-DOS), SEARCH (VAX/VMS) grep (UNIX) and XSEARCH (DECsystem20, TOPS-20) (see Chapter 5). Programs such as these will search a file or group of files for a string of characters or pattern of characters and report the lines and files where the string is found. If the file is a database index file, then the sequence identifier can be easily extracted and used to retrieve the sequence entry using the simple methods described in Section 6.2.1. If the index files are not available, text searching programs may be used to search through the database files themselves in order to locate entries of interest. However, this method is particularly inefficient because the programs have no information about the structure of the data entries. Thus a search for an author's name will scan all of the sequence characters in an entry looking for a match. A program with sufficient sophistication to avoid this pitfall is the QUEST program (ref 37, Chapter 5) available on the BIONET resource. The QUEST program can be instructed to search for character patterns only within certain contexts in a sequence entry (e.g. the annotations or the sequence data).

6.2.3 *Retrieval by sequence patterns*

The QUEST program has a particularly flexible and powerful language for describing patterns of sequence characters. Character patterns may be named and used as elements of increasingly complex patterns with defined separations between them. The efficient searching methods of the QUEST program enable these patterns to be searched for through an entire database. Chapter 5 discusses and illustrates the use of QUEST on the BIONET resource.

6.2.4 *Retrieval of sequence similarity*

Perhaps the most important type of retrieval activity is that which is intended to identify similarities between a query sequence and an entry in a database (a positive identification implying potentially common structure, biological function or ancestry). The importance of this problem requires a detailed description of the issues of appropriate sequence comparison algorithms for searching sequence databases, program efficiency and the criteria for determining what constitutes similarity. This is provided in Chapter 13.

6.3 **Analysis programs**

Any of the programs dicussed in this book could potentially be used to analyse data taken from a molecular sequence database. However, in the programs which have been discussed in this section (41 −46) there exists a particularly close association between a sequence database and sequence analysis methods. In general, this close association means that it is particularly easy for a user to move data from the database to the analysis software. In particular, the close links between database and analysis software permit analyses to be undertaken that treat the sequence databases as data worthy of study in their own right. These types of analysis are generally of a statistical nature, or are based on methods from information theory. In order to illustrate the types of analyses possible in these systems three examples have been chosen.

6.3.1 *Codon usage analysis*

Using the NAQ program (12) it was relatively simple to select all the recorded yeast mitochondrial coding regions from the NBRF nucleic acid sequence database and then analyse the codon usage patterns from those sequences. The resulting codon usage table could then be used directly to search an uncharacterized yeast mitochondrial sequence for potential protein-coding regions (Chapter 10).

6.3.2 *Classification of protein superfamilies*

Using the IDEAS system based on FRAMIS relational database management system (45,48) a statistical discrimination technique was used to classify proteins into superfamilies. The technique involved the identification of known sequences from the PIR database that can be classified into biological, chemical or physical groups and then the determination of the particular properties or combination of properties of those sequences that give the best discrimination between the classes. With this information it is then possible to classify uncharacterized sequences into the appropriate classes.

Subsets of the PIR database were constructed, each having a common superfamily classification. Each sequence was then analysed to determine the periodicity of nonpolar amino acids, the hydrophobicity, the net charge and sequence length. Using these parameters, the sequences were found to cluster into six large groups. The values of these parameters for each group could then be used in the discriminant analysis to classify new, uncharacterized sequences.

6.3.3 *Predicting regions of biological importance*

Using the SASIP system (42) an analysis of the GenBank and PIR databases (39) was executed to determine the frequency of occurrence of octanucleotides and tripeptides, respectively. The resulting frequency tables reveal some interesting observations at the extreme ends of the frequency distribution and information theory can be used to determine the relative importance of the oligomers. Simply stated, information theory predicts that in this type of analysis, the least frequently occurring oligomers carry the greatest amount of information and are thus most likely to be biologically important. Using the frequency tables as a reference, a new sequence may be analysed and the local information content calculated in a stepwise manner along the sequence. Peaks in the graph of information content for the sequence indicate regions of potential biological importance. Using some test sequences, this analysis method showed a correlation between the information peaks with regions of known biological importance such as the membrane-spanning and the cytoplasmic domains in the H-2K^b mouse histocompatibility protein.

6.4 **Molecular sequence databases as bibliography collections**

All of the molecular sequence data libraries and other data collections discussed in this chapter contain citations to the scientific literature used to complete each data entry. Therefore, in addition to their primary role as repositories for the original scientific data, the data libraries also constitute important specialist bibliographic data collections. For instance, GenBank release 45 (September 1986) contains over 7000 literature citations for its 9411 loci, 8 931 101 bases and 12 038 reported sequences.

The obvious use of the literature citation data is selecting subsets of the database entries in order to narrow down the number of candidates for subsequent analysis or inspection. Most data management systems for molecular sequence data permit the citation fields (e.g. the 'REFERENCE' lines; Section 3.4.7) and subfields (e.g. 'AUTHORS', 'TITLE' and 'JOURNAL') to be searched for the occurrence of key words and this is usually adequate for creating data subsets.

7. REFERENCES

1. Dayhoff,M.O. (ed) (1972) *Atlas of Protein Sequence and Structure.* National Biomedical Research Foundation, Washington, D.C.
2. Lewin,R. (1986) *Science,* **232**, 1598.
3. Martin,W.J. and Davies,R.W. (1986) *Bio/Technology,* **4**, 890.
4. Lesk,A. (1985) *Nature,* **314**, 318.
5. Fickett,J.W. (1986) *Trends Biochem. Sci.,* **11**, 382.
6. Anon (1986) *Nucleotide Sequences 1986.* IRL Press, Oxford.
7. Sprinzl,M. and Gauss,D.H. (1982) *Nucleic Acids Res.,* **10**, r1.
8. IUPAC-IUB Commission on Biological Nomenclature (1966) *J. Biol. Chem.,* **241**, 2491.
9. IUPAC-IUB Commission on Biological Nomenclature (1968) *J. Biol. Chem.,* **243**, 3557.
10. Kneale,G.G. and Bishop,M.J. (1985) *CABIOS,* **1**, 11.
11. Mount,D.W. (1985) *BioTechniques,* **3**, 102.
12. Orcutt,B.C., George,D.G., Fredrickson,J.A. and Dayhoff,M.O. (1982) *Nucleic Acids Res.,* **10**, 157.

13. Hamm,G.H. and Cameron,G.N. (1986) *Nucleic Acids Res.*, **14**, 5.
14. Burks,C., Fickett,J.W., Goad,W.B., Kanehisa,M., Lewitter,F.I., Rindone,W.P., Swindell,C.D., Tung,C. and Bilofsky,H.S. (1985) *CABIOS*, **1**, 225.
15. Bilofsky,H.S., Burks,C., Fickett,J.W., Goad,W.B., Lewitter,F.I., Rindone,W.P., Swindell,C.D. and Tung,C. (1986) *Nucleic Acids Res.*, **14**, 1.
16. Foley,B.T., Nelson,D., Smith,M.T. and Burks,C. (1986) *Trends Genet.*, **2**, 233.
17. Walker,J.R. and Willett,P. (1986) *CABIOS*, **2**, 89.
18. Devereux,J., Haeberli,P. and Smithies,O. (1984) *Nucleic Acids Res.*, **12**, 387.
19. Doolittle,R.F., Hunkapillar,M.W., Hood,L.E., Devare,S.G., Robbins,K.C., Aaronson,S.A. and Antoniades,H.N. (1983) *Science*, **221**, 149.
20. Waterfield,M.D., Scrace,G.T., Whittle,N., Stroobant,P., Johnsson,A., Wasteson,A., Westermark,B. and Heldin,C.H. (1983) *Nature*, **304**, 35.
21. Orcutt,B.C., George,D.G. and Dayhoff,M.O. (1983) *Annu. Rev. Biophys. Bioeng.*, **12**, 419.
22. Chen,H.R. and Barker,W.C. (1985) *Trends Genet.*, **1**, 221.
23. George,D.G., Barker,W.C. and Hunt,L. (1986) *Nucleic Acids Res.*, **14**, 11.
24. Doolittle,R.F. (1981) *Science*, **214**, 149.
25. Claverie,J. and Sauvaget,I. (1985) *Trends Biochem. Sci.*, **10**, (4) VIII.
26. Claverie,J. and Sauvaget,I. (1985) *Nature*, **318**, 19.
27. Fickett,J.W. (1986) *Trends Biochem. Sci.*, **11**, 190.
28. Claverie,J. and Sauvaget,I. (1986) *Trends Biochem. Sci.*, **11**, 381,382.
29. Claverie,J.M. and Bricault,L. (1986) *Proteins: Structure Function and Genetics*, **1**, 60.
30. Anon (1986) *Directory of Online Databases.* Cuadra/Elsevier, New York.
31. Rawlings,C.J. (1986) *Software Directory for Molecular Biologists.* Macmillan, London.
32. Bernstein,F.C., Koetzle,T.F., Williams,G.J.B., Meyer,E.F., Brice,M.D., Rodgers,J.R., Kennard,O., Shimanouchi,T. and Tasumi,M. (1977) *J. Mol. Biol.*, **112**, 537.
33. Roberts,R.J. (1985) *Nucleic Acids Res.*, **13**, suppl.
34. International Union of Biochemistry (1984) *Enzyme Nomenclature.* Academic Press, New York.
35. McKusick,V. (1983) *Mendelian Inheritance in Man.* Johns Hopkins Press, Baltimore.
36. Chapelle,A. de la (1985) *Cytogenet. Cell Genet.*, **49**, 1.
37. Nierman,W.C., Maglott,D.R., Benade,L.E., Blaine,L.D., Felix,J.S., Deaven,L.L., Badman,W.S., Van Dilla,M.A. and Dayton,D.A. (1986) *ATCC Q. Newslett.*, **6**, 1.
38. Blaine,L. (1984) *Bio/Technology*, **2**, 338.
39. Hay,R., Macy,M., Corman-Weinblatt,A., Chen,T.R. and McClintock,P. (1985) *Catalogue of Cell Lines and Hybridomas.* American Type Culture Collection, Rockville, Maryland.
40. Anon (1985) *MiCIS News*, **1**, (1). Laboratory of the Government Chemist, London.
41. Gouy,M., Milleret,F., Mugnier,C., Jacobzone,M. and Gautier,C. (1984) *Nucleic Acids Res.*, **12**, 121.
42. Claverie,J.-M. (1984) *Nucleic Acids Res.*, **12**, 397.
43. Schneider,T.D., Stormo,G.D., Yarus,M.A. and Gold,L. (1984) *Nucleic Acids Res.*, **12**, 129.
44. Harr,R., Fallman,P., Haggstrom,M., Wahlstrom,L. and Gustafsson,P. (1986) *Nucleic Acids Res.*, **14**, 273.
45. Kanehisa,M.I., Fickett,J.W. and Goad,W.B. (1984) *Nucleic Acids Res.*, **12**, 149.
46. Kuhara,S., Matsuo,F., Futamura,S., Fujita,A., Shinohara,T., Tagaki,T. and Sakaki,Y. (1984) *Nucleic Acids Res.*, **12**, 89.
47. Abarbanel,R.M., Wieneke,P.R., Mansfield,E., Jaffe,D.A. and Brutlag,D.L. (1984) *Nucleic Acids Res.*, **12**, 263.
48. Kanehisa,M.I., Klein,P., Grief,P. and DeLisi,C. (1984) *Nucleic Acids Res.*, **12**, 417.
49. Claverie,J.-M. and Bougueleret,L. (1986) *Nucleic Acids Res.*, **14**, 179.

CHAPTER 5

Online services

MICHELLE GINSBURG

"I, a stranger and afraid
In a world I never made."   A.E.Houseman

1. INTRODUCTION

Advances in all spheres of science have made it almost mandatory for laboratories to
have access to a computer. The rapid advances in recombinant DNA technology, and
the ease with which large DNA sequencing projects (1) can now be accomplished, is
a case in point: the use of computers to store, assemble and analyse sequences has greatly
increased the rate of data acquisition. In fact, large sequencing projects cannot now
be undertaken without computer assistance.

Many factors must be considered when deciding on the appropriate computing support for a molecular sequencing laboratory. Basically there are three alternatives:

(i)   Writing the necessary programs.
(ii) Obtaining one of the many packages available in the public or commercial do-
      mains and a compatible computer on which to run them (Chapters 2 and 3).
(iii) Finding a computer where the programs are already available.

For most laboratories, resources are limited and the cost of writing software can be
prohibitively high, especially in terms of the time needed to produce an error-free program (2). Commercially written software, although generally designed to be easy to
use (user friendly), internally documented, and powerful and flexible enough to perform many different functions (see Chapter 3) may be too expensive for a limited research
budget. The alternative, finding public domain software, also has its pitfalls since the
quality and friendliness is unpredictable although most authors will help if a user finds
himself in trouble. A commercial company should be willing to provide a higher level
of support to users of their software. Furthermore, it is possible that the amount of
sequence analysis required by the laboratory is unpredictable and is needed only infrequently or restricted to a limited set of methods. In this case it may be difficult to justify
the expense of buying the necessary computer hardware and software or it may be
premature to make such a commitment when the computing requirements of the
laboratory have not yet been fully determined. However, where it has been established
that a computer service exists that already has the necessary programs available then
the third alternative should be considered, since remote access to such a service can
be established at low cost and with very little extra hardware.

Finally, there is the issue of the molecular sequence databases which continue to pose a problem for many computer systems (see Chapter 3). One of the most frequently required sequence analyses is to search a sequence database for sequences similar to the users data (see Chapter 13) and 66% of available central processing unit (CPU) time on the BIONET resource is used for this purpose (3,4). The maintenance of large databases and the tools for handling them effectively imposes severe constraints on available resources (4). Supporting frequent database updates and accommodating new features as they become available is very time consuming and requires a considerable amount of disk storage especially when several databases are being maintained. Searching databases for sequence similarity occupies resources often for unacceptable periods of time and this is especially true for microcomputers (2,4). Therefore, an alternative approach to these problems would again be for users to connect to a central computer where the requisite resources are already resident. Recent developments in computer and telecommunications technology mean that communication between micro, mini and mainframe computers is now a very practical proposition (5) and a large number of research sites with a variety of machines, software, databases and management systems and electronic mail services are accessible by directly dialling over the telephone network or through the rapidly developing academic and commercial communication networks.

However, running software on a remote computer (an online system) brings its own problems and these may be compounded by trying to work with unfamiliar computers and operating systems. The difference between an online system and an online service is the quality of the help and support provided to ease the difficulties of outside users and ensure efficient use of time and resources. Whether the advice is by bulletin board to the user community at large, terminal links, telephone or electronic mail to a consultant, it is the criterion by which a service stands or falls. Since some of the services described here are still in their infancy, the quality of a service must also be judged by the continued effort put into improving facilities and keeping abreast of relevant scientific and technological developments.

The sharing of information, programs, data and other resources, without regard to the physical location of the machine or the user is an exciting concept that has been used to great advantage by the computer science and other research communities for some time. This chapter describes some of the basic principles and pitfalls of remote computer communications and illustrates a number of online services and systems currently available to the molecular biology and biotechnology research community.

2. ORGANIZED COMPUTER COMMUNICATIONS

When the need for computer communication over long distances was recognized the immediate solution was to utilize (6) a pre-existing network: the telephone system. However, this created certain problems and imposed certain restrictions. Firstly, the telephone system was designed to transmit *analogue* voice signals whereas computer signals are entirely *digital*. The latter must therefore be converted to an analogue form by connecting the computer to a device called a *modem* (MOdulator DEModulator) which converts digital signals into audio signals for transmission over telephone lines. However, the limited frequency response (*bandwidth*) of an analogue telephone system and degradation of the signal due to line noise restricts the rate at which data can be transmitted. These problems are further increased if, instead of connecting the modem electrically to the telephone system, an *acoustic coupler* is used to transmit the audio signals via the telephone handset. Using a standard grade analogue telephone line connected directly to a modem, the maximum transmission rate that might be expected is 1200 *baud* (bits per second) i.e. 150 characters per second. With an acoustic coupler the expected transmission rate would probably drop to 300 baud (38 characters per second). To obtain transmission rates greater than these, special grade communications lines must be leased from the telephone company.

The quality of the available communications methods has important implications for the types of use that will be made of the remote computer. Whereas 1200 baud might be considered as the minimum acceptable data rate if only simple *interactive* computing is envisaged, this would restrict other program functions which require high transmission speeds and low error rates such as data input using a digitizer (7) or high resolution graphics and unless special provision is made for them, they are unlikely to work satisfactorily.

In choosing an appropriate communications method, the distance between the computer and the user is an important consideration and whilst the telephone system is a simple and often adequate solution to communications with a nearby computer, a more satisfactory way of setting up a link to a distant computer is to use a computer *network*. An early formal definition describes a network as a set of autonomous computer systems interconnected so as to allow interactive resource sharing between any pair of systems (6). For true resource sharing to occur the network should be available on demand to users and their equipment, and the ability to exchange data should function easily. Generally, the way the network operates should be transparent, that is its operation should not be obvious to the user and failures at any particular point should not disable the whole network. However, in practice this is not always the case and users should be aware that failures may occur which can disable the network or prevent the desired host from being reached. This is especially so where not one but several networks may be used to reach a very remote host, for example, one in a different country.

Networks are described in terms of *links, nodes* and *hosts*. The links are the communication paths between the nodes which may either be endpoints or network junctions. Nodes generally play an active role in the network and are involved in the transmission of data, selecting alternative routes when faults occur and they can also be hosts, although not necessarily vice versa. The network hosts are computers connected to the network to which users (including other hosts) may connect. A local host is the user's computer, a remote host that to which he is connected. The number of alternative links between pairs of nodes will influence the reliability of the network.

Computer communications are usually sporadic in nature i.e. there will be frequent bursts of activity interspersed with periods of silence when no communication between the connected systems will take place. Therefore, if the connection between hosts was established by telephone, the line may be idle for much of the time and therefore unusable for other purposes. Furthermore, in some instances, establishing the connection may take longer than the total time the line is actually transmitting data. On a network, many communications are interleaved onto the same physical line to make optimum use of the available resources. Data may travel around a network by a variety of means. One of the most efficient methods is called *packet switching* (5) because the data is divided up into chunks called packets. Each packet sent onto the network may actually travel by a different route to its destination where it is reassembled correctly using information encoded in the packet. The device that assembles and disassembles packets is called a *PAD* (Packet Assembler Disassembler) and is logically equivalent to a modem. More efficient utilization of the communication links makes this mode of transmission relatively inexpensive. In order to ensure the orderly exchange of information on the network, sets of conventions called protocols have been devised. These may include such features as checking the data on arrival and asking for incorrectly received data to be retransmitted. Where two networks interconnect, or where it is required for incompatible systems to communicate, devices known as gateways perform the necessary interface. Whilst protocols do exist and standardization is theoretically possible, a user should be aware that for various reasons, the protocol may be only partially implemented. In practice this means that certain control sequences required by the remote computer may be filtered-out en route and certain features, for example the ability to use the DELETE key, may no longer function. If problems such as this occur it is advisable to consult with your local computer experts on how to deal with them.

2.1 Interactive computing and networks

Interactive computing implies a continuous dialogue between user and computer; the program prompting the user for information as it executes its assigned task, or the user issuing commands to be executed. Inevitably, there will be delays in this process if the computer has to perform particularly intensive numerical calculations or if the program requires frequent access to disk or tape storage. In addition to this, a *time-sharing* operating system may introduce delays while the *process* waits for its share of resources. If a computer is particularly heavily used, delays at peak times can lead to prolonged gaps in the interactive dialogue which can severely inhibit the effective use of interactive software. However, if the interactive dialogue occurs over a long distance network, yet more delays due to the length of the transmission route, error checking and retransmission of incorrectly received data further lengthen the time between command and response (5).

The user should therefore consider that it may well be more appropriate to run programs such as database searches overnight using *batch* processing. That is, a file containing all the input required by the program is created and submitted to the computer to run unattended after the user has ended his session. The use of batch processing can also be advantageous if the host is charging the user for computing resources. Costs are usually based on a combination of the time the user is connected to the computer and the amount of processing time consumed. Charges are generally higher at local peak times and therefore if it is possible to use the computer system outside peak times, considerable savings can be achieved. Batch processing may also help overcome another potential pitfall in the use of long distance remote computing: computer system and network failures.

Table 1. A brief description of some major networks

Network name	Country of origin	Description
ARPANET	United States	The prototype network, originally set up in the 1960s. Uses lines leased from the telephone system. Primary use appears to be mail.
TELENET	United States	A commercial value added network, using purchased lines which have been improved in terms of reliability.
TYMNET	United States	Operated by Tymshare Inc.
UNINET	United States	Public data network.
BITNET	United States	Network sponsored by IBM, mail and file transfer.
NORTHNET	Canada	IBM sponsored network.
DATAPAC	Canada	Public data network.
EARN	Europe	European Academic Research Network. IBM sponsored and designated to link with BITNET and NORTHNET.
TRANSPAC	France	Packet switch stream service.
PSS/IPSS	United Kingdom	Packet switch stream service offering connections to more than 30 countries.
JANET	United Kingdom	Joint Academic Network providing a single network with common standards. Anyone in the academic community may apply to join.
AGRENET	United Kingdom	Similar to JANET and linking units and institutes of the Agricultural and Food Research Council.
CSIRONET	Australia	Commonwealth Scientific Industrial and Research Organisation network.

2.2 System failures and networks

The complexity of long distance communications networks means that a user is more likely to experience unexpected breaks in a session than when using a local host. Failures can be attributed to either the CPU (back end) or communications hardware (front end). Generally, for the latter kind of failure the CPU will attempt to restore communications immediately. A back end failure is a much more serious problem and when this occurs the system may not be able to restart for some time.

Unfortunately, most software is written assuming that uninterrupted processing will take place and no allowances are made for lost network connections and other unexpected problems. If the network connection is broken for long enough, or the computer operating system is unforgiving of such problems, the user may lose the results of an entire session. At the very least this is frustrating and it may also be expensive if a lengthy database search must be repeated. It would be prudent for the user to consider before the start of the session whether interaction is essential and how the task may best be performed, so that data could be saved at frequent intervals and thus potential losses will be minimized.

3. ESTABLISHING THE CONNECTION

The route used to connect to a remote host should be established in consultation with the local computer department and those responsible for providing the remote service. A summary of the major national and international communications networks is given in *Table 1*. The network administration at the remote site will provide such informa-

Table 2. Example procedures for calling a remote computer using (A) a modem and (B) a packet assembler disassembler.

A.	A user equipped with microcomputer and modem will probably follow a similar procedure to the following.
1.	Set modem to ORIGINATE and requisite speed.
2.	Dial the number on the phone.
3.	The system answers with a high pitched tone.
4.	Switch the modem to ONLINE, replace the handset (or place in acoustic coupler).
5.	The carrier light should then come on and the system banner appears.
6.	The user logs on.
B.	Connecting through a packet assembler disassembler from a mainframe computer.
1.	The PAD is selected.
2.	A CALL command with appropriate DTE number is given to the PAD.
3.	If going via a service such as PSS the user is prompted for his authorization and the requisite address of the remote host.
4.	A 'call connected' message appears followed by the system banner.
5.	The user logs on.

tion as the relevant telephone number for access using a modem or *network address* when connection is intended via a network. The network address is the unique name or number used to identify a network host. The remote site may also provide information on special requirements such as terminal and modem types and the transmission speeds and may make recommendations on the types of file transfer protocols supported. For those users intending to establish the connection with a microcomputer and a modem it is generally recommended that a minimum speed of 1200 baud be used. Even at this speed it can take from 8 to 32 seconds to transmit one video screen full of text. *Table 2* outlines the important stages in establishing a connection with a remote host via a modem with acoustic coupler and a network PAD.

3.1 Terminal emulation and file transfer

Connecting to a remote computer using a microcomputer requires that the microcomputer be capable of emulating a display terminal of a type supported by the remote host. One of the most common standard terminal emulations is of the DEC VT100 (ANSI) terminal. The second requirement is for data files to be transferred reliably to and from the micro and the mainframe. Often the terminal emulator and file transfer utilities are parts of one program.

The essential features of a good terminal emulation program are that it sends all printing and control characters correctly to the remote host computer (including ASCII character 0), and it should not prevent transmission of the *escape* sequence necessary to switch the attention of the micro between itself and the remote host. The emulator should also be able to provide a full display of 80 columns for text or 132 for display terminal graphics, and it is essential that where programs rely on screen control (screen oriented editors; see Chapter 2) that the cursor movement commands be properly interpreted. More sophisticated programs may also provide features for user convenience such as storing telephone and login procedures. These may be used with *auto-dial modems* to automate the connection to the remote host.

The accurate transfer of data files between computers is an essential requirement for any computer communications link. When a microcomputer is used as an *intelligent terminal* to connect to the remote host a standard utility often provided is a program for recording all characters sent and received by the microcomputer on a disk file. These files are sometimes referred to as *typescript, dribble* or *log* files. There is usually also a complementary program that sends a text file to the remote host as if it were being typed at the microcomputer keyboard. These utilities, although useful for copying electronic mail or small sections of text, should not be confused with true file transfer programs using error-checking protocols that ensure that the data is accurately transferred. The error avoidance techniques used in file transfer programs check the received data so that spurious characters (e.g. due to line noise) may be detected and the data retransmitted correctly. The file transfer protocol also deals with synchronization of communication so that the data from one host is not sent faster than it can be received by the other and ensures that data is not lost even if short breaks occur in the communications link. Molecular sequence data should always be transferred using a file transfer program.

3.1.1 *The KERMIT file transfer program*

KERMIT is the name of a file transfer protocol implemented in programs of the same name on over 170 different computer systems, from the Apple II to Cray supercomputers. The success of KERMIT is largely due to the fact that the protocol specification is in the public domain and (most important from the users point of view) it is distributed free of charge, or for the cost of the distribution media. In order to use KERMIT, a version must be available on both systems.

The sequence of events in file transfer is relatively straight forward. Beginning with the local computer, the KERMIT program is started and after checking that communication speeds are all correctly set the user issues the CONNECT command to the remote host and login to his account. All KERMIT implementations have a minimum terminal emulation function to allow the user to login and run the KERMIT program on the remote computer system. However, in some versions the emulation is more sophisticated and may well be entirely adequate for running other programs. KERMIT is then started on the remote host and the systems made ready for file transfer. For example to download a file from the remote host, the KERMIT at that end is told to SEND the file. The user then escapes back (by typing a predefined combination of keys e.g. CONTROL and \ then C) to the local computer and tells the KERMIT there to receive a file. Once the file has been correctly received, the user then returns to the remote host, halts the KERMIT program there and logs out. Returning to his own computer he exits from KERMIT and the file is now available for use on the local system.

Most data is used and stored in the form of ASCII text files and these can be exchanged by file transfer programs without any significant problems. However, binary files (e.g. executable program files) are more difficult to exchange, particularly between computer systems with different storage methods and over networks which cannot transmit binary data. Most KERMIT implementations can send and receive binary files directly although this should be checked for your particular local and remote KERMIT versions. If your particular KERMIT (or other file transfer program) cannot exchange binary data then the responsibility for conversion of the binary codes to a coding based on ASCII characters (e.g. hexadecimal) is left to the user.

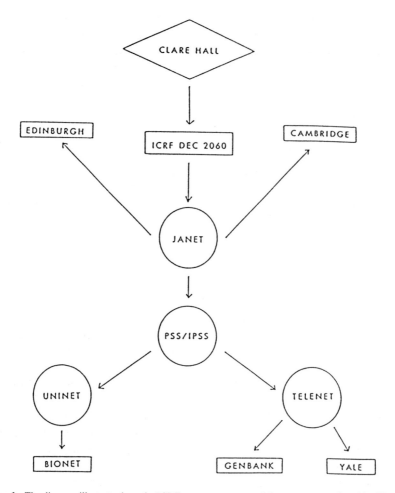

Figure 1. The diagram illustrates how the ICRF connects to some of the systems mentioned in this chapter.

3.2 Computer facilities at ICRF Clare Hall laboratories

Several of the systems described below have been used as part of the computing facilities available to scientists at the Imperial Cancer Research Fund (ICRF). The main computer is a Digital Equipment Corporation (DEC) 2060 with the TOPS-20 operating system. The majority of scientists have access to at least a VT100 compatible terminal for displaying text only.

Clare Hall is a newly opened research facility of the ICRF located about 20 miles from the central laboratories and the host computer. Access to the DEC-20 is provided via a system known as data over voice which utilizes the internal telephone system to carry the signals from the terminals and connects via *multiplexors* and modems at Clare Hall and the central site to complete the link to the computer (*Figure 1*). The communications lines operate at 9600 baud. The terminals used are CIFER T4 which offers DEC VT100, VT52 and Tektronix 4010 emulation, that is they provide text and monochrome graphics display.

Communication from the ICRF DEC-20 is through a PAD or PAD emulator software (X29PAD, DEC) providing terminal access to and from the UK JANET network. A gateway from JANET to the British Telecom national and international packet switch network PSS and IPSS provides routes to hosts in over 30 different countries.

4. THE MAJOR RESOURCES

4.1 **BIONET**

BIONET is funded by a 5 year co-operative agreement with the Division of Research Resources of the US National Institutes of Health. The IntelliGenetics division of IntelliCorp is providing the computer facilities, initial software and support. Users who wish to access BIONET must apply to the administrator at IntelliGenetics and pay a $ 400 subscription. Access is limited to academic and non profit-making organizations and applications from non-US users are accepted.

Approximately 1800 scientists now use the BIONET resource (3,8) which in the United States is principally accessed via the UNINET network. Direct access using the telephone system is also available. The system is based on a DECsystem 2060 running the TOPS-20 operating system. The resource provides a consultancy service which can be contacted by telephone, electronic mail and terminal links. The software available on the BIONET computer is essentially of three types.

(i)   The first comprises the core software based on nine programs provided by IntelliGenetics Inc. for manipulation and analysis of nucleic acid and protein sequence data.

(ii)   The second type comprises programs contributed by the user community.

(iii)   The third comprises a number of utilities including electronic mail, bulletin boards and file transfer programs.

For file transfer between microcomputers and the Bionet computer the public domain KERMIT (Section 3.1.1) program and other commercially available file transfer programs with error checking protocols are provided. To transfer a file from a microcomputer requires the user to run the file transfer program of choice on both the users' computer and the remote host. After initiating the dialogue between the two file transfer programs the data is sent and checked for errors as it is transferred.

4.1.1 *Organization of programs*

All programs in the core library share a similar layout. The program is started by giving its name to the system prompt, this is followed by the program banner and program prompt. Help is obtainable online at any time by typing a question mark to the prompt, which will return a list of what is expected in terms of commands, file names or parameter settings. More detailed help is available by typing HELP, HELP <topic name> or HELP <command name>. Training and reference manuals are also available and these include samples of typical sessions with the programs. Sample sessions are also available online. Typing 'DIR' to the system prompt shows the user a list of the files in his directory. For panic situations where the user wishes to stop the execution of a program pressing the CONTROL button and the letter C simultaneously a few times will achieve this, although it is not recommended for all programs.

```
Mail-From: BIONET created at 21-Mar-86 13:42:08
Date: Fri 21 Mar 86 13:42:08-PST
From: BIONET <BIONET@IC2060>
Subject: Re: GEN45 Gripe
To: VITEK.GINSBURG@IC2060
cc: BIONET@IC2060
In-Reply-To: Message from "VITEK.GINSBURG" of Fri 21 Mar 86 06:52:00-PST
Message-ID: <12192554487.19.BIONET@IC2060>

Dear Michelle,

Your discription of what happens if a person tries to use GENED with a
file that has been archived is perfectly accurate.  This is the way
the system works at the present time.  I have put in a suggestion of
an enhancement that would warn the user that the files have been archived.
When and if this gets done depends on how many other pressing problems
the system's operators have and also if the change is even possible.
It could be that there is no way for the operating system to distinguish
between an archived and a missing file.  In any case, I will enquire.

Marcia Allen
Bionet User Consultant
-------
R>
```

Figure 2. The Bionet user consultant responds to a GRIPE message sent by the author.

The programs are organized in a hierarchical structure; each level having its own prompt and set of commands. Returning to a higher level is accomplished by either QUITting that level or giving the '↑' command. Where appropriate the user can select a variety of parameters, influencing the exact operation of the program. This is done by giving the PARAMETER command. The options available are seen at the LIST command and are changed by typing the parameter name. The program then prompts for a new value. Generally the settings may be SAVEd after a session and reused at some future date. Results from the programs are saved by giving a SAVE, RECORD or COLLECT command, although this may depend on the program or command level. Other programs in the suite may sometimes be invoked from within a program by giving the new program name at the command level. QUITting that program returns the user to the command level of the original program.

Many of the programs operating on nucleic acid sequences recognize ambiguity codes to specify regions of uncertain or variable sequence and all programs contain a GRIPE command for sending complaints or suggestions to the IntelliGenetics staff. An example response from an IntelliGenetics user consultant to a GRIPE sent by the author is shown in *Figure 2*.

4.1.2 *Sequence databases*

The resource provides updated versions of EMBL, GenBank and PIR databases (see Chapter 4) as well as databases developed and maintained by IntelliGenetics. Vector-Bank contains maps of commonly used cloning vectors for use with cloning simulation programs. A library of restriction enzyme recognition sites contains data from all known restriction enzymes as well as lists of commercially available enzymes for use in several core programs (SEQ, PEP, CLONER etc). KeyBank contains predefined KEYs for use in the QUEST program.

The EMBL and GenBank databases are available in the original format or are specially formatted for use in the core programs. The nucleic acid databases are found in the sequence directories under <SEQUENCES>*.NIH for GenBank entries and

< SEQUENCES > *.MBL for EMBL entries, where * means any legal file name. Each file contains several sequences grouped according to some common feature (usually the source organism) which is reflected in the file name e.g. ECOA.NIH (*Figure 5*). During database searches it is more efficient to use indirect files to direct the search. These files contain the names of the files which actually hold the sequences, thus < IG > NIH-BACTERIAL.FLS is a list of all bacterial sequence files in GenBank. In programs this file name would be preceded by an @ to indicate that it is an indirect file. Protein sequences are found in the < PROTEINS > directory.

4.1.3 *MM*

This DEC program is concerned with the sending and receiving of messages and reading the bulletin boards. The program enables the user to organize a mail file. Options are specified at the top level to read the bulletin boards (of which there are over 30) on topics ranging from personal computer communications and software to protein engineering and yeast genetics. Mail is sent and read using the SEND and READ commands.

4.1.4 *GENED*

GENED is the sequence editor. Sequence data may be entered either using a keyboard or a digitizer. Amino acid sequences may be edited or entered providing the single letter code is used. GENED enables the sequence name, the sequence data, the comments and the sequence topology (linear or circular) to be modified and in the case of DNA the sequence can also be reversed.

Existing sequences are entered into the program by giving a LOAD command, the editing session begins by giving the subcommand EDIT or ESEQ, the former is a line oriented editor where commands and numerical argument are used to perform the editing changes. ESEQ is a display oriented editor (see Chapter 2), there is a continuous display of the sequence, and editing changes occur immediately at the cursor. Movement through the sequence is extremely rapid. The editing session ends by giving the QUIT command which automatically saves the session into user specified files.

4.1.5 *GEL*

GEL is the IntelliGenetics DNA sequencing management program (9). The manipulations are broadly divided into data entry, gel management and display and output of data. After the program banner has been displayed the program asks for a project name, it then asks if this is a new project. For new projects two files are created, one is the working file and records all alterations made to the data. The second file contains the individual gel readings and their associated comments. The program has a command level and several subcommand levels.

Contiguous stretches of sequences are called MELDs. The program has facilities for merging gels, joining sequences head to head, breaking up or removing sequences and adding comments. Scanning for the presence of vector or other specified sequences such as restriction sites can be done automatically. The program is designed for use with both dideoxy (chain terminator) and chemical (Maxam − Gilbert) sequencing. A valuable feature of the GEL program is that if required it can give a complete HISTORY of the edits, merges and disassemblies performed on the gels and melds.

4.1.6 *SIZER, MAP and CLONER*

These three programs enable the user to create restriction maps from enzyme digest data, display the maps, mark features on them and simulate cloning experiments. SIZER accepts fragment mobility data either input by digitizer or keyboard and compares them with the mobility of standard fragments to predict unknown fragment sizes. The data can then be saved and used in MAP to generate possible restriction maps. SIZER can also be called from within MAP.

CLONER can take previously created maps, including those from the VECTOR-BANK database and from MAP and insert one map within another, join previously existing maps, delete fragments, flip them and rotate circular maps. The maps can also be 'digested' to simulate a restriction digest on the final construction. Features such as origins of replication or antibiotic resistance genes can also be marked on the display.

4.1.7 *SEQ and PEP*

These are the DNA sequence (SEQ) and protein sequence (PEP) analysis (10) programs and certain features are common to both. Sequences can either be LOADed by specifying file names or created by entering GENED from within the program. Using the SITE option from within SEQ gives the user the option to produce restriction maps, list the sites recognized or predict the size and order of fragments that would be released from single or multiple enzyme digests. Within PEP a restriction site search can be performed on the reverse translation of a protein sequence, or a map of sites from enzymatic or chemical digestion of the protein can be produced.

Sequence comparisons are performed using either the SEARCH or ALIGN options. The latter aligns two separate sequences using the modified Needleman – Wunsch algorithm (see Chapter 13 and refs 11,12). Sequences are divided into stretches up to 200 nucleotides long by setting the SEGMENTSIZE parameter, and each segment is compared to nearby segments from the second sequence. The program then tries to identify perfect matches, ranking the lesser matches. The user can set the GAPPENAL-TY, GAPSIZEPENALTY and MISMATCHPENALTY, which introduces penalties for the number of gaps, their size and the number of mismatches in the sequence respectively. SEARCH is more suited to intrasequence comparisons and for locating repeats, complementary regions and dyad symmetries in DNA sequences.

4.1.8 *IFIND, QUEST*

These are the major database searching programs. IFIND is a similarity and alignment program and is run in two stages. Firstly, at the top level, the query sequence is compared to sequence files specified by the user and the matches located. The sequence file list can be DISPLAYed at this point and the distribution of matches examined or the results saved for later analysis. The alignments with the query sequence and individual sequences from the file list are produced by giving the ALIGN command. Parameters to produce the optimal alignment can be reset at this level. The output can be varied to show either sequence alignments or include comments from the original sequence file.

QUEST is a flexible high performance pattern matching program for locating se-

```
@quest
                    > O <
                    O: 10   IntelliGenetics
                    > O <

          QUEST - Biological Database Management System

      Version 4.5 for TOPS-20 - October 1985 - Timeshare version
                Copyright (C) 1985 by IntelliGenetics, Inc.
    IntelliGenetics and QUEST are trademarks of IntelliGenetics, Inc.


      For information enter one of these commands after "QUEST:":

      Introduction        - Gives an overview of this program
      ?                   - Lists commands
      Help                - Lists commands with short explanations
      Help topics         - Shows a list of informative topics
      Help help           - Explains how to use on-line assistance


  If you would like help setting up your search in the sequence databases,
  type GUIDE to the "QUEST:" prompt.

  QUEST [Comments]: guide

  Your current settings are:


          COLLECT = NO
          CONTEXT = NO
          PROMPT = YES
          OUTPUT = VERBOSE

          SCOPE = COMMENTS


  The GUIDE command will help you change these for your search.
  Do you want to change any of these settings? (Y or N):   (<CR>=Y)

  Will you be searching sequence files?:   (<CR>=Y)

  Will you be looking in the (C)omments or (S)equence portions of sequence
  files? (C or S): s

  SCOPE = SEQUENCE

  Do you want to collect the sequences into a file? (Y or N):   (<CR>=Y) Y

  File to write sequences to? ADASRCH.RES

      Opening ADASRCH.RES.


  Do you want to save a list of the names of the files containing
  sequence hits? (Y or N):   (<CR>=Y) N
```

Figure 3. Running a program on Bionet. The QUEST program is run using the GUIDE option to set up the search parameters.

quence patterns in molecular sequence databases (13). User defined patterns, called KEYs, can be simply constructed and combined into complex patterns using the logical operators AND, OR, THEN or & at the KEY subcommand level. Special symbols in the pattern description language enable ambiguous patterns to be described and QUEST can be instructed to narrow the scope of its search to specific regions of the data such as within lines, pages, files, comment fields or sequence data only. Various conventions exist for defining the KEY patterns and a database of predefined KEYS (KeyBank) can also be used. To help the user set up a search QUEST has a GUIDE option which engages the user in a dialogue to establish his requirements (*Figure 3*). *Figure 4* illustrates

```
KEY [Comments]: Q
QUEST [Comments]: HELP KEY
```

The Key command is used to bring you to the KEY: prompt. Here you can
execute a number of commands to define, change or manipulate KEYS, the
patterns of a search in QUEST.

You can Load keys from a file either at the top level at the QUEST:
prompt or at the KEY: prompt. Any editing changes, additions or
deletions must be made at the KEY: prompt.

If you add any new keys during a run in QUEST, you may wish to save
them into the file of keys that you have been using. You should first
List all of the keys with the List or Print command and Delete any
duplicates before saving the keys to a file. All of the keys
currently in QUEST will be saved to a single file when you issue the
Save Key command.

```
QUEST [Comments]: HELP CONJ
```

The conjunctions AND, OR, THEN, NOT, and & are used to connect simple
keys. Using parentheses with the conjunctions allows complex patterns
to be developed by appropriately grouping the simpler patterns.

QUEST uses conjunctions in the following manner:

THEN is used between patterns to make a match only
      when the second pattern follows the first.

&   acts similarly, but requires that the second pattern
      directly follow the first pattern. '&' may be used
      to split very long patterns.

OR   connected patterns will match when either the first or
      second pattern matches.

AND   connected patterns will match when both patterns match.

Figure 4. Obtaining HELP within a program. Here help is requested on KEYs and the CONJUNCTIONS
used to define patterns in QUEST.

how extensive online help can be used as an alternative to a manual to get an explanation of what type of information the prompt is soliciting and why.

The form taken by output from the program can be modified from a terse list of file
names to a more verbose output which includes comments from the original sequence
file (*Figure 5*). The names of sequences and files where the pattern was found can also
be stored for later analysis (e.g. with SEQ). Keys can also be saved for subsequent
searches. It is recommended, especially when a complex pattern is being constructed,
that the user test the KEYs on a few sample sequences to ensure that the desired match
will be found, before searching the database.

4.1.9 *Utility programs*

XSEARCH (an ancestor of QUEST) is a program for rapidly searching text files for
character strings. It is an efficient program but is only able to identify exact matches
with the target string. The output consists of the name of the file and the line where
the match was found. Targets can be combined using logical operators. By specifying
an equivalence file the program can be made to ignore line feeds, tabs, equate upper
and lower case letters, and T with U. The program can therefore be made to restrict
its search scope in a similar way to QUEST. The value of this program is its great
speed. It is ideal for pre-screening the database with parts of QUEST keys in order

```
searching IG:[SEQUENCES]ECOA.NIH
     ECOACE
     ECUALASYN
     ECOALKA


ECOALKA
: DEFINITION   E. coli alkA gene encoding 3-methyladenine DNA glycosylase II.
: LOCUS        ECOALKA     1954 bp ds-DNA              entered    09/03/85
:              complete cds.
: ACCESSION    K02498
: KEYWORDS     3-Methyladenine DNA glycosylase II; alkA gene.
: SOURCE       E. coli K12 (strain W363) DNA, clone pYN1000.
:   ORGANISM   Escherichia coli
:              Prokaryota; Bacteria; Gram-negative facultatively anaerobic rod


: COMMENT      Putative promoter region sites identified by [1] are: ribosome
:              binding site at position 506-566, -35 region at position 492-49
:              and -10 region at position 508-513.  Draft entry and printed
:              sequence kindy supplied by Y. Nakabeppu 02/22/85.
: FEATURES     from  to/span    description
:     pept     572    1420      3-methyladenine DNA glycosylase II
:   SITES
:     refnumbr     1        1      numbered 1 in [1]
:     ->mRNA     553        1      alkA mRNA start
:     ->pept     572        1      alkA cds start
:     pept<-    1420        1      alkA cds end
: BASE COUNT    458 a    493 c     547 g    456 t
: ORIGIN       EcoRI site; 43.2 min on K12 map.

       449         459         469         479         489         499         509
aaataatccc catgccggtg aagaaggggc gtgactttag cgaaatgttg ccgtcgcgac aaccggaata


: Hit starts at base 510 of ECOALKA and ends at 524
v
       519     524
tgaaagcaaa gcgca


       534         544         554         564         574         584         593
gcgtctgaat aacgtttatg ctgaaagcgg atgaataagg agatgcgatg tataccctga actggcagc
-------------

(S)ave-sequence, (Q)uit, (N)ext-sequence, (C)ontinue-this-sequence?  (<CR>=S)

     ECOAMPCFR
     ECOAMPCP
     ECOAMPIS1
     ECOAMPIS2
     ECOASD
     ECOAMPCR
     ECOALAS
1 hit found in IG:[SEQUENCES]ECOA.NIH
```

Figure 5. A 'hit' is found with QUEST, the key is called ADABOX. The *Escherichia coli* Ada gene positively controls its own expression (31) and that of several other genes including alkA. The sequence ADABOX is found within the promoter regions to which the Ada protein specifically binds.

to estimate the likely outcome of the more flexible but more costly QUEST search (*Figure 6*).

FINGER is a program for interrogating the status of other users of the Bionet DEC 2060 invoked by typing FINGER <argument>. The value of *argument* modifies the information that FINGER provides. For example if no argument is supplied then a terse alphabetical summary of all system users is displayed. If argument is a users account name then FINGER shows whether the user is currently logged in, if he is not then the program displays the last time he did so. It also shows when the user last read his mail file and whether there are any pending messages. It is also possible for a user to leave a PLAN of intended whereabouts in his absence. If he is FINGERed the plan

TARGET	EMBL DATABASE (August 1985)	GENBANK BACTERIAL SEQUENCES (November 1985)
AAA	103458	16596
GCGC	9532	3207
AAAGCGC	96	49
AAAGCGCA	30	12

Figure 6. Some idea of the quality of a pattern may be found by using subsets of a sequence in a program such as XSEARCH. The results shown here were derived using subsets of the ADABOX (31) sequence.

will be displayed to the FINGERer and perhaps suggest alternative means of communication.

Typing WHOIS < name > or WHOIS < topic > enables the user to find out about other users of the system by name, location or their particular research interest.

4.2 **Cambridge**

The Cambridge University mainframe IBM 3081 machine is located on the Cambridge University Data Network and is available to outside users through JANET or PSS. The machine runs under the MVS operating system, on top of which is a locally written command line processor called Phoenix. Charges for the use of the machine are based on a predefined resource unit, the cost of which varies with the category of user. Online help is available, and a variety of manuals and leaflets dealing with various aspects of the operating system are available. Queries may also be made by phone. Some programs require the use of a graphics terminal. Enquiries about access to the Cambridge resource should be addressed in the first instance to the External Receptionist, University of Cambridge Computer Service, at the Computer Laboratory.

4.2.1 *The programs*

The available programs operate on sequential files or on partitioned data sets. A sequential file is a single file, a partitioned data set may contain several members, each of which may be used as a single file. Files are examined by typing XFILE and members of a partitioned data set by typing XPDS.

The database files and associated programs are accessed by loading the appropriate library called EMBL.LIB:DNA. Typing HELP to the system READY prompt allows the user to see what online help is available. Having loaded the library, typing START displays a menu with eight options offering various kinds of help on the molecular biology software and databases. Using the 0 option takes the user back to the READY prompt where the programs can be run by simply typing their names and waiting for further prompts for input.

PACKAGE is an interactive set of programs designed for the retrieval and analysis of nucleic acid sequences from the EMBL database. A list of options are presented any of which may be specified by typing a single letter to the program prompt. The options include *DIRECT* retrieval using a sequence identifier, *SEARCHING* for keywords, *LISTING* the retrieved sequence or entering ZSEQ (a package of analysis

```
Ready
INDEX
GENBANK
Look-up facility for the sequence database indexes

search expression can be any combination of characters excluding spaces
If you are not reasonably specific, you will get too much output

name of database ( embl, genbank, pir, claverie or doolittl):
the indexes for GENBANK are as follows:

embl.GENBANK.index 13 Mar 86 11.43

ACCESSNO    AUTHOR      KEYWORD      SDIRECT

Directory blocks: used 1 free 49

name of index for searching: SDIRECT
enter search expression or sequence identifier:  POLYMERASE

searching embl.GENBANK.index:SDIRECT for "POLYMERASE"

HUMRSAHB1     HUMAN ALU FAMILY REPEAT CONTAINING RNA POLYMERASE III TEMPLATE.
HUMRSAHB2     HUMAN ALU FAMILY REPEAT CONTAINING RNA POLYMERASE III TEMPLATE.
XENRG40IN     XENOPUS LAEVIS POLYMERASE I PROMOTER REGION. 200BP
ECOPOLA       E.COLI K12 POLA GENE CODING FOR DNA POLYMERASE I. 4127BP
ECOPOLAMU     E.COLI B POLA GENE CODING FOR DNA POLYMERASE I. 201BP
ECORPOAPS     E.COLI RPOA AND RPSD GENES CODING FOR RNA POLYMERASE ALPHA-SUBUNIT
EBVL1         EPSTEIN-BARR VIRUS RNA POLYMERASE II PROMOTER REGION L1. 101BP
EBVL2         EPSTEIN-BARR VIRUS RNA POLYMERASE II PROMOTER REGION L2. 101BP
EBVR          EPSTEIN-BARR VIRUS RNA POLYMERASE II PROMOTER REGION R. 101BP
```

Figure 7. The INDEX program, here the user has opted for the GenBank database and short directory index.

programs). The relevant documentation may also be examined.

RETRIEVE will find sequence entries in the EMBL database by searching specified lines of the feature tables (see Chapter 4). The search may be conducted using single or multiple search expressions that can be linked using the logical operators AND, OR or NOT. Fourteen linecodes are allowable and the results of the search may be saved for later analysis.

INDEX automatically searches for keywords in an index of entries for the specified database (*Figure 7*). In some cases more than one index is available (see Chapter 4). The search expression may be any combination of characters excluding spaces. The program prompts the user to save the results. A new search can then be initiated with a different index or database if required.

Other utilities include NEWSEQ, a program to place sequences in EMBL format prior to submission to Heidelberg. EMBLSEND will send messages to other users of the system and EMBLMAIL will send mail to outside users. NEWS is a facility for obtaining information about new database releases and programs.

The remaining programs are divided into menus covering DNA sequence analysis, M13 shotgun sequencing projects, and protein sequence analysis. Programs are run by typing their name to the system prompt. The user is then prompted for more information as necessary.

4.2.2 *Databases*

The databases supported include those from EMBL, GenBank, the Protein Identification Resource (PIR) and a database, called CLAVERIE (also known as PGtrans in other systems), containing protein sequences translated from GenBank. The EMBL database is available as a partitioned data set or in four sequential files, split according to an alphabetical ordering of the entry identifier used in the indexes. GenBank is available as a partitioned data set with each partition corresponding to a group of source organisms and PIR is presented in EMBL format. Manuals for the databases are available online together with the various indexes for entry features such as sequence identifiers, accession numbers, keywords and authors (see Chapter 4).

4.2.3 *DNAMENU*

DNAMENU is the top level menu for selecting DNA sequence analyses. These include programs for converting sequences to conform with different database formats, programs for counting nucleotide frequencies and searching for direct and indirect repeats. Separate programs predict secondary structure, search for protein coding regions, oligonucleotide probes for a protein sequence and restriction sites. Comparison programs include sequence alignment, database searches and dot matrix comparison as well as phylogenetic analyses (see Chapter 14). The database searches can be performed with either a short target sequence or sequences from 50 bases to 20 kilobases long.

(i) *DIAGON.* This program (14) identifies regions of similarity between two sequences and the comparison can be displayed on a simple graphics terminal as a dot matrix (see Chapter 13). The program is interactive and the user may locate the position in each sequence of a region of interest in the dot matrix using the CROSSHAIR option to place the crosshair cursor. Online help is available and options within the program are selected by typing single characters selected from a menu. Scales may be drawn along the axes of plot by selecting the RULER option. The program will compare either DNA or protein sequences.

(ii) *ANALYSEQ/ANALYPEP.* These programs (15) are general DNA and protein sequence analysis programs which make extensive use of a graphics terminal to display results. Options are selected by number from at least two menu levels. The main menu options select submenus with options for GENERAL file manipulations and miscellaneous analyses, SCREEN CONTROL, GENE SEARCH BY SIGNAL and GENE SEARCH BY CONTENT for predicting protein coding regions (Chapter 10), TRANSLATION AND CODONS for translating and analysing a coding region, STRUCTURES AND REPEATS for intrasequence similarity search and STATISTICAL ANALYSIS OF CONTENT for compositional analysis.

When restriction analysis of a sequence is required the user may select subsets or individual enzymes and the output can be displayed either as a list of recognition sites or their positions represented graphically as a map. The estimations of coding potential of a stretch of DNA include searches for start or stop codons, ribosome binding sites, splice junctions or polyadenylation sites. All of these options as well as those searching for particular types of secondary structures (e.g. hairpins, regions of Z-DNA) provide a graphical display indicating the structural regions or sites in the sequence. The user may then select the CROSSHAIR or RULER options to move the crosshair to pinpoint the precise location of selected features or read the position from the numbering on the axis. It is also possible at any time to redefine the region of the sequence(s) used for analysis. Since the display area on the screen is a constant size, defining a shorter stretch of sequence effectively magnifies any features of interest.

```
                              Welcome to the
              University of Wisconsin Genetics Computer Group

                           Version 3.0, June 1985

                   Help is available with the command $ GenHelp

$ genhelp

HELP


                                   GENHELP

    You can get help on any topic in the UWGCG Program Manual by typing the
    topic(s) name(s) from the list below


    Additional information available:

    ASSEMBLE    BACKTRANSLATE      Batch_Queue           BESTFIT     ChouFas
    CODONFREQUENCY          CODONPREFERENCE    ComCheck    Command_Lines
    COMPARE     COMPOSITION        COMPRESS    COMPTABLE   Comp_Tables
    CONSENSUS   CORRESPOND Count   CRYPT       DETAB       DIVERGE     DOTPLOT
    Down        Enzymes    EXTRACT Fetch       FileCheck   FIND
    FINGERPRINT            FITCONSENSUS        FOLD        FRAMES      FROMIG
    FROMSTADEN GAP         GAPSHOW  GELINIT    GenHelp     Global_Switches
    HELP        HP7221     HP7475   Local_Files           Logical_Names
    MAP         MAPPLOT    MAPSORT  Moment     Names       Naming_Files
    Naming_Seqs            NoBeep   Nodoc      NoProt      OneCase     Over
    OVERPRINT   Partial    PEPDATA  PEPPLOT    PEPTIDEMAP PEPTIDESORT
    PlotChou    PlotTerm   POSTER   PRETTY     Private     Public      PUBLISH
    REFORMAT    REPEAT     REVERSE  SEARCHSET  SEGMENTS    SEQED       SeqFormat
    Shift       SHOWFILES  SHUFFLE  SIMPLIFY   SPEW        STATPLOT    Steal
    STEMLOOP    Store      Strings  Symbols    TERMINATOR TESTCODE     TOIG
    TOSTADEN    TRANSLATE  TypeData Up         VMS_Commands            VOLUME
    VT241       WINDOW     WORDSEARCH

Topic?
$
$
```

Figure 8. The figure shows the HELP menu from within the UWGCG program package exactly as it would appear on the VDU. The program asks the user for further information at the TOPIC prompt.

4.2.4 *GELMENU*

The programs provided here are for managing sequencing projects. Restriction site maps may be reconstructed from fragment length data. The majority of the menu is concerned with Staden's DB system (16 − 18, Chapter 8) for assembly and management of sequencing projects.

(i) *The Staden DB system.* The DB system is a complete management system for shotgun sequencing projects. The system stores the original sequencing gel readings, compares new readings to gels already incorporated into the project database and reports any that overlap. By aligning overlapping gel sequences a consensus contiguous sequence (contig) is built. Initially there may be many contig sequences but as the project progresses connections will be found and eventually a single sequence for the project will emerge. Extensive provisions are provided for editing the sequence data and the structure of the project. A detailed description of the DB system is provided in Chapter 8.

4.2.5 *PEPMENU*

PEPMENU is the top level menu for selecting protein sequence analyses. The options in this menu are equivalent to those provided under DNAMENU i.e. database search, dot matrix comparison and alignments, phylogenetic reconstruction and secondary structure prediction according to the method of Garnier, Osguthorpe and Robson (19, Chapter 12).

4.3 **Edinburgh**

The Edinburgh University system is based on a VAX 11/750 running the VMS operating system. The computer is connected to the Edinburgh University network and through this to JANET and PSS. Charges for CPU time are made by the University to external users of the service. The charge rate is determined by whether the user is from a commercial or non-commercial institute. Potential users should contact Dr Andrew Coulson in the first instance.

4.3.1 *The programs*

The programs available include the University of Wisconsin Genetics Computer Group (UWGCG) programs (20, *Figure 8*) based on a software tools approach to program design (21). Each program is designed to perform only one function and because minimal user input is required the programs are easy to use. Complex analyses are performed by using several programs in succession. The user selects the tools specific to the task he intends to perform.

The DB system of Rodger Staden (16 – 18, Chapter 8) is provided for the management of sequencing projects, and for extraction of data from protein and DNA sequence databases the programs PSQ and NAQ (22) are available. The collection also includes programs for protein secondary structure prediction. Certain programs may not be available to all network users because special output devices such as flat bed plotters and graphics terminals are required. These programs include DIAGON, ANALYSEQ, PLOTCHOU AND PEPPLOT.

4.3.2 *The databases*

The DNA sequence libraries supported are those of GenBank and EMBL. The protein sequence database comes from PIR (NBRF). Individual sequences can be identified in the databases by a variety of programs. The STRINGS program will locate sequences by keyword and NAMES will identify sequences by entry name. An entry can be located by using a short sequence pattern with the FIND program. Sequences can be copied from the libraries by using FETCH, whilst STORE will add locally generated sequences to the libraries. Files containing sequences (which may contain ambiguity codes) may be referred to by their names or as a group using a file of file names (an indirect file). Indirect files are processed as a file group by preceeding the file name with an @. The output from some programs is in the form of an indirect file.

4.3.3 *UWGCG programs*

The UWGCG programs are run by typing their names to the system prompt. The GENHELP command will take the user to the online help (*Figure 8*) and the full program documentation is also available online. When requesting user input with a prompt the programs indicate the default answer (the one used if the user just types a carriage return). The behaviour of some programs may also be modified when invoked from the operating system by using what are called command line switches. The selection of a switch will turn a certain behaviour on or off [see description of MAP, Section 4.3.3 (iii)]. Program output is written automatically to a file ready for use in other programs in the suite. A default file name will be supplied by the program if the user fails to do so.

The program are loosely divided into six groups: sequence comparisons, restriction enzyme and chemical and proteolytic cleavage site mapping and searching, prediction of nucleic acid secondary structure and protein secondary structure, pattern recognition and sequence manipulation. In the software tools style, certain UWGCG programs preprocess data for use by other programs. This has an important benefit when using these programs on a remote online service, since recovery from a mistake is simply a matter of typing a few commands to the preprocessing program. The more lengthy analysis need only be run when all the data are confirmed as correct. Where necessary a program will prompt the user to REFORMAT sequences not presented in a compatible format.

(i) *Sequence manipulation.* The sequence editor is called SEQED and is screen oriented (see Chapter 2). To facilitate sequence entry, the editing actions of individual keyboard keys may be redefined (e.g. moving the characters A, C, G, T to four adjacent keys) using the GELINIT option. DNA sequences can be TRANSLATEd, the user specifying the start and stop positions, and BACKTRANSLATE will generate a DNA sequence from a protein sequence using a codon frequency table. Nucleotide sequences may also be REVERSEd and complemented. The SHUFFLE program will produce a randomized copy of a sequence of the same composition as the original for use when estimating the statistical significance of a sequence comparison or pattern match.

(ii) *Sequence comparison.* Several programs are provided for sequence comparison. BESTFIT compares two sequences and produces an optimal alignment of the most similar segments. Where sequences must be aligned along their whole length GAP finds the optimal alignment by inserting spaces in either sequence as appropriate. Programs COMPARE and DOTPLOT compare sequences using the dot matrix method (see Chapter 13). Where the user wishes to compare a single sequence with a group of sequences WORDSEARCH and SEGMENT are used to identify and align similar segments and WORDSEARCH is fast enough to use in database searches. CONSENSUS generates a consensus matrix (Chapter 10, Section 2.2) from a set of pre-aligned sequences that summarizes the occurrence of each nucleotide present at each position in the pattern. FITCONSENSUS then uses this matrix to search for the consensus in a test sequence.

(iii) *Mapping.* MAP displays a restriction map of a sequence with translations beneath the DNA sequence. Using command line switches the user may opt for single cutters, specify whether the sequence is circular or linear and allow mismatches within the sites. MAPSORT will produce both the co-ordinates and sizes of fragments generated by single or multiple restriction enzyme digests. PEPTIDEMAP and PEPTIDESORT per-

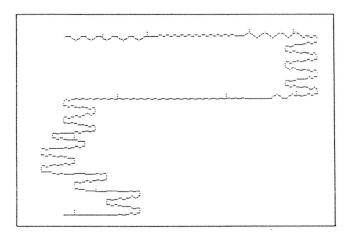

Figure 9. The diagram is the result of a 'screen dump' from the program EPLOTCHOU, where a secondary structure prediction of the first 100 residues of polyoma large T antigen has been requested. Here the display is generated on a terminal emulating a Tektronix 4010. (N.B. the remote host about 500 miles away.)

form analogous operations on peptide sequences where the products may be the result of chemical or enzymatic cleavage. PEPTIDESORT separates the peptide fragments based on weight, position in the sequence, and retention time in high pressure liquid chromatography at pH 2.1. The user may in any of these programs select individual enzymes or make his own special enzyme file.

(iv) *Secondary structure prediction.* STEMLOOP is used for finding inverted repeats in nucleotide sequences. The user may set minimum stem length and maximum loop size, as well as specifying a minimum number of bonds per stem. The program can be run repeatedly with the user resetting the parameters to find the most likely structures. The stems may be sorted by length, position or number of bonds. FOLD finds a structure of minimum free energy based on published tables of stacking and loop destabilizing energies (see Chapter 11). The program does take several minutes of CPU time and should be run in batch for sequences longer than a few hundred bases in length.

CHOUFAS creates a file containing the Garnier predictions (19) of local secondary structure where the measurements of alpha helix, beta sheet and turn potential are those of Chou and Fasman (23). The output file of CHOUFAS would normally be passed to the program PLOTCHOU for a plot of the protein secondary structure (*Figure 9*). Protein hydrophobicity (24) and secondary structure propensities can also be plotted using PEPPLOT. The output from these programs is normally passed to a *flat-bed plotter*. However, when a graphics terminal which conforms to the Tektronix T4010 standard is available, the programs EPLOTCHOU and EPEPPLOT will plot output on a display screen. This works satisfactorily over the JANET network.

(v) *Sequence composition and pattern recognition.* The program CODONPREFERENCE is used for locating coding regions in genes by comparing codon usage in each reading frame with codon usage in a known coding region (see Chapter 10). The user is asked to specify the start and stop base number for output of coding poten-

```
PSQ> FIND H-RAS
TVHUH    Transforming protein p21 (H-ras-1) - Human
VHV3H    Transforming protein (H-ras) - Harvey murine sarcoma virus
   2 entries found
PSQ> FIND /ADD
Find what: K-RAS
TVHUK    Transforming protein 1 (K-ras) - Human
TVHU2K   Transforming protein 2 (K-ras) - Human
TVHV2K   Transforming protein 1 (K-ras) - Kirsten murine sarcoma virus
   3 entries added
PSQ> USAGE /CURRENT


TVHUH    189 residues
Transforming protein p21 (H-ras-1) - Human

Composition  189 residues
   11  5.8%  Ala A    11  5.8%  Gln Q    14  7.4%  Leu L    11  5.8%  Ser S
   12  6.3%  Arg R    14  7.4%  Glu E    11  5.8%  Lys K    11  5.8%  Thr T
    5  2.6%  Asn N    13  6.9%  Gly G     5  2.6%  Met M     0  0.0%  Trp W
   15  7.9%  Asp D     3  1.6%  His H     5  2.6%  Phe F     9  4.8%  Tyr Y
    6  3.2%  Cys C    11  5.8%  Ile I     6  3.2%  Pro P    16  8.5%  Val V


TVHUK    189 residues
Transforming protein 1 (K-ras) - Human

Composition  189 residues
    9  4.8%  Ala A    10  5.3%  Gln Q    12  6.3%  Leu L     9  4.8%  Ser S
   12  6.3%  Arg R    15  7.9%  Glu E    16  8.5%  Lys K    13  6.9%  Thr T
    4  2.1%  Asn N    11  5.8%  Gly G     5  2.6%  Met M     0  0.0%  Trp W
   14  7.4%  Asp D     3  1.6%  His H     6  3.2%  Phe F     9  4.8%  Tyr Y
    5  2.6%  Cys C    15  7.9%  Ile I     5  2.6%  Pro P    16  8.5%  Val V


TVHU2K   188 residues
Transforming protein 2 (K-ras) - Human

Composition  188 residues
    9  4.8%  Ala A     9  4.8%  Gln Q    11  5.9%  Leu L    10  5.3%  Ser S
   10  5.3%  Arg R    13  6.9%  Glu E    21 11.2%  Lys K    13  6.9%  Thr T
    4  2.1%  Asn N    12  6.4%  Gly G     6  3.2%  Met M     0  0.0%  Trp W
   16  8.5%  Asp D     4  2.1%  His H     6  3.2%  Phe F     8  4.3%  Tyr Y
    4  2.1%  Cys C    12  6.4%  Ile I     4  2.1%  Pro P    16  8.5%  Val V


TVHV3H   241 residues
Transforming protein (H-ras) - Harvey murine sarcoma virus

Composition  241 residues
   27 11.2%  Ala A    11  4.6%  Gln Q    16  6.6%  Leu L    12  5.0%  Ser S
   18  7.5%  Arg R    16  6.6%  Glu E    11  4.6%  Lys K    12  5.0%  Thr T
    5  2.1%  Asn N    17  7.1%  Gly G     7  2.9%  Met M     0  0.0%  Trp W
   17  7.1%  Asp D     3  1.2%  His H     5  2.1%  Phe F     9  3.7%  Tyr Y
    6  2.5%  Cys C    11  4.6%  Ile I    18  7.5%  Pro P    20  8.3%  Val V


TVHV2K   189 residues
Transforming protein 1 (K-ras) - Kirsten murine sarcoma virus

Composition  189 residues
    8  4.2%  Ala A    11  5.8%  Gln Q    13  6.9%  Leu L    10  5.3%  Ser S
   12  6.3%  Arg R    15  7.9%  Glu E    16  8.5%  Lys K    14  7.4%  Thr T
    4  2.1%  Asn N    10  5.3%  Gly G     5  2.6%  Met M     0  0.0%  Trp W
   13  6.9%  Asp D     3  1.6%  His H     6  3.2%  Phe F     9  4.8%  Tyr Y
    5  2.6%  Cys C    13  6.9%  Ile I     5  2.6%  Pro P    17  9.0%  Val V


Cumulative frequencies from 5 entries

Composition  996 residues
```

Figure 10. Using the program PSQ (a) the user retrieves H-RAS sequences. (b) The FIND /ADD switch adds further sequences to the list. (c) The USAGE /CURRENT switch is used to display the amino acid content of the sequences in the current list.

tial to a plotter. CODONFREQUENCY is used to generate the codon usage tables by summing the codons in as many regions as the user requires. The output from this program is also used in several other programs. FRAMES marks the positions of the start and stop codons in all six reading frames. It will only place a mark where this would change the reading sense of the frame. A command switch /ALL will alter the program so that all the starts and stops will be marked. Using a set of dinucleotide frequency tables TERMINATOR will search for prokaryotic factor independent RNA polymerase terminators.

REPEAT is a program that allows the user to search for direct repeats. The user is asked for a minimum repeat length and stringency, and a range within which to search. The search may be repeated with refined parameters. FINGERPRINT simulates the result of a nucleotide sequence being digested with ribonuclease T1, the program identifies the labelled products of an RNA sequence fingerprint.

4.3.4 *PSQ and NAQ*

These programs (22) are concerned with retrieving and performing simple analyses of sequences retrieved from a database. After entering the program the user may give a variety of commands to the program prompt some of which may be qualified by modifiers (e.g. COMMAND /MODIFIER1 /MODIFIER2). Online help is available by typing HELP. Typing a ? to the program prompt displays a list of all the available commands together with their modifiers. The user may either enter the command and after the carriage return will be prompted for further action or he may give the whole command sequence on one line.

The databases consist of collections of entries which contain a title, text and a sequence. Generally the entries are located by using the FIND command (*Figure 10*) and the list retrieved can be modified in several ways. FIND /ADD < string > will search the database for the specified string and add any matches to the current list whilst FIND /SUBTRACT < string > will remove any matches from the current list. Entries may also be located by AUTHOR or by searching the entire text for a keyword or phrase, in this case the command FIND is modified by typing FIND /TEXT. Entries may also be SELECTed on the basis of such numeric information as length, molecular weight, or composition.

Four display commands are available. TYPE will display the title, sequence composition, references and feature table. USAGE displays the composition tables and accumulated composition tables for the current list of sequences (*Figure 10*). VIEW shows sequence features mapped from the feature table for the entry and REPORT will display numeric or taxonomic information. The COPY command retrieves an entry from the database using the entry code and copies it into a named file. EXTRACT enables the user to create new sequences by editing existing ones. The MATCH command searches for subsequences. Mismatches or ambiguous matches may occur between the search and target sequence but deletions or insertions are not admitted. ENZYME will perform restriction enzyme site searches on a DNA sequence or reverse translations of a protein sequence.

Certain commands are unique to the NAQ program. For example, HYPOTHETICAL will search a sequence and its complement for possible coding regions and TRANSLATE will display the sequence and its amino acid translation in all three reading frames. The user may search a database index for KEYWORDs or SEARCH for specified text strings in the titles, comments or feature tables. To leave the program the user types QUIT to the program prompt. Sequences retrieved using PSQ and NAQ must be edited slightly and then REFORMATted prior to use in the UWGCG suite.

4.3.5 *Utilities*

Electronic mail is read or sent using the MAIL program. Typing HELP to the MAIL prompt presents the user with a list of the available commands and also prompts for a topic title, after which the user is presented with a description of the command and any parameters that may modify its action. To read a message the user types READ

```
***** Log file opened at 06:35:34-GMT on March 28, 1986 *****

BREAK-CHARACTER (is) ~
ESCAPE-CHARACTER (is) `
Calling Remote DTE                    [OK]
Please enter your authorisation and address required in form:
(user,password).address
>                                   Call connected to remote address

Connecting to the PROPHET system.
NIH - PROPHET 7.02M.10 03:35:47 TTY35 system 1410/1359
Connected to Node DNA(3) Line C 10
Timesharing will cease in 1 hours 25 minutes, at 05:00.
          Welcome to the PROPHET system.

Please LOGIN


.LOGIN
JOB 16 NIH - PROPHET 7.02M.10 TTY35
[LGNTCI Timesharing will cease in 1 hours 25 minutes at  5:00:00]
C
Password:

03:36   28-Mar-86      Fri
Good morning, Michelle

Welcome  to  the  GenBank(R)  Genetic  Sequence  Data  Bank
You are using GenBank Release 41.0  (20 MARCH 1986)
This is PROPHET 71

*** System is scheduled to come down at 5:00 A.M. ***

CGENBANKMENUSS

Welcome to the GenBank Menu System!
(If you get lost, try typing ?<GO>)

MAIN MENU:

1.LEARN            about the GenBank System
2.ORDER            a Users' Guide
3.COMMUNICATE      with other users
4.GROUPS           of sequences
5.SEQUENCE         operations
6.SOFTWARE         Clearinghouse
7.HANDY            operations
8.DONE             with GenBank menus, for now

Choice:
```

Figure 11. Calling and logging onto a remote computer: GenBank. The entire procedure and welcome banner from GenBank is shown. Having logged on successfully the user has requested the GENBANKMENUS.

and to send a message he types SEND to the program prompt, he is then prompted for a user name to which the message should be sent, after which the message is typed. Finally pressing the CONTROL and Z keys will send the message. The text editor is EDT.

4.4 GenBank

GenBank (25) is a project sponsored by the United States Government and has primary responsibility for distributing the nucleic acid sequence library on floppy disk, magnetic tape or in a published compendium (26). The GenBank resource also supports online access in conjunction with the PROPHET system (27) through direct telephone dialling as well as the TELENET network. The PROPHET system is not, however, accessible from the GenBank computer. The data bank was created under the aegis of the National Institutes of Health, and the Computer and Information Sciences Division of Bolt, Berenek & Newman Inc. maintain the computerized data centre and distribute the database. The costs of access to GenBank are charged (by connect time) to the user according to the time of day and are lower during the night (US Eastern Time). Comprehensive manuals are provided as well as online help and a hot line telephone number. In order to abort any output the user must use CONTROL and W (not CONTROL

139

and C) as this will break the connection with the host. A rather unusual feature of the GenBank system is that the ESCAPE key is used to complete a command rather than RETURN.

4.4.1 *The programs*

The system tools are menu based, and sophisticated methods are provided for locating target entries, sequence patterns and specifying levels of base mismatching. The user can select to use the system in one of two ways: either using system commands or a menu based subsystem (GENBANKMENUS, *Figure 11*) which guides the user through the search process in a series of questions and choices, some of which present the user with new menus. The menus subsystem is user friendly and fully supported with online help which is available on any topic at any time.

One complete menu is dedicated to teaching the user about the system and another provides communication facilities with other users. Two menus contain the tools for searching and retrieving sequences using user specified conditions to locate the targets. The user may also tailor the presentation of sequences from within the menu system. One menu is dedicated to providing information about currently available software for sequence analysis. Results may be downloaded by direct data capture or file transfer for which the KERMIT program (Section 3.2.1) is available.

4.4.2 *Sequence retrieval*

To examine sequences through the menu system, the user selects the SEQUENCES menu and, if the index number or name is known, uses the SHOW option. There are 10 options in this particular menu, which also allow the user to examine the sequence or the feature tables. If the name or index is unknown the user must first select the GROUPS menu where sequences may be retrieved from the public sequence tables. Amongst the options in this menu the user may search for a new group of sequences using options including definition, keyword, organism, host, features, title, or journal. The results can then be seen by LISTing or SHOWing. Other options in the SE-QUENCE menu include analyses for single sequences. The user may opt for a translation, search for a pattern within the sequence (including restriction site analysis) or show all the fields relating to the sequence entry in the database.

4.4.3 *Software clearinghouse*

Information on more than 40 sequence analysis packages is available from this menu. The EXPAND option enables the user to see all the information on a selected package; this includes a description of the package's functions, contact numbers, charges, system requirements and methods of distribution (*Figure 12*). The user can SCAN the clearinghouse for keywords such as type of machine, name of the package or its author.

4.4.4 *The command level*

Using the command processor (rather than the menu system) a user can examine the database by using the DISPLAY command. The database is organized as if it were stored in tables and therefore a command is modified to display specific information

```
1.INFORMATION    about the Clearinghouse
2.DESCRIBE       fields of an entry
3.SCAN           all entries
4.SEARCH         the Clearinghouse
5.REFINE         the previous search
6.EXPAND         an entry
7.MAIN           menu

Choice:6S


Index of clearinghouse entry to expand:20S


[20] Protein Comparison          (Posted: 2/1/86)

Authors: D.J. Lipman and W.R. Pearson

Description:
      FASTP is a computer program for searching the NBRF
(National Biomedical Research Foundation) protein sequence
database for proteins homologous to a query sequence.  The
program uses a rapid and sensitive algorithm which first
finds regions of amino-acid identity and then scores the
aligned identical and differing residues in those regions
using an amino-acid replaceability matrix (the  PAM250
matrix).  The program is very fast, comparison of a 200
residue protein to the 600,000 NBRF library takes less than
2 minutes on a VAX, and less than 10 minutes on the IBM PC.

      In addition, a second program, RDF, is available for
testing the significance of sequence similarities found with
FASTP.  RDF uses the same scoring algorithm as FASTP,  but
compares two sequences by first measuring their similarity,
and comparing the score to scores generated when the second
sequence is randomly shuffled.


Functions:
compare; search; sequence alignment; sequence similarity measure;
protein sequence comparison;
search for subsequence with mismatches

Distributors:
Dept. of Biochemistry
Box 440 Jordan Hall
U. of Virginia
Charlottesville, VA 22908

Contact: W. Pearson          (804-924-2818)

References:
Science 227:1435-1441(1985)

Cost: VAX/VMS tape no charge; IBM PC disks with programs and
      library 6S0

Computer System Requirements:
      Hardware/OS: IBM PC, XT, AT MS DOS 2.0+; DEC VAX VMS; DEC
                   VAX UNIX
      Memory: 128 K
      Disk Storage: 100 K program; 900 K library
      Color and/or Graphics: None
      Required Peripherals: None
      Optional Peripherals: None
      Other Required Software: None
```

Figure 12. Utilizing the menu system, the user has requested information on a particular software package, from information contained in the SOFTWARE CLEARINGHOUSE.

by the addition of ROW and COLUMN, where COLUMN refers to LOCUS, DEFINI-TION, SEQUENCE, TYPE, DATE, ACCESSION NUMBER, KEYWORDS, ORGANISM, COMMENT, FEATURES, AUTHORS etc. (see Chapter 4). Thus a command sequence might contain DISPLAY PHAGESEQUENCES ROWS n TO m COL-UMNS y TO z. Information can also be displayed for a specific organism by using a command sequence such as DISPLAY VERTSEQUENCES WHERE LOCUS CON-TAINS *nnn*, this will display information for every entry in the vertebrate sequence data that contains *nnn* in the LOCUS field of the database.

SEARCHGENBANK enables the user to search for entries combining several sets of criteria using data such as authors names, subsequence patterns, and the source organism by extending the DISPLAY command with the WHERE subcommand. The user is prompted for the type of search he wishes to perform. A summary is also available in table form and may be examined using the ACCESS command.

```
* Your password has been accepted.  Welcome to the Yale Gene Mapping Library.
* Working...
[intro]
You are in the first menu of a newly-released (April 15, 1986) menu and help
system.  The old menu system (previously accessed by typing $HELP) will
remain available through June 15, 1986, although its use is not recommended.
Would you like to:
1    try the new menu and help system
2    use the old ($HELP) system
3    leave the Library (end this session completely)
4    leave the menu system, to give raw SPIRES commands
5    give a suggestion/comment/question before leaving the Library
Please enter the number of your selection:
  1

[db0]
No database is selected at this time.  Please select one of the following:
1.  LIT        literature citations and abstracted information
2   MAP        summary information on gene/locus mapping
3   RFLP       restriction fragment length polymorphisms
4   PROBE      DNA probes
5   CONTACT    people to contact for probes (names and addresses)
6              quit (leave $MENU or the Library)
Selection:
  1

[db.lit]
The LIT database has been selected.  There is no search result.
Would you like to:
1    start a search
2    display entries in the database (by key)
3    change databases or leave the Library
Selection:
  1

[db.lit.srch.index]
Starting a new search in the LIT database.  What index do you want
to search in:
1    authors
2    gene or locus symbols, from HGM Workshops if possible
3    keywords from the title, comments, etc.
4    year of publication
Selection:
  3

What KEYWORD do you want to search for? VIRUS

* Found 96 item(s).
[db.lit.srch.done]
Found 96 literature citation(s).  Do you want to:
1    display the results of the search on your terminal
```

Figure 13. The user has successfully logged onto the computer and now uses the menu system to perform a search.

4.5 The Howard Hughes Medical Institute human gene mapping library

The human gene mapping library consists of several databases managed by a system called SPIRES. The project was originally funded by a grant from the National Institutes of Health. Extensive help facilities are provided and these are immediately available to the user after successfully logging onto the system. The library may be accessed by direct dial and through the TELENET network. The user can work through the help system, which is presented as a series of MENUs called by number, or work directly at the command level of the management system. The user can return to the Help/Menu level at any time by typing a single command. Searches are initiated by looking through the indexes associated with a database selected through the menu system. The user may then ask to have the results displayed on the screen.

The available databases include literature citations, information on mapped genes, restriction fragment length polymorphisms, DNA probes and their source (*Figure 13*).

4.5.1 *The programs*

At the SPIRES command level a complex but flexible set of commands is available for interrogating the databases. Within the SPIRES command level the collections of data are known as subfiles which may be SELECTed. The searches look through indexes for the keywords named in the FIND command which may use logical operators

```
-> HELP

*
* You have subfile 'LIT' selected.
* The 'ID+CITE' format is set.
*
-No stack exists
-Result: 5 REFERENCES
*
* Your search result was obtained as follows:
FIND YEAR AFTER 1980
AND KEYWORD VIRUS
AND KEYWORD BURKITT
*
* To find out how to display your results, issue the commands:
*
*     EXPLAIN TYPE COMMAND
-> AND KEYWORD EBV

-Zero result, previous result retained
-Result: 5 REFERENCES
-> AND KEYWORD EPSTEIN

-Result: 2 REFERENCES




* You have subfile 'LIT' selected.
* The 'ID+CITE' format is set.
*
-No stack exists
-Result: 2 REFERENCES
*
* Your search result was obtained as follows:
FIND YEAR AFTER 1980
AND KEYWORD VIRUS
AND KEYWORD BURKITT
AND KEYWORD EPSTEIN
*
* To find out how to display your results, issue the commands:
*
*     EXPLAIN TYPE COMMAND
-> TYPE

H0351
Petit, P.
On the chromosomal sites of Epstein-Barr virus in Burkitt tumor cell lines.
Am. J. Hum. Genet. 36:480-481,1984.
```

Figure 14. Using HELP to check on the progress of a literature retrieval search.

such as AND, OR, or NOT to extend the search specification (*Figure 14*). Results may be viewed at any time by using the TYPE command. A BACKUP command enables the user to go back one step in his search and try again with a different specification.

An estimate of the size of each subfile is obtained by using the SHOW SUBFILE SIZE command and a brief description of its contents is obtained using SHOW SUBFILE DESCRIPTION. Information on a variety of topics is available with the EXPLAIN command and the user is prompted to do this after giving a HELP command. This command shows the user the current state of the search and the allowable search terms for any subfile can be displayed with SHOW INDEXES.

The output of a search can be modified in several ways. Firstly the amount of output is established with SET FORMAT. SHOW FORMAT displays the formats relevant to the subfile as well as the current setting. The records can then be ordered using the SEQUENCE command, and several modifiers known as elements can be added to this command. Single entries may be examined by giving the DISPLAY command, where the entry is identified by a key unique to that entry.

4.6 Protein identification resource

The Protein Identification Resource (PIR) (28) is cosponsored by the US National Institutes of Health and the National Biomedical Research Foundation (NBRF). The NBRF protein sequence database contains all completely sequenced proteins and sequences translated from nucleic acids as well as amino terminal sequences and substantial numbers of protein fragments with relevant annotations (see Chapter 4). In addition to the NBRF protein and nucleic acid sequence databases PIR also supports the GenBank and EMBL sequence libraries. The online resource on a DEC VAX 11/780 is accessible through direct telephone connection and the TYMNET network with charges made according to use for commercial users and as a flat rate subscription to non-profit organizations, others may pay an hourly rate for access. The principal support on the PIR system is for database retrieval and searching the protein database for similar sequences.

4.6.1 *The programs*

The core PIR software are the database management and retrieval programs PSQ and NAQ (22) for protein and nucleic acid sequence databases (as described in Section 4.3). Sequence entries in the PIR databases may be retrieved using the entry title, bibliographic source, keywords or other fields in the entry annotations (see Chapter 4). The results from a PSQ or NAQ session can be saved as database subsets and used as input data for other programs in the PIR collection. These include methods for searching nucleic acid sequences for open reading frames, sequence translation or back translation, general data manipulation and analyses of codon usage. The PIR online system also provides the fast and sensitive FASTP and FASTN (4) programs for finding sequence similarities in protein and DNA sequence databases. Protein secondary structure predictions are available using the program PRPLOT based on the Chou and Fasman method (23) which also estimates protein hydrophilicity and hydrophobicity. A program called ALIGN, based on the Needleman – Wunsch algorithm (11) can be used to determine the best alignment between two sequences and program ISEARCH searches a sequence for small protein fragments which may include ambiguous residue codes. For moving data from the PIR system file transfer programs are also available.

4.7 CITI2

The Centre Interuniversitaire de Traitment de l'Information 2 (CITI2) has been entrusted as the French national computing resource for molecular sequence analysis. It presently has online three databases and software for retrieval, and approximately 100 programs for analysing sequence data, developed both at CITI2 and other research centres. The system is based on a CII-Honeywell Bull computer running under the GCOS8 operating system and can be used interactively or using batch processing (see Chapter 1). CITI2 can be accessed via direct telephone connections or through the

TRANSPAC network. Some of the available programs require a Tektronix-compatible terminal. Currently some 90 research groups are connected to the system.

4.7.1 *The databases*

Both GenBank and EMBL nucleic acid databases are available with database access provided by the ACNUC program (29). ACNUC provides a flexible and powerful approach to data retrieval based on a hierarchical network of keywords that can select data entries on all aspects of the sequence annotations (see Chapter 4). The protein sequence database is from PIR and the Brookhaven crystallographic database of protein structure (30) is also available.

4.7.2 *The programs*

The programs available include packages by Rodger Staden, the University of Wisconsin Genetics Computer Group, as well as programs developed at CITI2. The system is organized around a principal menu and secondary menus, although for experienced users it is also possible to short-circuit the menus and go directly from one function to another.

4.7.3 *Database searches*

The database searches can also be performed using a menu system. The user is taken through a simple dialogue generally picking options from a numbered list or answering yes or no. A search may be made selecting entries on many criteria combining keywords with the standard logical operators and extracting subsequences from the data using the database feature tables (see Chapter 4).

5. REFERENCES

1. Baer,R., Bankier,A.T., Biggin,M.D., Deininger,P.L., Farrell,P.J., Gibson,T.J., Hatfull,G., Hudson,G.S., Satchwell,S.C., Seguin,C., Tuffnell,P.S. and Barrell,B.G. (1984) *Nature*, **310**, 207.
2. Bishop,M.J. (1984) *Bioessays*, **1**, 29.
3. Bionet National Computer Resource for Molecular Biology (1986) Annual Progress Report.
4. Lipman,D.J. and Pearson,W.R. (1985) *Science*, **227**, 1435.
5. Newell,A. and Sproull,R.F. (1982) *Science*, **215**, 843.
6. Bacon,M.D., Stokes,A.V. and Bacon,J. (1984) *Computer Networks: Fundamentals and Practice.* Chartwell-Bratt (Publishing and Training) Ltd, UK.
7. Coulson,A. (1985) *Binary*, **4**, 22.
8. Smith,D.H., Brutlag,D., Friedland,P. and Kedes,L.H. (1986) *Nucleic Acids Res.*, **14**, 17.
9. Grymes,R.A., Travers,P. and Engelberg,A. (1986) *Nucleic Acids Res.*, **14**, 87.
10. Brutlag,D., Clayton,J., Friedland,P. and Kedes,L.H. (1982) *Nucleic Acids Res.*, **10**, 279.
11. Needleman,S.B. and Wunsch,C.D. (1970) *J. Mol. Biol.*, **48**, 443.
12. Wilbur,W.J. and Lipman,D.J. (1983) *Proc. Natl. Acad. Sci. USA*, **80**, 726.
13. Abarbanel,R.M., Wieneke,P.R., Mansfield,E., Jaffe,D.A. and Brutlag,D.L. (1984) *Nucleic Acids Res.*, **12**, 263.
14. Staden,R. (1982) *Nucleic Acids Res.*, **10**, 2951.
15. Staden,R. (1982) *Nucleic Acids Res.*, **12**, 521.
16. Staden,R. (1980) *Nucleic Acids Res.*, **8**, 3673.
17. Staden,R. (1982) *Nucleic Acids Res.*, **10**, 4731.
18. Staden,R. (1984) *Nucleic Acids Res.*, **12**, 499.
19. Garnier,J., Osguthorpe,D.J. and Robson,B. (1978) *J. Mol. Biol.*, **120**, 97.
20. Devereux,J., Haeberli,P. and Smithies,O. (1984) *Nucleic Acids Res.*, **12**, 387.

21. Kernighan,B.W. and Plauger,P.J. (1976) *Software Tools*, Addison-Wesley Publishing Company, Reading, Massachusetts.
22. Orcutt,B.C., George,D.G., Fredrickson,J.A. and Dayhoff,M.O. (1982) *Nucleic Acids Res.*, **10**, 157.
23. Chou,P.Y. and Fasman,G.D. (1978) *Ann. Rev. Biochem.*, **47**, 251.
24. Kyte,J. and Doolittle,R.F. (1982) *J. Mol. Biol.*, **157**, 105.
25. Bilofsky,H.S., Burks,C., Fickett,J.W., Goad,W.B., Lewitter,F.I., Rindone,W.P., Swindell,C.D. and Tung,C.-S. (1986) *Nucleic Acids Res.*, **14**, 1.
26. *Nucleotide Sequences 1985*, IRL Press, Oxford, UK.
27. BBN Laboratories Ltd. (1986) *Nucleic Acids Res.*, **14**, 21.
28. George,D.G., Barker,W.C. and Hunt,L.T. (1986) *Nucleic Acids Res.*, **14**, 11.
29. Gouy,M., Milleret,F., Mugnier,C., Jacobzone,M. and Gautier,C. (1984) *Nucleic Acids Res.*, **12**, 121.
30. Bernstein,F.C., Koetzle,T., William,G.J.B., Meyer,E.,Jr, Brice,M.D., Rodgers,J.R., Kennard,O., Shimanouchi,T. and Tasumi,M. (1977) *J. Mol. Biol.*, **112**, 535.
31. Teo,I., Sedgwick,B., Kilpatrick,M.W., McCarthy,T.V. and Lindahl,T. (1986) *Cell*, **45**, 315.

6. APPENDIX

Relevant addresses

BIONET, Intelligenetics Inc., 1975 El Camino Real West, Mountain View, CA 94040, USA

P.Le Beux, CITI2, 45 rue des Saints-Peres, 75006 Paris, France

Dr A.Coulson, Department of Molecular Biology, King's Building, Mayfield Road, Edinburgh, EH9 3JR, UK

GenBank, BBN Laboratories Inc., 10 Moulton Street, Cambridge, MA 002238, USA

Howard Hughes Medical Institute Human Gene Mapping Library, 25 Science Park, New Haven, CT 06511, USA

JANET, University of London Computing Centre, University College London, Gower Street, London WC1E 6BT, UK

KERMIT Distribution, Columbia University, Centre for Computing Activities, 612 West 115th Street, New York, NY 10025, USA

Protein Identification Resource, National Biomedical Research Foundation, Georgetown University Medical Centre, 3900 Reservoir Road NW, Washington DC 20007, USA

PSS, 1 Swan Lane, London EC4, UK

University of Cambridge, Computer Laboratory, Corn Exchange Street, Cambridge, CB2 3QG, UK

CHAPTER 6

Approaches to restriction map determination

G. ZEHETNER, A. FRISCHAUF and H. LEHRACH

1. INTRODUCTION

Restriction mapping is usually one of the first steps in the characterization of a cloned DNA fragment and is frequently a prerequisite for planning a sequencing strategy and selecting appropriate fragments for subcloning. Overlaps between different clones are usually first established on the basis of the restriction maps, either by eye, or with the help of computer programs. There are several ways to establish a restriction map that vary in effort and time spent on experimental work and on data analysis. The choice of the optimal strategy should depend on the expected complexity (number of enzymes and sites), on existing information about the DNA fragment and on the availability of suitable computer programs.

Restriction maps can be determined either directly using partial digestion techniques, or by indirect approaches such as those based on the analysis of fragments created by single and double digestions with a set of restriction enzymes.

In the first strategy a common end point allows the direct determination of the positions of restriction sites along the linear map from the fragment sizes. The second method arrives at a map compatible with the measured fragment sizes by a process of exclusion. This is usually quite straightforward if the number of sites for each enzyme is small. However, the number of possible arrangements and the difficulties in assigning fragments increase dramatically with the number of sites. The indirect approaches are therefore better suited to the analysis of relatively simple problems involving enzymes with few sites.

This chapter will first discuss the general strategies used in the two kinds of approaches, then go through the most commonly used techniques and describe some experimental protocols.

1.1 Mapping strategies

1.1.1 Indirect approaches

The indirect methods are based on the analysis of fragment sizes arising from complete digestion of the DNA with each enzyme separately as well as with combinations of enzymes (in two-dimensional gels partial digestion can be used in one dimension). The complexity of the problem increases rapidly with increasing numbers of sites and enzymes and may be impossible to solve, even if computer programs are used in the analysis. If high resolution maps of long DNA molecules have to be established by such a protocol then in most cases subfragments or subclones have to be analysed separately to determine the entire map.

The indirect strategy most commonly used for the mapping of a small number of sites on DNA molecules is the analysis of the results of single and double (and occasionally triple) digestion patterns. One modification involves the cleavage of separated fragments derived from single digests with a second enzyme, so as to unambiguously assign the source of each double digestion product. If a gel strip containing the separated products of the single enzyme digest is incubated in a reaction solution with the second enzyme, this step can be carried out in parallel on all fragments in a two-dimensional procedure (see Section 2.1.2). Other two-dimensional restriction mapping procedures use separation of partial digestion products in one dimension, followed by redigestion within the gel with either the same or different enzymes (see Section 2.1.3). Additional information can also be gained by hybridization techniques. Isolated fragments originating from a complete digestion can be labelled and used as probes to identify fragments from a different digest. Two-dimensional versions capable of identifying in parallel all cross-hybridizing bands have been described (see Section 2.1.4). In conjunction with the indirect mapping approaches it is useful to obtain information identifying the end fragments (derived from, for example, end labelling) or information identifying fragments hybridizing (with different intensity) to common probes (e.g. repetitive DNA for mammalian clones). The order of fragments can also be deduced from the loss of sequences during the course of exonucleolytic degradation [e.g. digestion of the linear DNA with *Bal*31 (1) or exonucleases III and IV of *Escherichia coli* (2), followed by recutting with a restriction enzyme].

1.1.2 *Direct approaches*

By analogy with DNA sequencing techniques, the direct mapping approaches rely on the identification of partial digestion products extending from a unique point (usually one end) to all restriction sites in the DNA molecules.

The steps used in these approaches are given below.

(i) *Linearization (for circular DNAs)*. Circular molecules have to be cut (at one or more positions) to allow analysis by this technique. While for small plasmids unique restriction sites are usually available, finding unique sites can be rather difficult for large circular DNAs (e.g. cosmids). Cosmids can be linearized by specific cleavage of the *cos* sequence by lambda terminase *in vitro* or *in vivo* (3–5).

(ii) *Asymmetric labelling*. To identify all partial digestion products containing one specific end, direct or indirect end labelling techniques are used. If the DNA is initially labelled at both ends, the ends are separated by recutting with a second enzyme (6). If the second site is close to one of the labelled ends, e.g. by using two sites within a polylinker sequence, the short radioactive fragment can be left in the reaction. If not, the two labelled fragments have to be separated and processed independently unless a two-dimensional analysis technique is used (7). Alternatively, one of the two ends may be labelled selectively. This is easy for DNA fragments ending in two different restriction sites, either by taking advantage of the preference of enzymes for 3′ or 5′ protruding termini, or by filling in protruding 5′ termini by DNA polymerase (Klenow fragment) with the different bases to be incorporated into different sites. Similarly, fragments ending in asymmetric sites (created for example by cleavage with a restriction nuclease with one or more undefined bases in the protruding end sequence) can be selectively labelled on one or other end by choosing the appropriate radioactive triphosphate in a copying reaction with Klenow DNA polymerase. Most indirect labelling procedures are inherently asymmetric if appropriate probes are chosen. For both lambda and cosmid (linearized at the *cos* site) clones we use hybridization to a radiolabelled oligonucleotide complementary to either the right or the left cohesive end of lambda to selectively label partial digestion products originating from the corresponding end of the molecule. Similarly, labelled probes can detect fragments originating from one or other end of the molecule by hybridization to the separated partial digestion products.

(iii) *Partial digestion.* In most cases a satisfactory partial digestion is achieved by varying the time of incubation and/or the amount of enzyme. Special strategies to standardize the degree of digestion have been developed. One procedure, applicable to the mapping of restriction sites containing TT sequences in the recognition sequence, involves u.v.-induced formation of TT dimers to partially block the cleavage reaction (8) (see Section 2.2). An alternative method is the digestion of circular DNA molecules in the presence of ethidium bromide, which should block further reaction after the first cut has occurred (9).

2. THE DIFFERENT APPROACHES TO RESTRICTION MAP DETERMINATION

For comparison of the different approaches a map of phage lambda cI857S7 DNA (10) for the enzymes *Eco*RI and/or *Bam*HI is constructed using different methods. *Figure 1* shows the map for both enzymes and all possible fragments generated by partial *Eco*RI digestion.

2.1 Indirect methods

2.1.1 *Products of single and double digestions*

This approach is experimentally simple (for digestion protocols see *Table 1*). Small problems, involving only few fragments, can essentially be solved by trial and error. Larger problems require a more systematic approach in the analysis of the data, often involving computer programs.

A systematic approach allowing the manual analysis of fairly complex mapping problems is given by a Branch and Bound technique (11). In this procedure fragments created by double digestion are assigned to single digestion products under a set of rules, allowing the elimination of most or all alternative arrangements. Other information, gained, for example, by the direct determination of double digestion products created by redigestion of one or more single digestion products (see Section 2.1.2) or from other experiments, can be incorporated. If each double digestion fragment can be unambiguously assigned to the single digestion products, the restriction map can be determined by identifying the pairs of single digest fragments sharing one double digestion product. The solution matrix and the resulting map for the example mentioned above are shown in *Figure 2*.

In many cases a computer program (12−17) will be used in the analysis. To illustrate the difficulties and strengths of these programs, program MAP from IntelliGenetics (version 4.5) has been used to analyse the *Eco*RI−*Bam*HI digestion dataset. The program is based on the algorithm of W.R.Pearson (12) where for each pair of restriction enzymes, the fragment order for the two single digests are permutated in all possible

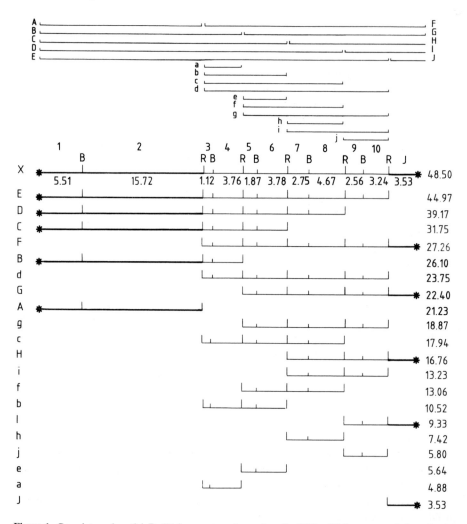

Figure 1. Complete and partial *Eco*RI fragments and complete *Eco*RI/*Bam*HI fragments of phage lambda cI857S7. Above, the origin of all partial *Eco*RI fragments is shown, and below the fragments are ordered by decreasing size. Each *Eco*RI fragment is identified by a letter (upper case letters for fragments containing a labelled DNA end, lower case letters for internal fragments), and its size [in kb, taken from the sequence (10)] is given on the right. Within partial *Eco*RI fragments *Bam*HI and *Eco*RI sites are marked by a line. On the full length DNA line (X) the *Eco*RI/*Bam*HI double digestion fragments are indicated by numbers. Complete *Bam*HI fragments are indicated by the numbers of the two *Eco*RI/*Bam*HI double digest fragments they contain. Thick lines on partial fragments show the labelled fragment created after complete re-cleavage with *Eco*RI.

ways. The calculated double digest fragment sizes resulting from each permutation are compared with the actual double digest data, and a numerical measure of the fit is computed. If the match lies within a user-defined error limit it will be saved and presented to the user as one of the possible maps. To increase efficiency the program is able to use heuristic rules to avoid calculating a major part of the possible fragment permutations.

150

Table 1. Complete single and double digestions.

Restriction enzyme buffers:

10 × low	100 mM Tris-HCl pH 7.5
	100 mM MgCl$_2$
	10 mM dithiothreitol (DTT)
10 × medium	10 × low buffer containing 500 mM NaCl
10 × high	10 × low buffer containing 1 M NaCl
10 × Sma	100 mM Tris-HCl pH 8
	100 mM MgCl$_2$
	10 mM DTT
	200 mM KCl

1. Add water to a final volume of 20 μl to an Eppendorf tube.
2. Add 2 μl of 10 × buffer (high, medium, low, Sma buffer)[a]
3. Add 1−2 μg of DNA (enough to give more than 20 ng in the lowest molecular weight band you want to see)[b], and mix.
4. Add a 2-fold excess of enzyme(s)[a] according to units (taking into account the average number of sites per length of DNA)[c]
5. Mix and incubate at the appropriate temperature for 1−2 h.
6. Heat for 5 min to 70°C. Mix 12 μl with 3 μl of loading buffer (Loening buffer[d] containing 50% glycerol, 60 mM EDTA pH 8 and 0.1% bromophenol blue) and load on the gel.

[a]Double digestions can be carried out simultaneously by adding both enzymes to the reaction mix if both require the same enzyme buffer. Otherwise digest first with the enzyme requiring lower salt conditions and then add 1 M NaCl stock as required for the second digestion step.
[b]If small fragments are not visible, treatment of the digests with RNase can help. In some cases it might be useful to label the ends of all fragments to detect fragments not visible in ethidium bromide staining.
[c]Do not add more than 10% enzyme solution to the total volume as otherwise the glycerol concentration becomes too high.
[d]36 mM Tris, 30 mM NaH$_2$PO$_4$, 1 mM EDTA, pH 7.7

Even with this (unrealistically) good dataset the identification of a completely unrestricted solution requires a large amount of computer time (42 minutes CPU time on a VAX 11/780) when no additional information is given. Defining one of the two *Eco*RI or *Bam*HI end fragments decreases the CPU time to 7 min. The program shows how long the calculation is expected to take and prompts the user as to whether he wishes to proceed.

If the positions of more than one end fragment are fixed in advance, both computational effort and uncertainty of the solutions decrease rapidly. *Figure 3* shows a run of the program (taking a few seconds) and the calculated maps with the locations of the left and right *Eco*RI end fragments specified as known. Even so, eight possible solutions are generated but the true map is given the highest probability of being correct. Additional information about the *Bam*HI end fragments enables the program to exclude all incorrect solutions.

Similarly, other programs are capable of taking into account information on fragment positions (e.g. vector or end fragments or fragments known to be adjacent to each other), or fragments known not to be cleaved by the second enzyme. Since in general all possible permutations of single or double digestion products are tested, leading to very large computation times for large problems, any such reduction in the complexity of the problem will contribute to making the analysis feasible. Even if a program is not able directly to take advantage of additional information, this data will usually be

Fragment (kb)	1.1	1.9	2.6	2.8	3.2	3.5	3.8	3.8	4.7	5.5	15.7	
3.5						S						
4.9	A						A					
5.6	~~A~~	~~A~~B	~~A~~				B			~~S~~		
5.8	~~A~~	~~B~~	C		C		~~B~~	~~A~~				
7.4	~~BE~~	~~A~~	~~DF~~	~~BE~~		~~CE~~	~~CD~~	~~BF~~	~~A~~			
21.2		~~C~~	~~B~~	~~B~~		~~C~~			A			A~~BC~~
(order)	2	4	5	6	5	1	2	4	6	3	3	
(kb)	1.1	1.9	2.6	2.8	3.2	3.5	3.8	3.8	4.7	5.5	15.7	
(order)	1	5	3	6	2	2	5	6	3	4	1	
5.5	~~A~~	~~A~~B	~~AC~~	~~C~~		~~B~~				S		
5.6	~~A~~	~~A~~B	~~A~~				B			~~S~~		
6.5	~~AE~~	~~BE~~	~~C~~	D		~~E~~		~~C~~D	~~B~~	~~A~~		
6.8	~~B~~	~~B~~			A	A	~~B~~					
7.2	~~CEF~~	~~D~~	A~~CDF~~	~~DE~~	~~E~~	~~BCF~~		~~B~~	A			
16.8	A											A

```
                 1        5         6         4         2          3
        EcoR1   3.5   3.2-2.6   4.7-2.8   3.8-1.9   3.8-1.1   15.7-5.5
                  \   /   \   /   \   /   \   /   \   /    \
        BamH1   3.5-3.2  2.6-4.7  2.8-3.8  1.9-3.8  1.1-15.7  5.5
                    2        3         6         5         1        4
```

Figure 2. Solution matrix of the Branch and Bound analysis (11). The first column lists the lengths (in kb) of lambda *Eco*RI fragments (upper set) and *Bam*HI fragments (lower set). In the middle row the *Eco*RI/*Bam*HI double fragment lengths are shown. In the first step of the analysis all possible combinations of double digest fragments which add up to the length of any single digest fragment (within the allowed error of ± 0.1 kb) are marked in the matrix by letters (A,B,C etc). The same letters in any row indicate double digestion fragment members of one such combination for the single digestion fragment in the first column of that row. 'S' stands for fragments presumably not cut by the second enzyme. Next, starting with a single digestion fragment from each enzyme which shows just one (and therefore unambiguous) combination of double digest fragments, (e.g. the 3.5-kb and the 4.9-kb *Eco*RI fragment and the 16.8-kb *Bam*HI fragment) it is necessary to find definite assignments of double digest fragments for all fragments from either single digest set by excluding the majority of possible combinations using several rules (e.g. double digest fragments already assigned to a single digest fragment cannot be used in a combination involving another single digest fragment). A slash through letters indicates that this combination of fragments has been ruled out. The order of definite assignments is shown by upper and lower numbers in the boxes containing the respective double digest fragment lengths. If all possible combinations except one can be discounted for each single digest fragment, it is easy to determine the order of the double digest fragments by linking assigned fragment pairs (or triplets) alternating between the two enzymes as shown below the matrix. Sometimes more than one consistent combination remains for single digest fragments and additional analysis has to be performed by checking all possible restriction maps created by the ambiguous combinations to find a valid solution. For a detailed description of the rules for this method see ref. 11.

required to exclude incorrect restriction maps compatible with the fragment sizes.

In many programs error limits have to be specified in advance, with the number and quality of the solutions depending critically on the error limit. In this case the analysis should be started with small error values, which can then be increased until solutions compatible with the given error are found. An algorithm described by Wulkan and Lott (16) avoids the definition of such an error limit by the user. The efficiency is increased further, since in this program both vector fragments and fragments not recleaved by the second enzyme can be specified.

Circular maps can be calculated with the same algorithm as linear maps but, in addition to the fragment order, the 'offsets' of the digests relative to one another must be determined. Usually the length of a double digest fragment not present in either of the single digests (each end has therefore been created by a different enzyme) is used to

```
MAP: New
Enter tolerance: 0.2E-02
Topology: Linear
Enter all of the enzymes (? for help): EcoR1 BamH1

Ready for SINGLE-digest results.

EcoR1-Unknown-Fragments: 21.2 7.4 5.8 5.7 4.9 3.5
BamH1-Unknown-Fragments: 16.8 7.2 6.8 6.6 5.6 5.5

Ready for DOUBLE-digest results.

EcoR1/BamH1 fragments: 15.7 5.5 4.7 3.8 3.8 3.5 3.2 2.8 2.6 1.9 1.1

MAP: Edit
FragEdit: Single
The enzymes in the single digests are:
 1) EcoR1
 2) BamH1
Digest to be edited? (<CR> to quit): 1
DigestEdit(EcoR1): List
 1) 21.2    2) 7.4    3) 5.8    4) 5.7    5) 4.9    6) 3.5

DigestEdit(EcoR1): Known
Which end are these fragments on? (L or R): L
Number(s) of the fragment(s) to be marked as known: 1
DigestEdit(EcoR1): Known
Which end are these fragments on? (L or R): R
Number(s) of the fragment(s) to be marked as known: 6

MAP: Generate
Entering data checking mode...
Data checking is complete.
Average Length is 48533
the best solution with an rms error of .140998E-02 is:
Map # 1    Linear      Length 48533 base pairs      RMS Error is .140998E-02

                              22315   27919      34523      41728
               5504          21214  26118   31822     39227    45031
|-----------------------------------------------------------------------|
0              ^              ^ ^    ^  ^    ^  ^      ^  ^    ^    48533
            BamH1            EcoR1  EcoR1  EcoR1     EcoR1  EcoR1
                            BamH1  BamH1   BamH1    BamH1

Fragment Order Reference Table -- Map #1
EcoR1    A 21.21    E 4.9      D 5.7      B 7.41    C 5.8    F 3.5
BamH1    F 5.5      A 16.81    E 5.6      D 6.6     B 7.2    C 6.8

There are 7 other solutions within the current tolerance.
The error and changes from the best map are given for each.

Solution 2  RMS Error is .157009E-02      Solution 3  RMS Error is .162339E-02
BamH1   E A C B D F                        EcoR1    A D E C B F
                                           BamH1    F A D E C B

Solution 4  RMS Error is .162484E-02      Solution 5  RMS Error is .163802E-02
EcoR1   A E B D C F                        EcoR1    A C E D B F
BamH1   F A D B C E                        BamH1    F A D E C B

Solution 6  RMS Error is .164580E-02      Solution 7  RMS Error is .165252E-02
EcoR1   A E C B D F                        EcoR1    A E C B D F
BamH1   F A D E C B                        BamH1    F A E C B D

Solution 8  RMS Error is .179420E-02
EcoR1   A E C B D F
BamH1   E A C F D B
```

Figure 3. Sample run of program MAP (IntelliGenetics) to determine the *Eco*RI and *Bam*HI restriction map of phage lambda DNA. User input is underlined. First error limit (0.002 means that fragment sizes may not differ by more than 0.2% to be considered equal) and DNA topology (linear or circular) are specified, then the single and double digest fragment lengths are entered. (In this example sizes are taken from the sequence and rounded to the next hundred bases). Then the left and right *Eco*RI end fragments are defined as known. Eight possible solutions within the defined error limit are presented. The correct map has the lowest error value.

Table 2. DNA digestion in agarose gels (18).

1. Cut the required portion out of the gel[a] (either a single fragment or a whole lane).
2. Soak the gel slice in 50 vol of the appropriate reaction buffer containing 100 μg/ml of nuclease-free bovine serum albumin for 6−8 h at 4°C with buffer changes every 2 h.
3. Soak the gel slice in 5 vol of reaction buffer containing 20−100 units/ml enzyme[b] for about 12 h at 4°C.
4. Incubate for 16 h at 37°C.
5. Equilibrate the gel slice in electrophoresis buffer for 1 h.
6. Seal the slice into a well with hot agarose. Run the gel.

Using low melting point agarose (19).

1. Cut the band out of the gel and move to a 1.5 ml Eppendorf tube.
2. Add 40−60 μl of 1 M NaCl (depending on the enzyme).
3. Heat for 50−70 sec to 70°C.
4. Shake twice and move to a 37°C water bath.
5. If necessary reduce the agarose concentration to 0.5% by adding an appropriate volume of water.
6. Add 0.1 vol of 100 mM $MgCl_2$ and 0.05 vol of 100 mM 2-mercaptoethanol.
7. Add 1−2 units of enzyme.
8. Incubate for 30 min at the appropriate temperature.

[a]If possible, agarose with low electroendosmosis should be used since some enzymes are inhibited in agarose with medium or high electroendosmosis
[b]Boehm and Drahovsky (18) have tested the following 10 enzymes for their ability to cut DNA in agarose and polyacrylamide gels to completion. Numbers in brackets give the used enzyme units per ml for low electroendosmosis agarose: HpaII (20), MspI (20), HaeIII (20), HindIII (50), TaqI (20), HhaI (20), AluI (25), BamHI (100), EcoRI (50), SalI (25)

define this offset.

The advantage of the single and double digestion approach to restriction mapping is the simplicity of the experimental steps. The analysis, however, can be rather complicated or impossible. Further information may be necessary, and can be incorporated both in the manual analysis (carried out for example by the Branch and Bound procedure) or in the analysis by some computer programs.

2.1.2 *Restriction maps from fragment redigestion in a two-dimensional analysis*

To discriminate between alternative assignments of double digestion products to the single digestion products, fragments separated in a first dimension are recleaved within the gel and the products separated in a second dimension. *Table 2* shows two protocols for digestion in gels (18,19). The analysis of the resulting fragment pattern can be carried out essentially as in the Branch and Bound technique mentioned above (11). This protocol can be used either by itself (20) if all single digest gels are analysed by redigestion, or, probably more commonly, to support the standard analysis.

2.1.3 *Restriction maps from redigestion of partial digestion fragments in a second dimension*

Another type of restriction mapping is based on the redigestion of partial digestion fragments created by one enzyme either with the same enzyme or with a second enzyme (see partial mapping protocols in Section 2.2.2 for the case of end-labelled DNA). In the first case the analysis is straightforward. The second strategy requires complex analysis if the end fragments are not labelled but it also gives information on the sites for the second enzyme. *Figure 4* shows such a dataset: a partial *Eco*RI digest separated

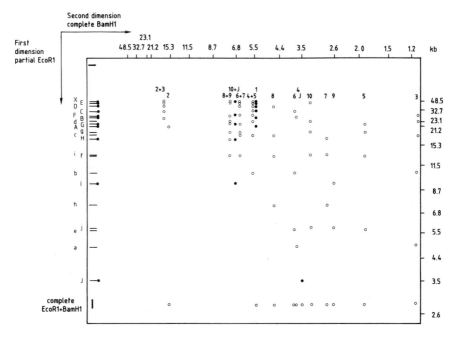

Figure 4. Two-dimensional gel of phage lambda DNA partially digested with *Eco*RI, run in the first dimension and then cut to completion with *Bam*HI and run in the second dimension. Letters and numbers identifying fragments are described in the legend to *Figure 1*. The lane on the bottom of the gel shows complete *Eco*RI/*Bam*HI digestion fragments. Black stars indicate fragments containing (labelled) ends. The derivation of restriction maps from this fragment pattern is described in Section 2.2.3 for end-labelled DNA.

in the first dimension, recleaved *in situ* with *Bam*HI to completion, and separated in the second dimension. In addition, a complete *Eco*RI/*Bam*HI digestion was loaded in the second dimension. The dark dots identify fragments that would carry a radioactive end if end-labelled DNA had been used. The mobilities of all fragments (sizes are taken from the sequence) are calculated by a computer program (20) using the parameters of a gel (0.5% agarose, 20 × 20 × 0.5 cm) with indicated size marker fragments.

To obtain a map from this dataset, arguments have to be based on a series of conclusions (and/or exclusions). Since a detailed discussion of the analysis would go beyond the scope of this chapter the reader should consult references 2, 22 and 23.

2.1.4 Southern cross-hybridization

This method is based on the hybridization of overlapping fragments resulting from digestion with different restriction enzymes (24). Fragments of a digest with one enzyme are radioactively labelled. The labelled fragments and fragments of a different digest are then separated on gels using long slots across the entire gel width. The unlabelled fragments are transfered to a nitrocellulose filter. The filter is then placed at right angles on the second gel containing the labelled fragments. After blotting the filter is washed, dried and put on X-ray film. Labelled fragments which have hybridized to homologous sequences on unlabelled fragments will lead to a two-dimensional pattern of spots. Analysis of these spots will give the restriction map for the two enzymes.

A reaction kit offered by NEN (New England Nuclear, Postfach 401240,

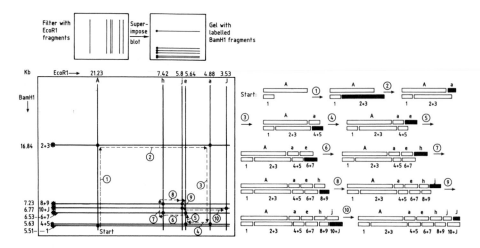

Figure 5. Above, a schematic drawing of a filter containing *Eco*RI fragments of lambda DNA and a gel with labelled *Bam*HI fragments of the same DNA is shown at right angles for Southern cross-hybridization, below the resulting autoradiogram after blot-hybridization. Black dots indicate hybridization spots. The identification of fragments is as described in the legend to *Figure 1*. On the right, the 10 analysis steps according to the numbers in the autoradiogram are shown. Starting at the cross of *Bam*HI fragment 1 and *Eco*RI fragment A each new overlapping fragment as it is assigned in each step is drawn in black.

D-6072 Dreieich, FRG) uses a modified procedure where the labelled fragments are blotted on GeneScreen membrane (which allows fragments to be eluted under defined conditions) and the unlabelled fragments are blotted on GeneScreen-Plus membrane which does not release the DNA. The advantage of this method is that several filters with unlabelled fragments can be processed at once.

The analysis of a Southern cross-autoradiogram of *Eco*RI- and *Bam*HI-digested phage lambda DNA is shown in *Figure 5*. Ten steps are necessary in 'jumping' from one hybridization spot to the next to determine the whole *Bam*HI−*Eco*RI map. Next to the schematic autoradiogram the stepwise assembly of the map by addition of overlapping fragments is shown. The map starts at the overlap spot of the *Bam*HI fragment 1 and *Eco*RI fragment A. The next point is the hybridization spot between *Eco*RI fragment A and *Bam*HI fragment 2+3 [step (1) on *Figure 5*]. This process is continued until all fragments are ordered.

This type of protocol is experimentally rather complex but does lend itself to a general analysis strategy. A special advantage of this method is that the filter hybridization protocol can be carried out on many digests in parallel, and can therefore make efficient use of time and materials. In general, we prefer partial mapping protocols, possibly combined with supporting single and double digestion data, as requiring less work and being easier to analyse. Both complete digestion and partial digestion protocols will however have limitations in special cases.

2.2 Direct approach: partial mapping protocols

Table 3 gives examples for protocols to obtain partial rather than complete digestion of DNA. Two different types of mapping protocols have been described.

156

Table 3. Partial digestion procedures.

A. *Partial digestion of linear DNA molecules*

Analytical

1. Dilute the enzyme to approximately 0.1 units/µl.
2. Set up a reaction mixture with 2 µg of DNA and 0.1 unit of enzyme in a total volume of 20 µl at the desired temperature.
3. Take 4 µl aliquots at 5, 10, 20, 40 and 80 min. Immediately add 1 µl of 0.1 M EDTA.
4. Check the samples on 0.7% gel and select the optimal digestion time.

Protocol

1. Set up a reaction mixture with approximately 2 µg of DNA and 0.1 unit of enzyme in a total volume of 20 µl.
2. Take 6 µl aliquots around the previously determined optimal time (e.g. at 2, 5 and 10 min or at 15, 30 and 60 min). Add 1 µl of 0.1 M EDTA immediately after digestion.
3. Pool the samples of all time points, add loading dye and load on a gel.

B. *Partial digestion after u.v. irradiation of DNA (8)*

The activity of restriction enzymes with a recognition sequence containing TT is inhibited after u.v. irradiation due to formation of thymine dimers. The method was tested with five enzymes: *Dra*I, *Eco*RI, *Hind*III, *Hpa*I and *Xmn*I. Restriction endonucleases with other potential dimerization sites in the recognition sequence (*Bam*HI, *Bgl*II, *Cla*I, *Eco*RV, *Hpa*II, *Kpn*I, *Pst*I, *Pvu*I and *Sac*I) have been found unsuitable for this procedure. Since the electrophoretic mobility of DNA fragments is increased by u.v. irradiation, the size marker fragments have to be irradiated with the same u.v. dosage to allow correct size determination.

1. Take $0.1-1$ µg of DNA in 20 µl of restriction buffer in an Eppendorf tube (the presence of Mg^{2+} during u.v. irradiation is an absolute requirement, otherwise the DNA will be degraded).
2. Place the open tube directly under a u.v. lamp (distance 6 cm, fluence rate 30 W/m^2, intensity maximum 254 nm) for 30 min.
3. Add the restriction enzyme.

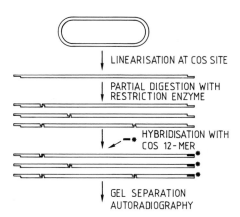

Figure 6. Strategy of procedure for restriction mapping of cosmids.

2.2.1 *Partial mapping of DNA labelled at one end*

Linear DNA labelled selectively at one end is partially cut with a restriction enzyme, the fragments are separated on a gel and the sizes of the partial digestion products are determined after autoradiography (3,6,25).

157

Table 4. Restriction analysis of *in vitro* linearized cosmid clones.

A. *Preparation of terminase extract*

Strain: AZ760 or AZ1069 (3)

DPA buffer :   20 mM Tris-HCl pH 8.0
                3 mM $MgCl_2$
                1 mM EDTA pH 8.0

DPA−DTT buffer : DPA buffer containing 5 mM dithiothreitol (DTT)

DPB buffer :   6 mM Tris-HCl pH 7.5
                18 mM $MgCl_2$
                30 mM Spermidine
                60 mM Putrescine

DPC buffer :   0.6 M NaCl
                0.2 M Tris-HCl pH 8.0

1.   Inoculate 400 ml of LB medium containing 20 μg of ampicillin/ml with a temperature-sensitive colony or streak of AZ760 or AZ1069 and grow at 30°C under vigorous aeration to OD_{600} = 0.3.
2.   Transfer the culture to a 44°C water bath for 15 min (to induce terminase production) and incubate for 45 min at 38°C with vigorous aeration.
3.   Chill in ice water and centrifuge at 6000 r.p.m. for 10 min in a pre-cooled GSA rotor. Pour off the supernatant.
4.   Resuspend the pellet in 0.5 ml of DPA−DTT buffer. Add 3.5 ml of the same buffer and transfer to a plastic tube.
5.   Sonicate the suspension in a NaCl−ice bath until the mixture becomes translucent (~ 20 times for 2 sec). Take care to keep the mixture ice cold and avoid foaming.
6.   Centrifuge the extract at 4000 r.p.m. for 30 min at 4°C. The resulting pellet should be very small.
7.   To the supernatant add phenylmethylsulphonyl fluoride to a final concentration of 0.1 mM and Aprotinin to 50 μg/ml.
8.   Divide into 50 μl aliquots, freeze in liquid nitrogen and store at −70°C. Under these conditions the extracts retain their full activity for more than a year.

B. *In vitro linearization by cos cleavage*

Sufficient linearized cosmid DNA for five partial digestions is produced by the following protocol.

1.   Remove salt from the cosmid DNA (as it inhibits the terminase cleavage) by ethanol precipitation and washing the pellet with 70% ethanol. In addition cosmid DNA can be dialysed against DPA buffer. This is very important.
2.   Dissolve 5−10 μg of cosmid DNA in 30 μl of DPA−DTT buffer. Add 8.5 μl of DPB buffer, 1.5 μl of 100 mM ATP and 10 μl of terminase extract.
3.   Leave for 30 min at room temperature and then add 50 μl of DPC buffer.
4.   Extract twice with 100 μl of phenol/chloroform/isoamylalcohol (25:24:1) equilibrated with 1 M NaCl. Extract once with 500 μl of ether.
5.   Remove the ether by heating the open tubes to 75°C, then precipitate the DNA with 200 μl of ethanol for 10 min in dry ice. Wash the pellet with 70% ethanol and dissolve in 25 μl of TE (10 mM Tris-HCl pH 8.0, 1 mM EDTA). Typically 10−20% of the circular cosmid molecules are linearized. This can be difficult to visualize on an ethidium bromide-stained gel. It can be verified by hybridization to labelled oligonucleotides. Alternatively, to check the quality of the terminase extract and the reaction conditions, a test substrate (cosmid vector) can be used which on re-cutting with a restriction enzyme will give a characteristic fragment for molecules linearized at the *cos* site.
6.   For partial digestion 5 μl of linearized cosmid DNA in 10 μl total volume is used.

C. *5' End-labelling of oligonucleotides*

Two dodecamers ON-L (5'-dAGGTCGCCGCCC-3') and ON-R (5'-dGGGCGGCGACCT-3') complementary to the left and right cohesive lambda DNA teminus, respectively, are used.

1.   Incubate 1 μl of ON-R or ON-L oligonucleotide [1 OD/ml (260nm)], 20 μCi [$\gamma-^{32}$P] ATP (5000 Ci/mmol) and 8 units of polynucleotide kinase in 10 μl of 70 mM Tris-HCl pH 7.6, 10 mM $MgCl_2$ and 5 mM DTT at 37°C for 1 h.

2. After 0 and 60 min spot an aliquot on a polyethyleneimine t.l.c. foil (Machary-Nagel, Dueren). Develop the chromatograph in 0.75 M potassium phosphate buffer pH 3.5. An autoradiogram should show an incorporation of at least 75% of the [^{32}P]phosphate.
3. Stop the reaction by heating to 100°C for 1 min. The oligonucleotides are stored at −20°C and used without further purification.

D. *Hybridization of partial DNA digests with oligonucleotide probes*

Hybridization mixes:

    1 μl of 5'-[^{32}P]oligonucleotide (ON-L or ON-R)
    50 μl of 1 M NaCl
    200 μl of gel loading buffer (see *Table 1*)

1. Divide each partial digest reaction containing about 0.01 pmol of DNA into two equal parts and mix one with 5 μl of hybridization mix ON-L and the other with 5 μl of hybridization mix ON-R.
2. Incubate the mixture for 2 min at 75°C. Transfer immediately to a 45°C water bath for 30 min.

E. *Electrophoresis*

1. Pour a 0.5% agarose gel (20 × 20 × 0.5 cm) with up to 28 slots in Loening buffer (see *Table 1*).
2. Load the marker in the first, middle and last slot. Load the samples labelled on the same end in adjacent lanes.
3. Run the gel at 1 V/cm for 18−24 h (at least until the free labelled nucleotides have run out of the gel; check the gel bottom with a hand monitor). Do not re-circulate the buffer because the free oligonucleotides and radioactive nucleoside triphosphates might re-enter the top part of the gel. Change the buffer after 8−10 h. The aim is to separate the large fragments as well as possible without loosing too many small fragments that run out of the gel. Frequently the smallest bands represent sites within the vector.
4. Transfer the gel onto DE-81 cellulose paper (Whatman). Dry on a gel dryer.
5. Expose on X-ray film overnight.

The strategy of partial mapping to determine the restriction maps of cosmids is outlined in *Figure 6*. *Table 4* contains a complete protocol including the preparation of terminase extract. *Figure 8* gives an example of an autoradiogram used for the analysis of the restriction site positions.

2.2.2 *Partial mapping of DNA labelled at both ends in a two-dimensional gel*

A linear DNA molecule labelled at both ends is partially digested with one enzyme, separated on a gel and then redigested to completion within the gel by a second enzyme generating labelled fragments of different sizes from both ends (7). Partial digestion products originating from one or the other end can therefore be independently identified and analysed on the basis of radiolabelled product present in the second dimension. An example is shown in *Figure 4* where the labelled fragments are indicated by a black star. The partial fragments can be easily divided into two sets by the two different labelled end fragments (5.5 kb and 6.8 kb) they generate after complete digestion with the second enzyme. The analysis is described in Section 2.2.3.

2.2.3 *Analysis of partial mapping results*

All methods to determine a restriction map include a step where the sizes of digestion fragments on a gel photograph or on an autoradiogram have to be calculated. This is done by comparing the mobility of unknown fragments with the mobilities of marker fragments of known length. Several algorithms are available to convert mobilities to

Table 5. Size marker for partial mapping experiments.

1. Cut aliquots of lambda cI857S7 DNA separately with the following enzymes: *Bam*HI, *Eco*RI, *Hind*III, *Hpa*I, *Pst*I, *Sal*I (Marker-1).
2. Alternatively use *Sph*I instead of *Pst*I and *Hpa*I (Marker-2)[a].
3. To check the digestion take 1 μg of DNA from each single digestion mix, hybridize to 2.5 μl of each hybridization mix (see *Table 4*), run on a 0.5% gel, dry the gel and expose overnight. Only two bands in each lane should be visible. If not, continue the digestions until all are complete.
4. Mix the samples in equal proportions, add uncut lambda DNA and dilute with water so that 5 μl of mix will give visible bands after overnight exposure.

[a]Marker fragment sizes in kilobases :
Marker-1: 48.5, 32.8, 23.1, 21.2, 15.3, 11.5, 8.7, 6.8, 5.5, 4.4, 3.5, 2.6, 0.7
Marker-2: 48.5, 32.8, 23.1, 21.2, 15.3, 9.1, 6.8, 5.5, 4.4, 3.5, 2.2

sizes. We have found an algorithm described by Southern (26,27) to be most reliable. For an account of the accuracy attainable see Chapter 7.

Since the correct determination of the final map depends on the accuracy of fragment sizes it is important to use a size marker with an optimal distribution of fragments in the range of interest. *Table 5* lists two lambda size marker mixtures. The marker mixture should be applied in at least two, preferably three, slots (in the first, middle and last) on the gel. The fragment size determination method should use the information of both flanking marker lanes for the calculations. It is essential that the set of marker fragments should cover the entire molecular weight range to be analysed, since the accuracy of all methods of size calculation decreases significantly if extrapolation outside the range covered by the marker fragments is used.

By measuring fragment lengths from either end of the DNA two restriction maps are established. Each becomes increasingly inaccurate towards the non-labelled end due to the decrease in accuracy of size determination with increasing length of DNA fragments. The combination of the two maps (taking the more accurate first half of each one) alleviates this problem and leaves only moderate uncertainty in the middle. The exact combination of the two maps, however, requires the knowledge of the total DNA length which is usually unknown. Therefore, it has to be first determined (without additional experiments) from the partial mapping data. Two theoretical methods see applicable:

(i) Each lane normally contains a band representing the undigested linearized DNA, but due to its large size determination of its length can be inaccurate.
(ii) The lengths of each pair of corresponding fragments (cut at the same site within the DNA but labelled at the right or left DNA end) will sum up to the total DNA length [Equation (1)].

$$S[i] = LF[j] = TL - RF[k] \qquad \text{Equation 1}$$

S = site position
LF = length of fragment labelled on left DNA end
RF = length of fragment labelled on right DNA end
TL = total DNA length

Unfortunately it is never possible to be sure which of the bands on the autoradiogram

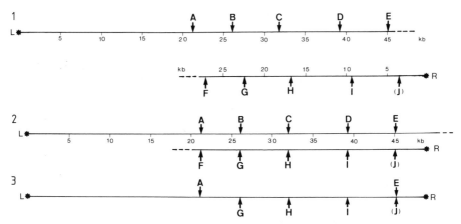

Figure 7. Restriction map determination from 'partial mapping' data. An autoradiogram of the gel in *Figure 4* is used as the starting point. Two different labelled complete *Bam*HI digestion fragments appear in the second dimension (6.8 kb and 5.5 kb). Accordingly, the partial *Eco*RI fragments are classified as originating from either the left or the right DNA end (marked L or R). Note that fragment J cannot be assigned since it is not cut by *Bam*HI and has to be omitted from the analysis. (If *Eco*RI instead of *Bam*HI had been used for the complete digestion in the second dimension, 21.2-kb and 3.5-kb labelled fragments would have been produced, see *Figure 1*, allowing unambiguous assignment of fragment J). For each end a map of the *Eco*RI sites based on the lengths of the partial digestion fragments is drawn (**1**) : set L = fragments A 21.23 kb, B 26.10 kb, C 31.75 kb, 39.17 kb, E 44.97 kb; set R = fragments J 3.53 kb (only if *Eco*RI was used for complete digestion), I 9.33 kb, H 16.76 kb, G 22.40 kb, F 27.26 kb. Then the two maps are shifted relative to each other until a clear overlap of restriction sites is found (**2**). The final map and the total DNA length is calculated using the site positions from the smaller partial fragments since they should be more accurate (**3**). If fragment J could not be assigned to a specific end, fragment E has to be used to calculate the site at position 44.97 (measured from the left end).

belong together since some small bands may have run out of the gel or large ones remain unresolved. There are empirical ways, however, to determine the total DNA length.

(i) *Analysis by hand.* The two maps (from the right and left DNA end) are drawn on two sheets of graph paper. The graph papers are then shifted relative to each other until an optimal overlap of two or three restriction sites in the middle of both maps is found. This method can be rather unsatisfactory if it is not possible to assign well defined restriction sites in one map clearly to corresponding sites in the second map (within a tolerable error limit). Once an unambiguous overlap is found a resulting map (by taking the more accurate site positions of either map) can be drawn and the approximate total DNA length can be determined from the distance between the beginning and the end of the map.

Figure 7 shows such an analysis using the data of an autoradiogram of the gel shown in *Figure 4*. Fragment J is an *Eco*RI fragment not cut by *Bam*HI and therefore it is not clear from the gel from which DNA end it originated. Such ambiguities can be avoided if the same enzyme is used for the partial and the complete digestion in both dimensions.

(ii) *Analysis using a computer program.* The main advantage of the program (21) is the simultaneous consideration of all available restriction sites measured from both sides during the calculation of the total DNA length and the resulting restriction map.

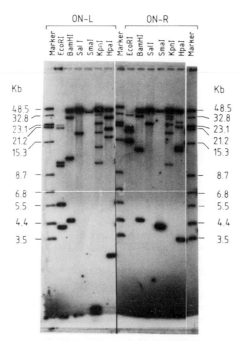

Figure 8. Partial digestion products of a linearized cosmid clone labelled either with an oligonucleotide complementary to the left (ON-L) or the right (ON-R) *cos* sequence of phage lambda and analysed on a 0.5% agarose gel as described in *Table 4*.

The program starts with two assumptions.

(i) The majority of restriction sites are found in both maps, generated from the right and from the left DNA end ('DOUBLE sites') with few found in only one map ('SINGLE sites').

(ii) In the 'optimal map' the number of DOUBLE sites is a maximum and the two single positions (calculated from either DNA end) for each such site show a minimum of discrepancy.

To define the reliability of a restriction map a measure G is evaluated using Equation 2. A smaller value of G indicates a better fit of the data to the calculated map.

$$G = \text{difsum} * (\text{single} / \text{double})^4 \qquad \text{Equation 2}$$

G = measure of reliability of the tested map

difsum = sum of all differences between site positions found from both DNA ends for one site

single = number of sites found only from one DNA end

double = number of sites found from both DNA ends

The weighting of the different factors was found empirically and shows that the most important parameter is the number of DOUBLE sites.

First, the optimal assignment of the restriction sites measured from both ends is determined for each total length value (tested stepwise within a range) by evaluating Equation (2) with the actual map parameters. For the final map both the resulting optimal total length and optimal assigned site pairs are used.

```
*****  Predicted MAP of DNA 66B   ***  File 1W66B.PAM   ***  10-APR-1986  ***  12:49  *****

Used FILE(S): 13W66B.PAR

Individual site limits used ('END-SEARCH' with LIMITS * 2.0)  Total length = 40.90 kb   Used CPU time: 0.60 seconds

4 BAMHI  L   4.52  11.25         29.09                    38.44
         R   4.52  11.25         28.35                    36.52
             6.06  11.96         28.35                    36.52

7 ECORI  L   4.20   5.59   9.91  10.45        20.83  22.27        26.19
         R   4.20   5.59   9.91  10.45        20.88  22.38        26.19
             4.20          11.76              20.94  22.48        26.19

6 HPAI   L   2.63  16.30  19.67         27.50  27.94               38.00
         R   2.63  16.30  19.80         27.50  28.19               37.58
                   17.05  19.93                28.19               37.58

5 KPNI   L  10.36  13.61  18.46                23.81        32.95
         R  10.36  13.61  18.46                22.65        30.97
            10.31  13.77  18.30                22.65        30.97

1 SALI   L  18.61
         R  18.53
            18.45

2 SMAI   L   1.18                                                  38.08
         R   1.18                                                  36.90
                                                                   36.90

L = from LEFT DNA end, R = from RIGHT DNA end

        .----1----.----2----.----3----.----4----.----5----.----6----.----7----.----8
4 BAMHI >---I---I------------------------------------------------I-------I-----------<  BAMHI
7 ECORI >---I---I--I----II-------------I--I----------I-------I----------I------------<  ECORI
6 HPAI  >---I----------I---------I---------I--------II-------------------------I-----<  HPAI
5 KPNI  >---I-------I--------I----------I--------------------I--------I-------------I <  KPNI
1 SALI  >------------------------I------------------------------------------------- I <  SALI
2 SMAI  >--I---------------------------------------------------------------------I--.--<  SMAI
        .----1----.----2----.----3----.----4----.----5----.----6----.----7----.----8
     5.00     10.00     15.00     20.00     25.00     30.00     35.00     40.00 kb

One step = 500 bases    I = one site    2,3 .. 9 = two to nine sites    X = ten or more sites
```

Figure 9. Calculated restriction map using the data from *Figure 8* and the computer programs described in ref. 21 which are available from the authors. For each enzyme the first and third line contain the calculated site positions from the left (L) or right (R) DNA end. The second line shows the final selected values (site positions determined from the closer DNA end and except in an interval of 5% around the centre of the map where the mean values are taken). A graphic representation of the restriction map is shown below.

163

If the map is established from one end of the DNA only, the program uses the average size of the linear uncut DNA bands as the total length value and site positions are derived from the single set of partial fragment sizes. This simplification gives satisfactory results only if the second half of the analysed DNA contains mainly vector DNA with known site positions (e.g. using phage vectors with one long and one short vector arm).

To determine the restriction map from the autoradiogram of such a mapping experiment the mobilities of the marker fragments and the labelled partial fragments visible on the X-ray film are entered into the first program using either a digitizer or the keyboard. The fragment sizes are then calculated and stored in a file. Next, the name of the DNA to be mapped and the name(s) of the file(s) containing partial fragment data for this DNA are specified to the mapping program. Using the partial fragment data of the autoradiogram shown in *Figure 8* a restriction map has been calculated and a part of the output file is presented in *Figure 9*. The programs, available from the authors, are written in Pascal and run on a DEC VAX under the VMS operating system.

3. REFERENCES

1. Legerski,R.J., Hodnett,J.L. and Gray,H.B.,Jr. (1978) *Nucleic Acids Res.*, **5**, 1445.
2. McDonell,M.W., Simon,M.N. and Studier,F.W. (1977) *J. Mol. Biol.*, **110**, 119.
3. Rackwitz,H.-R., Zehetner,G., Murialdo,H., Delius,H., Chai,J.H., Poustka,A., Frischauf,A. and Lehrach,H. (1985) *Gene*, **40**, 259.
4. Poustka,A., Rackwitz,H.-R., Frischauf,A.-M., Hohn,B. and Lehrach,H. (1984) *Proc. Natl. Acad. Sci. USA*, **81**, 4129.
5. Little,P. and Cross,A. (1985) *Proc. Natl. Acad. Sci. USA*, **82**, 3159.
6. Smith,H.O. and Birnstiel,M.L. (1976) *Nucleic Acids Res.*, **3**, 2387.
7. Kovacic,R.T. and Wang,J.C. (1979) *Plasmid*, **2**, 394.
8. Whittaker,P.A. and Southern,E.M. (1986) *Gene*, **41**, 129.
9. Parker,R.C., Watson,R.M. and Vinograd,J. (1977) *Proc. Natl. Acad. Sci. USA*, **74**, 851.
10. Daniels,D.L., Schroeder,J.L., Blattner,F.R., Szybalski,W. and Sanger,F. (1983) In *Lambda II.* Hendrix,R.W., Roberts,J.W., Stahl,F.W. and Weisberg,R.A. (eds), Cold Spring Harbor Laboratory Press, New York, p. 519.
11. Fitch,W.M., Smith,T.F. and Ralph,W.W. (1983) *Gene*, **22**, 19.
12. Pearson,W.R. (1982) *Nucleic Acids Res.*, **10**, 217.
13. Polner,G., Dorgai,L. and Orosz,L. (1984) *Nucleic Acids Res.*, **12**, 227.
14. Durand,R. and Bregegere,F. (1984) *Nucleic Acids Res.*, **12**, 703.
15. Nolan,G.P., Maina,C.V. and Szalay,A.A. (1984) *Nucleic Acids Res.*,**12**, 717.
16. Wulkan,M. and Lott,T.J. (1985) *CABIOS*, **1**, 235.
17. Smith,D.H., Brutlag,D., Friedland,P. and Kedes,L.H. (1986) *Nucleic Acids Res.*,**14**, 17.
18. Boehm,T.L.J. and Drahovsky,D. (1984) *J. Biochem. Biophys. Methods*, **9**, 153.
19. Herrmann,R.G., Whitefeld,P.R. and Bottomley,W. (1980) *Gene*, **8**, 179.
20. Hildebrand,M., Jurgenson,J.E., Ramage,R.T. and Bourque,D.P. (1985) *Plasmid*, **14**, 64.
21. Zehetner,G. and Lehrach,H. (1986) *Nucleic Acids Res.*,**14**, 335.
22. Villems,R., Duggleby,C.J. and Broda,P. (1978) *FEBS Lett.*, **89**, 267.
23. Chen,C.W., Braun,R. and Thomas,C.A. (1984) *Experientia*, **40**, 921.
24. Sato,S., Hutchinson,C.A. and Harris,J.I. (1977) *Proc. Natl. Acad. Sci. USA*, **74**, 542.
25. Rackwitz,H.-R., Zehetner,G., Frischauf,A.-M. and Lehrach,H. (1984) *Gene*, **30**, 195.
26. Southern,E.M. (1979) *Anal. Biochem.*, **100**, 319.
27. Elder,K.J. and Southern,E.M. (1983) *Anal. Biochem.*, **128**, 227.

CHAPTER 7

Computer-aided analysis of one-dimensional restriction fragment gels

J.K.ELDER and E.M.SOUTHERN

1. INTRODUCTION

Gel electrophoresis is the standard technique for measuring the size of DNA restriction fragments. The estimated lengths are often used in constructing DNA restriction maps (Chapter 6), and it is therefore important that the fragment lengths are estimated as accurately as possible. As with large sequencing projects, large mapping projects become unwieldy, and to tackle the mapping of genomes such as those of *Escherichia coli* or yeast, or substantial regions of mammalian chromosomes, it will be necessary to automate much of the analysis, particularly in assembling complex restriction maps when the size of restriction fragments is the only information used to detect overlaps. When other information is available, such as shared sequence homology between fragments, or fragment order in the cloned sequences (Chapter 6), estimation of the fragment length is less critical. But when sequence length is the only information used in comparing two fragments, the quality of the matching process is crucially dependent on the accuracy of size measurement.

In this chapter we discuss computer-aided methods for measuring the mobility of restriction fragments and for converting mobilities to DNA length, and compare them with manual methods. The assembly of fragment lengths into a map is dealt with in Chapter 6.

2. MEASUREMENT OF ELECTROPHORETIC MOBILITY

2.1 Sources of error

Mobility measurements are usually made from a positive or negative photograph of the gel. There are four main sources of error which affect the accuracy with which band positions are measured:

(i)   limited accuracy of the method of measurement,
(ii)   limited ability to determine the centres of bands,
(iii)   loss of registration between tracks,
(iv)   transcription errors.

We discuss these types as they affect the methods described below.

2.2 Traditional methods

The simplest method of measuring mobility, with a ruler or a micrometer, suffers from all the sources of error listed. The use of a digitizing tablet (1) eliminates transcription errors, and also inter-track registration errors when a two-dimensional frame of reference is used. In addition, it is less tedious. However, both these methods suffer from low accuracy because they depend on the ability of the human eye to determine the centre of each band, and this ability is limited, especially for diffuse bands or bands which overlap. Positional accuracy can be improved by photographic enlargement, but the problem of estimating the centres of bands remains.

The use of an analogue densitometer to produce density profile plots brings two advantages. The centres of bands, taken as the tops of the peaks in the profiles, can be found more easily than the centres of bands on photographs; furthermore, many densitometers provide facilities for expanding the linear scale. This method is time-consuming however, since a separate profile has to be produced for each track in the gel, and without special provision it is difficult to ensure proper registration of tracks.

2.3 Computer-based methods

Computer-controlled scanning devices are being used increasingly for measuring electrophoretic mobility. As with analogue densitometers, the scanner records a density profile for each track, but this time in a digital form which can be analysed by computer.

Two types of digital scanner can be used for this purpose: those which scan one track at a time, and those which scan all tracks in a single pass of the gel. Like analogue scanners, the former may suffer from loss of registration between tracks, leading to loss of accuracy in length estimation when external markers are used. Most scanners which scan all tracks together use a linear diode array (2,3) or a laser beam (4), (Chapter 9, Section 2.4). The array or laser beam scans over the gel in a direction parallel or perpendicular to the tracks; in both cases however, a single frame of reference is used so that there is no loss of registration between tracks.

A common photographic negative size used for electrophoresis gels is 125×100 mm ($5'' \times 4''$). For these dimensions, a pixel size of 50 μm corresponds to 2500 readings down the length of a track and provides sufficient resolution.

The agarose gels used to analyse restriction fragments differ from sequencing gels in that the tracks are usually straighter and run more nearly parallel to each other. If the gel has been run with care the bands are straight over much of their length and are perpendicular to the track. This means that the scan need not cover the entire area of the gel, as is necessary for sequencing gels. Instead only those areas of the gel containing tracks need to be scanned and a one-dimensional density profile can be produced directly for each track.

The track areas to be scanned can be specified by performing a fast, low resolution scan of the entire gel photograph, displaying the digital image on a graphics screen and interactively defining the positions of tracks. Since the tracks are usually equispaced and of constant width, this can be done rapidly by using a cursor or a light pen to specify the centres of the first and last tracks to be scanned and the number of tracks between them. It is best to avoid the edges of tracks where bands may be curved, and so the track width to be scanned can be adjusted interactively so that it covers, say, the central two-thirds of the total track width. The superimposition of the scan boundaries on the gel image (*Figure 1*) also allows the operator to verify that the photograph is correctly aligned.

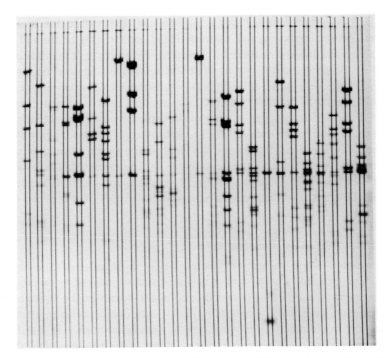

Figure 1. A coarse resolution scan of a restriction fragment gel with boundaries superimposed indicating the sections to be scanned at high resolution.

The quality of the density profiles can be improved by applying digital filtering to the pixel data to reduce noise, both from the specimen and from the scanner. In a typical analysis, 20–50 pixels will lie across the central width of a track. Our practice is to split these into five adjacent sets of consecutive pixels, sum the pixel readings within each set, and take the median of the five sums as the final reading for that point in the track. This operation is performed in real time for each track and for each reading down the length of the gel.

2.4 Mobility measurement from digital profiles

The digital form of each profile allows band positions to be measured more accurately than by eye. A peak finding procedure such as that described in *Chapter 9*, Section 3.6 can be used to find an initial set of peaks. However, the results should be reviewed by an operator, to eliminate obvious artefacts not dealt with by the computer program. The initial position of a peak can be taken as the top of the peak, but this can be improved by finding the position that corresponds to the peak's centre of mass, or by fitting a Gaussian to the upper part of the peak and taking its mean. In this way information from the whole peak, rather than just its maximum height, is used in determining its position.

It is important for the accurate construction of a restriction site map that unresolved multiple peaks are detected. In theory, the amount of DNA in a restriction fragment will be proportional to the area of the corresponding peak in the profile. Therefore, if all the peaks in a track represent single fragments, an approximate linear relationship should exist between peak area and fragment length. Since the digital form of the track profiles allows the areas of peaks to be calculated, the presence of multiple peaks can be tested by fitting a straight line by least squares to the fragment lengths and the corresponding peak areas and, for each peak, examining the deviation of expected area from actual area. The possible presence of a multiple peak is signalled by a large deviation. For each suspect peak, the test can be refined by removing the suspect from the set of peaks used for the least squares fit, performing a new fit and again examining the deviation in peak area. If the peak is a compound one, the actual area should be an integral multiple of the area predicted by the fit. In practice, this assumption will only be true if the linear relationship between peak area and fragment length holds in the gel and if there are no other multiple fragments in the track. The test must therefore be used with caution.

2.5 Correction for inter-track variation

DNA standards of known length are always present in the gel, and typically several adjacent tracks containing fragments of unknown size are flanked by two tracks each containing identical standards. The mobilities of the standards can be projected to each track containing unknown fragments by linear interpolation between each pair of standards; this corresponds to drawing straight lines between pairs of standards and using the points of intersection with the track of interest as the mobilities of the standards. This procedure, which requires that all band positions in all tracks are recorded in a single two-dimensional frame of reference, has the advantage of being insensitive to distortions in the gel and to misalignment of the gel photograph with respect to the scanning direction (2).

An alternative, which avoids variation between tracks, is to use standards in the same track as the fragments of unknown size. A minor disadvantage of this approach is that the standards and unknowns have then to be separated by the analytical program.

When either the semi-logarithmic or reciprocal method (Sections 3.1 and 3.2) is used to convert mobility to length, the choice of origin for mobility measurements is arbitrary since it only alters the value of a constant in the relationship. The positions of the gel slots, which are often difficult to define accurately, can therefore be ignored.

2.6 Relative accuracy of manual and computer-based methods for measuring mobility

In a comparison of the accuracy of different methods for measuring mobility (5), we used a digitizing tablet, an analogue densitometer and a digital densitometer to measure the mobilities of a polymeric series of DNA fragments. The polymer mobilities were converted to lengths by the reciprocal method (Section 3.2) using other polymers in the series as standards, and the estimated lengths were compared with the true lengths. The tablet gave a maximum error of approximately 0.6%. The analogue densitometer with an expansion ratio of 2:1 gave a similar maximum error, which was reduced to 0.3% with an expansion ratio of 5:1. The digital densitometer with subsequent computer processing gave a maximum error of less than 0.1%. The digital method was therefore several times more accurate than all of the other methods. This improvement was entirely due to the greater accuracy of measurement of relative band positions.

3. METHODS FOR CONVERTING MOBILITY TO FRAGMENT LENGTH

Mobility in the gel can be used to compare two fragments and test them for identity, but to perform the arithmetic used in building a restriction map it is necessary to convert mobility to DNA size. For example, to decide whether two fragments are produced from cleavage of a larger one, it is necessary to know the sizes of all three, and again, the power of this test depends critically on the accuracy of the molecular weight estimate. This depends on both the accuracy of mobility measurement, and the method used to convert this measurement to DNA size.

To estimate the lengths of unknown fragments, a relationship must be established between the mobilities and lengths of standard fragments so that it can be used to convert the mobilities of the unknown fragments to lengths. Many relationships have been used (6−11) and we discuss two of the most commonly used ones. Whatever method is used, it is important that the standards span the entire size range of the unknown fragments, since extrapolation beyond the range of the standards gives rise to large errors in length estimates. Also, there should be no large gaps in the size distribution of standards, since no relationship between mobility and length holds accurately over a wide size range.

3.1 Manual and semi-logarithmic methods

A graph of length versus mobility can be constructed manually and the lengths of unknown fragments read off by eye, but accuracy is limited because of the marked curvature of the graph. The most commonly used method is to plot log(length) against mobility (6). This relationship holds approximately over a narrow size range which depends on the gel concentration and running conditions, but has marked curvature in the high and low ranges, which makes estimates inaccurate unless the unknown is very close to a standard.

3.2 Reciprocal method

The reciprocal relationship

$$(m - m_0)(L - L_0) = c$$

where m is mobility, L is length, and m_0, L_0 and c are constants has been shown to hold over a wider molecular weight range than the semi-logarithmic method (7). Since the reciprocal relationship has three constants, three standards are needed to determine them. If the unknown is in the middle of four standards, a convenient way to apply this method in estimating the length of a fragment is to use these standards in two sets of three. The two closest standards of higher mobility and the closest of lower mobility are used to determine one set of constants, and the two closest standards of lower mobility and the closest of higher mobility are used to determine a second set of constants. The two sets are used with the fragment mobility to calculate two length estimates as described in the Appendix, and the mean of these is taken as the fragment length. When the unknown has only one standard to one side of it, then only one set of three can be used. When the unknown is outside the range of the standards, again only one set of three can be used, and in this case the estimate is likely to be inaccurate.

An alternative to this local method of estimation is to perform a least squares fit of all the standard mobilities and lengths (8), and to use a single set of constants m_0, L_0 and c. It might be thought that estimates derived from a global fit would be more accurate than those from local fits, since the global fit uses more information. In practice however the local estimates are usually better since none of the relationships between size and mobility that have been proposed holds accurately over a wide size range (9) (Section 3.3), and the use of standards distant from the unknown introduces errors. The use of a global fit does however have the advantage that the contribution of a faulty or misestimated standard is less than with local fits. Our practice is to calculate both local and global estimates; the local estimates are normally used, but the global ones serve as a check.

We also use the reciprocal relationship to test the internal consistency of each set of standards, by regarding each standard in turn as an unknown fragment, using other standards to estimate its length, and comparing the estimated and true lengths.

The reciprocal method is described in more detail in the Appendix to this chapter.

3.3 Relative accuracy of semi-logarithmic and reciprocal methods

We used the first 11 fragments in the polymer series of ligated pAT153 *Bam*HI DNA to compare the accuracy of the semi-logarithmic and reciprocal methods (9). The fragments ranged in size from 3.7 to 40.2 kb. Mobilities were measured with a digital densitometer (Section 2.6). The length of each of the fragments 3 to 9 in the series was estimated by using other fragments in the series as standards. The methods were used in local and global forms: in the local form the two closest fragments on each side of the unknown were used as standards; in the global form all the other fragments were used as standards. Both relationships were fitted by least squares in the global form, as was the semi-logarithmic method in the local form; the local version of the reciprocal method was applied as described (Section 3.2).

Using the semi-logarithmic method, errors in length estimates ranged from -7.6 to 16% in the global form, and from 1.4 to 4.6% in the local form. The reciprocal method gave errors ranging from -1.6 to 1.0% in the global form, and from -0.07 to 0.06% in the local form. These results show the improvement in accuracy to be gained by using only standards close to the unknown, and the greater accuracy of the reciprocal method over the semi-logarithmic method.

3.4 Effect of base composition on gel mobility

We have shown (9), (Section 3.3) that DNA size can be estimated with an accuracy of one part per thousand. However, as all fragments in a polymer series have the same base composition, this analysis masks any effect of base composition on gel mobility. In analyses of restriction fragment digests of phage λ DNA, in which the fragments have a wide range of base compositions, we found that the errors in fragment sizes compared with those known from the sequence were much larger, up to 3%. For mapping, it is important to know whether the effects of base composition are consistent when large fragments are broken into smaller ones, or more precisely, whether the size of a large fragment estimated from its gel mobility is equal to the sum of component fragments estimated in the same way. We have not investigated this.

4. REFERENCES

1. Kieser,T. (1984) *Nucleic Acids Res.*, **12**, 679.
2. Gray,A.J., Beecher,D.E. and Olson,M.V. (1984) *Nucleic Acids Res.*, **12**, 473.
3. Elder,J.K., Green,D.K. and Southern,E.M. (1986) *Nucleic Acids Res.*, **14**, 417.
4. Marsman,H. and Wijnaendts van Resandt,R. (1985) *Electrophoresis*, **6**, 242.
5. Elder,J.K., Amos,A., Southern,E.M. and Shippey,G.A. (1983) *Anal. Biochem.*, **128**, 223.
6. Fisher,M.P. and Dingman,C.W. (1971) *Biochemistry*, **10**, 1895.
7. Southern,E.M. (1979) *Anal. Biochem.*, **100**, 319.
8. Schaffer,H.E. and Sederoff,R.R. (1981) *Anal. Biochem.*, **115**, 113.
9. Elder,J.K. and Southern,E.M. (1983) *Anal. Biochem.*, **128**, 227.
10. Stellwagen,N.C. (1983) *Biochemistry*, **22**, 6180.
11. Plikaytis,B.D., Carlone,G.M., Edmonds,P. and Mayer,L.W. (1986) *Anal. Biochem.*, **152**, 346.

5. APPENDIX: ESTIMATION OF FRAGMENT LENGTH USING THE RECIPROCAL METHOD

We give here formulae for using the reciprocal relationship

$$(m - m_0)(L - L_0) = c$$

in both its local and global forms. The formulae are easily programmed for a programmable calculator or a computer.

5.1 Local form

Three pairs of standard mobilities and lengths (m_1, L_1), (m_1, L_2), (m_3, L_3) are needed to determine the constants m_0, L_0 and c in the equation. The constants are found by the following formulae:

$$A = \frac{(m_3 - m_2)(L_2 - L_1)}{(m_2 - m_1)(L_3 - L_2)}$$

$$m_0 = \frac{m_3 - m_1 A}{1 - A}$$

$$L_0 = \frac{(m_3 - m_0)L_3 - (m_1 - m_0)L_1}{m_3 - m_1}$$

$$c = (m_1 - m_0)(L_1 - L_0)$$

Lengths of unknown fragments of mobility m are then given by:

$$L = \frac{c}{m - m_0} + L_0$$

5.2 Global form

A single set of constants m_0, L_0 and c is calculated from n pairs of standard mobilities and lengths (m_1, L_1), ..., (m_n, L_n), $(n \geq 3)$ by performing a least squares fit. This is the form of the method used by Schaffer and Sederoff (8), who also give a FORTRAN program listing. All summations are over the range $i = 1, ..., n$.

$$p_i = m_i L_i \qquad (i = 1, ..., n)$$

Restriction fragment gels

$$\bar{m} = \frac{1}{n}\Sigma m_i \qquad\qquad \bar{L} = \frac{1}{n}\Sigma L_i \qquad\qquad \bar{p} = \frac{1}{n}\Sigma p_i$$

$$CS_m = \Sigma(m_i - \bar{m})^2 \qquad\qquad CS_L = \Sigma(L_i - \bar{L})^2$$

$$CP_{mL} = \Sigma(m_i - \bar{m})(L_i - \bar{L})$$

$$CP_{mp} = \Sigma(m_i - \bar{m})(p_i - \bar{p})$$

$$CP_{Lp} = \Sigma(L_i - \bar{L})(p_i - \bar{p})$$

$$\Delta = CS_m CS_L - (CP_{mL})^2$$

$$m_0 = (CS_m CP_{Lp} - CP_{mL} CP_{mp})/\Delta$$

$$L_0 = (CS_L CP_{mp} - CP_{mL} CP_{Lp})/\Delta$$

$$c = \frac{1}{n}\Sigma(m_i - m_0)(L_i - L_0)$$

As in the local form, lengths of unknown fragments of mobility m are given by:

$$L = \frac{c}{m - m_0} + L_0$$

CHAPTER 8

Computer handling of DNA sequencing projects

R.STADEN

1. INTRODUCTION

For large scale DNA sequencing projects the shotgun strategy (1,2) is currently regarded as the fastest technique. Every base will be sequenced many times and the data processing is more complicated than for other methods, but if the data handling preserves all the evidence used to derive the final sequence, we can be confident of its accuracy. In this chapter I describe a data handling method for shotgun sequencing projects, although many of the techniques and programs will also be applicable to more directed sequencing strategies.

1.1 The shotgun sequencing strategy

In the shotgun sequencing procedure the sequence to be determined is randomly broken into fragments of about 400 nucleotides in length. These fragments are cloned and then selected randomly and their sequences determined. The relationship between any pair of fragments is not known beforehand but is found by comparing their sequences. If one sequence is found to overlap that of another for sufficient length then those two fragments can be joined. The process of select, sequence, and compare is continued until the whole of the DNA to be sequenced is in one continuous, well determined, piece.

1.2 Definition of a contig

It is useful to have a word to describe the overlapping sets of gel readings that are manipulated during a sequencing project. A *contig* is a set of gel readings that are related to one another by overlap of their sequences. All gel readings belong to one and only one contig and each contig contains at least one gel reading. The gel readings in a contig can be summed to produce a continuous consensus sequence and the length of this sequence is the length of the contig. At any stage of a sequencing project the data will comprise a number of contigs; when a project is complete there should only be one contig and its consensus will be the finished sequence.

1.3 Symbols for uncertainty in gel readings

In order to record uncertainties when reading gels the codes shown in *Figure 1* can be used. Use of these codes permits us to extract the maximum amount of data from each gel and yet record any doubts by choice of code. The programs can deal with all of these codes and any other characters in a sequence are treated as dash (-) characters.

```
        SYMBOL                    MEANING
        ------                    -------
          1          PROBABLY       C
          2             "           T
          3             "           A
          4             "           G
          D             "           C      POSSIBLY   CC
          V             "           T         "       TT
          B             "           A         "       AA
          H             "           G         "       GG
          K             "           C         "       C-
          L             "           T         "       T-
          M             "           A         "       A-
          N             "           G         "       G-
          R          A OR G
          Y          C OR T
          5          A OR C
          6          G OR T
          7          A OR T
          8          G OR C
          -          A OR G OR C OR T
            else = -
```

Figure 1. Uncertainty codes for gel readings.

1.4 List of programs

There are a number of sets of programs for handling such data (for example ref. 3), but here I describe only the ones I have developed (4), and which have been used on a number of the very largest sequencing projects such as the complete sequence of Epstein−Barr virus·(5). The following programs are described.

BATIN, GELIN, GELINB	— data input programs.
BSPLIT	— splits files created by GELINB.
SCREENV, SCREENR, DBCOMP	— screening programs.
DBUTIL, DBAUTO	— database management programs.
PREDIT, POSTED, DBMERGE	— ancillary programs for facilitating editing.
ENDSOUT	— copies gel readings out of a database.
HIGH	— used to highlight errors in contigs.

1.5 Equipment

The methods described require particular hardware and software if the problems of converting programs are to be avoided. If a different computer is chosen, a machine with at least 512 kbytes of random access memory, a hard disk and a FORTRAN 77 compiler would be needed, as well as a knowledge of the FORTRAN language.

1.5.1 Computers

The programs described here were written to run on a DEC VAX computer or on an IBM Personal Computer AT or XT with a hard disk (see Chapter 1). Other machines could be used as the software is easily *ported*. The program that uses a digitizer for

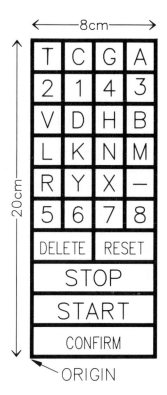

Figure 2. The menu for GELIN.

reading sequences from autoradiographs runs on both the VAX and the IBM micro-computers, and also on a BBC micro made by Acorn Computers, Cambridge, UK.

1.5.2 *Digitizer*

A digitizer enables the spatial coordinates of a special pointing device (e.g. a pen) placed on a two dimensional surface to be recorded by a computer. These coordinates can be interpreted by a program.

The digitizing device we used has a stylus which emits sparks giving a high frequency sound which is picked up by two microphones positioned at the rear of the working area. The pen position is determined by triangulation and the digitizing device sends the coordinates to the computer. As no special surface is required the device can conveniently be positioned on a light box giving the sequencer an unobscured view of the autoradiographs.

The digitizer used is called a GRAPHBAR MODEL GP7 (see Chapter 1).

The program uses a menu to allow the user to select commands or to enter the uncertainty codes for areas of the gel that are difficult to interpret. A menu is simply a series of boxes drawn on the digitizing surface that each contain a command or uncertainty code. When the user puts the pen down in these special regions the program interprets the coordinates as commands and acts appropriately. The menu is shown in *Figure 2*. It should be drawn on a piece of paper, exactly as shown, and struck down

175

on the surface of the light box in the digitizing area. For convenience it is best to position it to the right of the digitizing area, but in practice as long as its top edge is parallel to the digitizer box, it can be put anywhere in the active region.

1.5.3 *Software*

The programs described can be obtained by writing to me at MRC Laboratory of Molecular Biology, University Medical School, Hills Road, Cambridge CB2 2QH, UK. For users of VAX computers the programs will be sent with command procedures to install them ready for use. The IBM PC versions will be available commercially. For users of other machines notes on how to port the programs are provided, and a lot of effort has gone into making this as easy as possible.

2. INTRODUCTION TO THE COMPUTER METHOD

It is useful to consider the objectives of a sequencing project before outlining how we use the computer to help achieve them.

2.1 **Objectives of a sequencing project**

The aim of a shotgun sequencing project is to produce an accurate consensus sequence from many overlapping gel readings. It is necessary to know, at all stages of the project,

```
              10        20        30        40        50
 -6  HINW.010  GCGACGGTCTCGGCACACAAAGCCGCTGCGGCGCACCTACCCTTCTCTTATA
     CONSENSUS  GCGACGGTCTCGGCACACAAAGCCGCTGCGGCGCACCTACCCTTCTCTTATA

              60        70        80        90        100
 -6  HINW.010  CACAAGCGAGCGAGTGGGGCACGGTGACGTGGTCACGCCGCGGACACGTC
 -3  HINW.007                                        GGCACA*GTC
     CONSENSUS  CACAAGCGAGCGAGTGGGGCACGGTGACGTGGTCACGCCG-G-ACACGTC

              110       120       130       140       150
 -6  HINW.010  GATTAGGAGACGAACTGGGGCG3CGCC*GCTGCTGTGGCAGCGACCGTCG
 -3  HINW.007  GATTAG4AGACGAACTGGGGCGACGCCCG*TGCTGTGGCAGCGACCGTCG
 -5  HINW.009                               GGCAGCGACCGTCG
 17  HINW.999                               AGCGACCGTCG
     CONSENSUS  GATTAGGAGACGAACTGGGGCGACGCCCGCTGCTGTGGCAGCGACCGTCG

              160       170       180       190       200
 -6  HINW.010  TCT*GAGCAGTGTGGGCGCTG*CGGGCTCGGAGGGCATGAAGTAGAGC*
 -3  HINW.007  TCT*GAGCAGTGTGGGCGCTGC*CGGGCTCGGAGGGCATGAAGTAGAGC*
 -5  HINW.009  TCT*GAGCAGTGTGGGCG*T*G*CGGGCTCGGAGGGCATGAAGTAGAGC*
 17  HINW.999  TCTCGAGCAGTGTGGGCGCTG**CGGGCTCGGAGGGCATGAAGTAGAGCG
 12  HINW.017                                    GTAGAGC*
     CONSENSUS  TCTCGAGCAGTGTGGGCGCTG-CGGGCTCGGAGGGCATGAAGTAGAGCG
```

Figure 3. A display of a contig produced by the program DBUTIL. This shows the left end of a contig from position 1 to 200. Overlapping this region are gel readings numbered 6, 3, 5, 17 and 12; 6, 3 and 5 are in reverse orientation to their original reading (denoted by a minus sign). Each gel reading also has a name (e.g. HINW.010). It can be seen that in a number of places the sequences contain characters other than A, C, G and T. Some of these extra characters have been used by the sequencer to indicate regions of uncertainty in the initial interpretation of the gel reading, but the asterisks (*) have been inserted by the automatic assembly program in order to align the sequences. Underneath each 50 character block of gel reading sequences is the consensus derived from the sequences aligned above (the line labelled CONSEN-SUS). For most of its length the consensus has a definite nucleotide assignment but in a few positions there is insufficient agreement between the gel readings and so a dash (−) appears in the sequence.

how accurate the consensus sequence is. This enables us to know which regions of the sequence require further work and also to know when the project is finished. To show the quality of the consensus, the programs described here produce displays like that shown in *Figure 3.*

This shows the left end of a contig from position 1 to 200. Overlapping this region are gel readings numbered 6, 3, 5, 17 and 12; 6, 3 and 5 are in reverse orientation to their original reading (denoted by a minus sign). Each gel reading also has a name (e.g. HINW.010). It can be seen that in a number of places the sequences contain characters other than A, C, G and T. Some of these extra characters have been used by the sequencer to indicate regions of uncertainty in the initial interpretation of the gel reading, but the asterisks (*) have been inserted by the automatic assembly program in order to align the sequences. Underneath each 50 character block of gel reading sequences is the consensus derived from the sequences aligned above (the line labelled CONSENSUS). For most of its length the consensus has a definite nucleotide assignment but in a few positions there is insufficient agreement between the gel readings and so a dash (-) appears in the sequence. This display contains all the evidence needed to assess the quality of the consensus; the number of times the sequence has been determined on each strand of the DNA, and the individual nucleotide assignments given for each gel reading.

So the aim is to produce the consensus sequence, and, equally important, a display of the experimental results from which it was derived.

2.2 Required operations

In order to obtain an accurate final sequence the following operations need to be performed.

(i)   Interpret autoradiographs and put individual gel readings into the computer.
(ii)   Check each gel reading to make sure it is not simply part of one of the vectors used to clone the sequence.
(iii)   Check each gel reading to make sure that those fragments that span the ligation point used prior to sonication are not assembled as single sequences (see Section 5).
(iv)   Compare all the remaining gel readings with one another to assemble them to produce the consensus sequence.
(v)   Check the quality of the consensus and edit the sequences.
(vi)   When all the consensus is sufficiently well determined, produce a copy of it for processing by other analysis programs.

It is very unlikely that this procedure will only be passed through once. Usually steps (i) − (v) are cycled through repeatedly, with step (iv) just adding new sequences to those already assembled. Generally step (vi) is also used in order to analyse an imperfect sequence to check if it is the one intended to sequence, or to look for interesting features. Analysis of the consensus, such as searches for protein coding regions (Chapter 10), can also help to find errors in determination of the sequence. The display of the overlapping gel readings shown above can be used to indicate not only the poorly determined regions but also which clones should be resequenced to resolve ambiguities, or those which can usefully be extended or sequenced in the reverse direction.

2.3 **Gel reading files and files of file names**

The individual gel readings for a sequencing project are each stored in separate files which we call *archives*. As the gel readings are entered into the computer (usually in batches, say 10 from a film), the file names they are given are stored in a further file, called a file of file names. Files of file names enable gel readings to be processed in batches.

2.4 **Project databases**

The assembled sequences for a sequencing project are stored in a database that has a structure specifically designed for dealing with shotgun sequence data.

In order to arrive at the final consensus sequence, many operations will be performed on the sequence data. Individual fragments must be sequenced and compared in both senses (i.e. both orientations) with all the other sequences. When an overlap between a new gel reading and a contig are found they must be aligned and the new gel reading added to the contig. If a new gel reading overlaps two contigs they must be aligned and joined. Before the two contigs are joined one of them may need to be turned around (reversed and complemented) so that they are both in the same orientation.

Clearly, keeping track of all these manipulations is quite complicated, and to be able to perform the operations quickly requires careful choice of data structure and algorithms. For these reasons it is not practicable to store the gel readings aligned as shown displayed in *Figure 3*. Rather, it is more convenient to store the sequences unassembled, and to record sufficient information for the programs to assemble them during processing. The data used to assemble the sequences is called relational information.

Note that the size of the database is not automatically increased as required by the addition of new data: when the database is created a fixed amount of space is reserved on the disk. Users must decide what their requirements are going to be, and the database should be large enough to contain all the gel reading and contig information. One line is needed for each gel reading and one for each contig; the database size is the sum of these two numbers. The maximum permitted size for a database is 1000. To minimize the amount of disk space required in the short term, the size can be set to less than the expected final requirements, and the database expanded using the copying option (option 13 in the main menu of the program DBUTIL, see Section 9.9) when necessary. We call the actual size of the database IDBSIZ. The database comprises three files and they are described below.

2.4.1 *Working versions of gel readings*

After each new gel reading has been compared with the contents of the database it is copied into the database and aligned with any sequences it overlaps. Gel readings stored in the database are called the *working versions* of gel readings.

The working versions of gel readings are stored in a file with IDBSIZ records. Each record is 512 characters in length and so no gel reading can exceed 512 nucleotides. The first gel reading entered into the database is stored in record number 1, the second in record number 2 and so on.

2.4.2 *The file of relationships*

This file contains all of the information that is required to assemble the working versions into contigs during processing; any manipulations on the data use this file and it is automatically updated at any time that the relationships are changed. The information in this file is as follows.

Gel descriptor lines. These contain information about each gel and its relationship to others.

(i)   The number of the gel (each gel is given a number as it is entered into the database).

(ii)   The length of the sequence from this gel.

(iii) The position of the left end of this gel relative to the left end of the contig of which it is a member.

(iv)   The number of the next gel to the left of this gel.

(v)   The number of the next gel to the right.

(vi)   The relative strandedness of this gel, i.e. whether it is in the same sense or the complementary sense as its archive.

Contig descriptor lines. These contain information about each contig.

(i)   The length of this contig.

(ii)   The number of the leftmost gel of this contig.

(iii) The number of the rightmost gel of this contig.

General data

(i)   The number of gels in the database.

(ii)   The number of contigs in the database.

The file contains IDBSIZ lines of data: the general data are stored on line IDBSIZ; data about gels are stored from line 1 downwards; data about contigs are stored from line IDBSIZ-1 upwards. A database of 500 lines containing 25 gel readings and 4 contigs would have a file of relationshps as is shown in *Figure 4*.

As each new gel is added into the database a new line is added to the end of the list of gel lines. If this new gel does not overlap with any gels already in the database a new contig line is added to the top of the list of contig lines. If it overlaps with one contig then no new contig line need be added but if it overlaps with two contigs then these two contigs must be joined and the number of contig lines will be reduced by one. Then the list of contig lines is compressed to leave the empty line at the top of the list. Initially the two types of line will move towards one another but eventually, as contigs are joined, the contig descriptor lines will move in the same direction as the gel lines. At the end of a project there should be only one contig line. A fullsize database is thus capable of handling a project of 998 gels.

2.4.3 *The file of archive names*

This is simply a list of the names of each of the archive files (original gel reading files) in the database but on line number 1000 we also store the size of the database, i.e.

```
-------------------------------------------------
    1    Gel descriptor record
    2     "         "         "
    3     "         "         "
    4     "         "         "
    5     "         "         "
    '     '         '         '
    '     '         '         '
   25     "         "         "
   26    Empty record
    '     '         '
    '     '         '
  495     '         '
  496    Contig descriptor record
  497     "         "         "
  498     "         "         "
  499     "         "         "
  500     NUMBER OF GELS=25, NUMBER OF CONTIGS=4
-------------------------------------------------
```

Figure 4. The arrangement of the data in the file of relationships. Here we see a database of 500 lines containing 25 gel readings and 4 contigs.

the number of lines of information allowed in the database files. This file always has 1000 lines but the length of the file of relationships and the file of working versions can be set by the user when creating a database or when copying from one to another.

2.5 **File and project names**

There are some restrictions on file and project names and they are given below.

2.5.1 *Project names*

Project names must be exactly six characters in length; each character must be a legal file name character but must not be a point (.). The files for the database are created using the program DBUTIL. If we choose, say LAMBDA, for the project name then DBUTIL will automatically name the file of relationships LAMBDA.RL0, the file containing the archived gel reading names LAMBDA.AR0 and the working versions of gel readings LAMBDA.SQ0.

Database copy numbers. The last symbol in the database file names is called the *copy number*. When the database is initially created it will always be set (as above) to 0, but if the database is copied using the program DBUTIL to create a backup on the disk, any other legal file name symbol can be used. As a rule we use copy 0 for the active copy of the database and copy 1 as a backup.

2.5.2 *Gel reading file names*

The maximum permitted length for gel reading file names is 10 characters. I would recommend using names like LAMBDA.001, LAMBDA.002, ...LAMBDA.999 where LAMBDA is the project name.

2.5.3 *Files of file names*

There are no restrictions on the names of files of file names, but I would recommend the use of LAMBDA.NAM where LAMBDA is the project name.

2.6 **How contigs are named**

Generally users need not concern themselves with how the relational information is used by the programs, but it is necessary to know how contigs are named. As contigs are constantly being changed and reordered the programs identify them by the numbers of the gel readings they contain. Whenever users need to identify a contig they need only know the number of one of the gel readings it contains. The program will always ask users for the number of the left gel reading in a contig, but if the gel number the user types is not a left end, the program finds the correct number.

2.7 **Searching for overlaps**

Before it is entered into the project database each new gel reading must be searched to look for overlaps with all the data already contained within the database. This last point is important: all searching for overlaps is between individual new gel readings and the data already in the database. There is no searching for overlaps between sequences within the database and overlaps must be found when new gel readings are entered into the database. The algorithm used to find the overlaps searches for a minimum number of consecutive matching characters − typically 15.

2.8 **Overview of computer processing**

Below I give an introduction to how the sequences are processed as they are passed from one program to the next.

DBUTIL is used to start a database for the project and then the following procedure is used.

Data in the form of individual gel readings are entered into the computer and stored in separate files using either program GELIN or BATIN. Batches of these gel readings are passed to the screening programs SCREENV and SCREENR via files of file names. SCREENV searches for overlaps with vector sequences and SCREENR for matches with restriction enzyme recognition sequences that should not be present in the gel readings. Each run of SCREENV or SCREENR passes on only those gel readings that do not contain unwanted sequences. Sequences are passed via files of file names and eventually are processed by DBAUTO. This program compares each gel reading with a consensus of all the previous gel readings stored in a database for the sequencing project. If it finds any overlaps it aligns the overlapping sequences by inserting padding characters, and then adds the new gel reading to the database. Gels that overlap are added to existing contigs and gels that do not overlap any sequence in the database start contigs of their own. If a new gel overlaps two contigs they are joined. Any gel readings that appear to overlap but which cannot be aligned sufficiently well are not entered and have their names written to a file of failed gel reading names. The screening and assembly steps are usually run as a single batch procedure.

Generally data is entered into the database in batches as just described. DBUTIL is used to examine the data in the database, to enter gel readings that DBAUTO cannot align, and to make final edits. Edits to whole contigs can also be made using system editors (e.g. the full-screen editor EDT on a VAX or WORDSTAR on the IBM PC AT computer). This is made possible by the programs PREDIT, POSTED and DBMERGE. Disagreements between gel readings in contigs and their consensus sequences can be highlighted by use of the program HIGH. If it is thought that some overlaps may have been missed the programs ENDSOUT and DBCOMP can be used to compare the ends of contigs.

3. ENTERING GEL READINGS INTO THE COMPUTER

There are two methods of entering gel readings into the computer. The original method requires the user to interpret the autoradiograph and write the sequence on a piece of paper before typing the sequence into the computer using program BATIN. A faster, and more accurate method uses a digitizing device to simultaneously interpret the autoradiograph, and enter the sequences into the computer. Two programs are available for digitizing, one, GELIN, works on the machines that can also assemble the sequences (such as the VAX), and the other, GELINB, works on a BBC microcomputer. If the BBC microcomputer program is used the gel readings must be transferred to a more powerful machine for assembly.

All the subsequent screening and assembly programs operate on batches of sequences. To facilitate this, the input programs write the individual gel readings into separate files, but also record the names of these files in a file of file names (see Section 2.3). As was noted in the overview (Section 2.8, see also Section 8) a command procedure is usually used to pass batches of gel readings through the screening and assembly procedures. To avoid having to change this command procedure it is best to always use the same file of file names for a project. It is therefore also necessary to process all the gel readings from one batch before deleting the file of file names ready for the next round of film reading and data entry.

3.1 **Typing gel readings into the computer**

The program to use is called BATIN. It writes each gel reading into a separate file and records the names of all the files created in a file of file names.

(i)   Start the program by typing BATIN and the program will ask the first question:
    " ? FILE OF FILE NAMES = "

(ii)   Type a name for the file of file names and the main loop of the program procedes with the next question.

(iii)   " ? FILE NAME FOR NEXT GEL READING = "
    Type the name for the next gel reading (see Section 2.5.2 about file names).

(iv)   " TYPE THE SEQUENCE, FINISH WITH @"
    Type in the sequence on lines that are less than 81 characters in length and when finished type the special symbol @. To make checking easier it is helpful to type lines that are the same length as those you are copying from.

(v)   " TO ENTER ANOTHER GEL READING TYPE 1"
    To enter another gel reading type 1 and the program will ask for its file name etc; otherwise type only carriage return.

(vi) *"* TO GET GEL READINGS PRINTED TYPE 1*"*

To get all the gel readings that have just been entered printed out on the terminal screen type 1; otherwise type only carriage return and the program will stop.

3.2 **Entering gel readings using a digitizer**

For our purposes, the digitizing device is placed at the rear of a light box and the autoradiograph is stuck to the light box. In order to read an autoradiograph the user need only define the positions of the four sequencing lanes and the bases to which they correspond and then use the pen to point to each successive band progressing up the gel. The program examines the coordinates of each pen position to see in which of the four lanes it lies and assigns the corresponding base to be stored in the computer. Each time the pen tip is depressed to point to a position on the surface of the digitizer the program sounds the bell on the terminal (a different sound for each of the four bases on the microcomputer version of the program) to indicate to the user that a point has been recorded. As the sequence is read the program displays it on the screen.

The program uses a menu (*Figure 2*) to allow the user to select commands or to enter the uncertainty codes for areas of the gel that are difficult to interpret (see Section 1.3). As well as the uncertainty codes the following commands are included in the menu: DELETE removes the last character from the sequence; RESET allows the lane centres to be redefined; START means begin the next stage of the procedure; STOP means stop the current stage in the procedure; CONFIRM means confirm that the last command or set of coordinates are correct.

The digitizing device also has a menu of its own. This lies in a two inch wide strip immediately in front of the digitizing box. Pen positions within this two inch strip are interpreted as commands to the digitizer and are not sent to the GELIN program. In general the only time users will need to use the device menu is when they tell GELIN where the program menu lies in the digitizing area. This is done by first touching ORIGIN in the device menu and then touching the bottom left hand corner of the program menu. The program menu can hence be positioned anywhere in the active region but should be arranged parallel to the digitizer.

The user should try to touch the bands as near as possible to the centre of the lanes because the program tracks the lanes up the film using the pen positions. By using this tracking strategy the user only has to define the centres of the bottom of the lanes before starting to read the film. The program can correctly follow quite curved lanes and constantly checks that its lane centre coordinates look sensible. If the lane centres appear to be getting too close the program stops responding to the pen positions of bands and hence does not ring the bell. If this occurs users must hit the reset box in the menu and the program will request them to redefine the lane centres at the current reading position. Then they can continue reading. As a further safeguard the program will only respond to pen positions either in the menu or very close to the current reading position.

3.3.1 *Running the gel reading program*

(i) The autoradiograph should be firmly stuck down on the light box and the program started by typing GELIN. It will ask the first question.
" ? FILE OF FILE NAMES = *"*

(ii)   Type the name for the file of file names and then follow the instructions.
(iii)   " HIT DIGITIZER MENU ORIGIN"
(iv)   " THEN PROGRAM MENU ORIGIN"
(v)   " THEN HIT START IN PROGRAM MENU"
         If the bell does not sound after you touch START try touching metric in the
         device menu (the program uses metric units, and some digitizers are set to default
         to use inches; touching metric switches between the two).
(vi)   After the bell has sounded the program will give the default lane order.
(vii)   " LANE ORDER IS T C A G"
(viii)   " IF CORRECT HIT CONFIRM, ELSE HIT RESET"
         If the lane order, reading from left to right is correct touch CONFIRM in the
         program menu. If you are using a different order touch RESET and you will
         be asked to define the lane order from left to right using the program menu (as
         follows).
(ix)   " DEFINE LANE ORDER (LEFT TO RIGHT) USING MENU"
         Touch the boxes in the menu that contain the symbols A, C, G, T in the left-
         right order of the lanes. The program will respond with the lane order as above
         and ask for confirmation. When this is received, the next task is to define the
         start positions of the next four lanes.
(x)   " HIT START, THEN HIT (LEFT TO RIGHT)"
(xi)   " THE START POSITIONS FOR THE NEXT FOUR LANES"
         Touch the centres of the four lanes at a height level with the first band that is
         going to be read. The program will report the coordinates of the lane centres
         and asks for confirmation that they are correct.
(xii)   " LANE CENTRES ARE A B C D"
(xiii)   " HIT CONFIRM TO CONTINUE"
         Asking for confirmation allows users to try again if their hand had slipped;
         generally the lane centre coordinate values can be ignored. Touch CONFIRM,
         and the program will give the mesage.
(xiv)   " HIT START WHEN READY TO BEGIN READING"
         Touch START and the program will give the message.
(xv)   " HIT BANDS, UNCERTAINTY CODES, RESET OR STOP".
         Touch the bands, interpreting the sequence progressing up the film. If necessary
         use the uncertainty codes. If the pen stops responding touch RESET and follow
         the instructions as above. When the sequence becomes unreadable touch STOP
         and the program will ask for a file name for the gel reading just read.
(xvi)   " ? FILE NAME FOR THIS GEL READING="
         Type the file name observing the rules given in Section 2.5.2. The program will
         display the length of the last reading and ask if you wish to read another sequence.
(xvii)   " TO ENTER ANOTHER GEL READING TYPE 1"
         To enter another type 1 and you will be back to the step of defining the lane
         order. Typing anything else will stop the program.

3.3.2 *Running the microcomputer version of the gel reading program*

The microcomputer version of GELIN is slightly different and is called GELINB. The
BBC micro does not have the capacity to process the gel readings beyond the reading

stage, and some users of the IBM PC AT version might also prefer to process their gel readings on a larger machine. This means that users of these programs would need to transfer their gel readings from the micro to another machine using a file transfer program (see Chapter 1). Transferring many files is tedious and so the microcomputer versions of the gel reading programs store all the gel readings for each run of the program in a single file. This special file contains both sequences and file names and can be moved in a single transfer to another machine. Once on the other machine the single file must be split into separate gel reading files and a file of file names. This is done using the program BSPLIT. The only difference in using the microcomputer versions of GELIN is that the first file name the program requests is not a file of file names, but a name for the single file to contain all the gel readings and their names.

3.3.3 *Running the program to split concatenated gel reading files*

(i)   Having, if necessary, transferred the file of concatenated gel readings, start the splitting program by typing BSPLIT. You will be asked the first question.

(ii)   " ? INPUT FILE FROM GELINB="

Type the name of the concatenated file created by GELINB.

(iii)   " ? FILE OF FILE NAMES FOR THIS BATCH="

Type the name of the file of file names for the project and the program will create all the files passed from the gel reading program.

4. SCREENING GEL READINGS AGAINST VECTOR SEQUENCES

It is possible that gel readings may contain sequences from any of the vectors used in the production of the clones from which they were obtained. Before they are assembled the gel readings must be compared against all the vectors with which they could be contaminated. As usual, gel readings are processed in batches using files of file names. The program reads a file of file names containing the names of all the gel readings to compare against the vector sequence. It writes a new file of file names containing the names of the files of all those gel readings that do not match the vector. This new file of file names can be used by the next program in the procedure. The algorithm the program uses looks for a perfect uninterrupted match of some minimum length such as 15 nucleotides. All the matches found are displayed.

4.1 **Format of vector and consensus sequence files**

For convenience, the vector sequences have exactly the same format as the consensus sequences calculated for shotgun project databases. They are text files with records less than 81 characters in length. The consensus of the project database may contain many contigs and these need to be both named and separated from one another. This is achieved by preceding each contig consensus with a title of exactly 20 characters. A typical title would be < ---LAMBDA.076-----> where LAMBDA is the project name and gel 76 is the leftmost gel reading to contribute to this consensus sequence. The angle brackets < > and the position of the gel number are important as they are used by programs to identify the contigs. This means that vector sequences used for screening should also contain similar titles with a six character name followed by a point and a three digit number positioned exactly as shown above. For example a title of the form

$<$---M13MP7.001-----$>$ must precede the nucleotides. The only parts that may vary are the six characters M13MP7 (but not their position or number), and the number 001 can be any three digit number.

4.2 **Running the vector screening program**

(i)   Start the program by typing SCREENV and it will ask the first question.

(ii)   " ? INPUT FILE OF FILE NAMES = "

Type the name of the file of file names. It will either be the one created by GELIN or BATIN or one passed on from a previous screening step.

(iii)   " ? OUTPUT FILE OF FILE NAMES = "

Type the name of the file of file names to be passed on to the next stage of processing. When the program has finished running this file will contain the names of all the gel readings that did not match the vector sequence.

(iv)   " ? FILE NAME OF VECTOR SEQUENCE = "

Type the name of the file containing the vector sequence.

(v)   " ? MINIMUM MATCH LENGTH ($>$11) = "

Type the minimum match length, which must be at least 12. If a number less than 12 is typed the program will use 12 as its minimum. Lengths of 14 or 15 are suitable.

Each gel reading whose name appears in the input file of file names will be compared with the vector sequence. Both strands are compared and any matches found are displayed. Once the comparison procedure is started it will take of the order of 1 second per gel reading.

5. SCREENING GEL READINGS AGAINST RESTRICTION ENZYME SITES

If sonication is used to produce the shotgun fragments the DNA is first circularized. Some of the fragments will therefore span the ligation point and must be prevented from being entered into the project database as single sequences. The program SCREENR is used to screen a batch of gel readings identified by a file of file names against any restriction sites that should not be present in single fragments. It writes a new file of file names that contains the names of all the gel readings that do not match with the restriction enzyme recognition sequences. The recognition sequences are stored in a file.

5.1 **Format of restriction enzyme recognition sequence files**

The file containing the restriction enzyme recognition sequences that should not be present in the gel readings for a project is a simple text file that can be written by a system text editor. The file should contain only the recognition sequences, and they should be written one recognition sequence to a line, in upper case letters.

5.2 **Running the restriction enzyme screening program**

(i)   Start the program by typing SCREENR and it will ask the first question.

(ii)   " ? INPUT FILE OF FILE NAMES = "

Type the name of the input file of file names, probably passed on from SCREENV.

(iii)   *" ? OUTPUT FILE OF FILE NAMES = "*
Type the name of the file of file names to contain the names of all those gel
readings that do not match the restriction enzyme recognition sequences.

(iv)   *" ? NAME OF RESTRICTION ENZYME FILE = "*
Type the name of the file containing the restriction enzyme recognition sequences
that should not be present.

6. STARTING A NEW PROJECT DATABASE

At the start of each new project a database is created to contain all the assembled data.
The three files that make up the database are described in Section 2.4. They are created
by program DBUTIL which will need to be given a six character name for the project
and an initial size for the project.

(i)   Start the program by typing DBUTIL and it will begin by offering 18 different
options:

        1 = HELP
        2 = STOP
        3 = OPEN A PROJECT DATABASE
        4 = EDIT CONTIG
        5 = DISPLAY A CONTIG
        6 = LIST A TEXT FILE
        7 = DIRECT OUTPUT TO DISK
        8 = CALCULATE A CONSENSUS
        9 = SHOW RELATIONSHIPS
        10 = ENTER NEW GEL READING
        11 = COMPLEMENT A CONTIG
        12 = JOIN CONTIGS
        13 = COPY THE DATABASE
        14 = CHECK THE DATABASE FOR LOGICAL CONSISTENCY
        15 = EXAMINE CONTIGS FOR QUALITY OF DATA
        16 = ALTER RELATIONSHIPS
        17 = SEARCH FOR A GEL READING BY NAME
        18 = SET DISPLAY PARAMETERS

(ii)   Select option 3 and the program will ask the question.

(iii)   *" TO START A NEW DATABASE TYPE 1, TO OPEN AN OLD ONE TYPE
0"*
Type 1, and the program will ask the next question.

(iv)   *" ? SIZE FOR NEW PROJECT DATABASE (<1001, DEFAULT=50) = "*
the database should be large enough to contain all the gel reading and contig
information. One line is needed for each gel reading and one for each contig;
the database size is the sum of these two numbers. To minimize the amount of
disk space required in the short term, the size can be set to less than the expected
final requirement, and the database expanded using the copying option (option
13 in the main menu, see Section 9.9), when necessary. Type the database size
chosen and the program will ask for a project name.

(v) " ? PROJECT NAME (EXACTLY SIX SYMBOLS) FOR NEW DATABASE="

The project name must contain exactly six letters and no point (.). The program will create the three files setting the version number to 0 and the number of gels and contigs to zero. If this is done successfully the message "DATABASE FOR PROJECT ?????? COPY 0 SUCCESSFULLY STARTED" will be given, the main menu will reappear and the program should be stopped using option 2. If there is a problem such as insufficient disk space or if any of the files already existed the program will give the message "ERROR OPENING PROJECT DATABASE" and the main menu will reappear. If so return to the operating system and check for the two suggested difficulties.

7. AUTOMATIC ASSEMBLY OF THE GEL READINGS

The automatic assembly program DBAUTO takes a batch of gel readings (probably passed on as a file of file names from SCREENV) and enters them into the database for a sequencing project. It takes each gel in turn, and compares it with the current consensus for the database. It then produces an alignment for any regions of the consensus it overlaps; if this alignment is sufficiently good it then edits both the new gel reading and the gels it overlaps and adds the new gel to the database. The program then updates the consensus accordingly and carries on to the next gel reading. If a new gel reading overlaps two contigs they should be joined. To achieve this DBAUTO enters the gel reading into one contig, and then recalculates its consensus, before recomparing it with the other contig. If the alignment is still good enough the contigs are edited and joined, if not, the name of the gel reading is added to an error file. If a new gel reading overlaps more than two contigs, it is not entered into the database and its name is written to the error file.

Both strands of each gel reading are compared with the consensus. Overlaps are found by searching for exact matches of some minimum length (typically 15 bases). Using this matching procedure the number of contigs the new gel reading overlaps is determined, and then a more careful search is done to find the best alignments. Alignments are achieved by putting padding characters (*'s) in both the new gel reading and the contig. No data is deleted by DBAUTO.

The user controls the editing performed by the program by specifying three parameters:

(i) the maximum number of padding characters that the alignment routines may enter into the contigs for each gel reading,

(ii) the maximum number of padding characters the alignment routines may place in the sequence of the new gel reading, and

(iii) the maximum percentage mismatch allowed between the new gel and the consensus after alignment has been achieved.

The program types a commentary of the processing. An example of a typical run of DBAUTO is given in *Figure 5* which shows the beginning of a run starting on an empty database. The first two gel readings do not match, the third does overlap one of the first two, and the fourth overlaps two contigs.

All alignments are displayed and any gels that do match but that cannot be aligned sufficiently well have their names written to a file of failed gel reading names. The program works without any user intervention and can process any number of gel readings in a single run. Those gel readings that fail can be recompared using program DBCOMP and the user can enter them into the database manually using the program DBUTIL. Earlier versions of the program had a number of limitations on the types of joins that could be made, but in the latest version these have been removed.

(a)

```
$DBAUTO
    DBAUTO V5.4
 OPEN A PROJECT DATABASE
 ? PROJECT NAME =LAMBDA
 ? COPY NUMBER(DEF=0)=0
  NUMBER OF GELS=    0 NUMBER OF CONTIGS=    0 DB SIZE=    50
 CHECKING DATABASE FOR LOGICAL CONSISTENCY
 DATABASE IS LOGICAL CONSISTENT
 ? FILE OF GEL READING NAMES=1.NAM
 ? FILE FOR NAMES OF FAILED GELS=FAILED.OUT
 ? MINIMUM MATCH (DEFAULT= 12)=15
 ? MINIMUM ALIGNMENT BLOCK (DEFAULT=3)=0
 ? MAXIMUM PADS ALLOWED IN GEL (DEFAULT=5)=6
 ? MAXIMUM PADS ALLOWED IN CONTIG (DEFAULT=5)=6
 ? MAXIMUM PERCENT MISMATCH BETWEEN NEW GEL AND CONTIG(DEFAULT=5.)=_
  TO PERMIT JOINING OF CONTIGS TYPE 1 1
 >>>>>>>>>>>>>>>>>>>>>>>>>>>>>>>>>>>>>>>>>>>>>>>>>>>>>>>>>>>>>>>>>>>>>>>>>>>>
 NAME OF NEW GEL READING= A1E1.SEQ
 LENGTH OF NEW GEL READING=    195
 NEW GEL READING DOES NOT OVERLAP. START A NEW CONTIG
 TRYING TO ENTER NEW GEL READING INTO DATABASE
 THIS GEL READING HAS BEEN GIVEN THE NUMBER        1
 ADDING NEW GEL READING TO CONSENSUS
 >>>>>>>>>>>>>>>>>>>>>>>>>>>>>>>>>>>>>>>>>>>>>>>>>>>>>>>>>>>>>>>>>>>>>>>>>>>>
 NAME OF NEW GEL READING= A1B7.SEQ
 LENGTH OF NEW GEL READING=    120
 SEARCHING FOR OVERLAPS
 STRAND        1
 NO MATCHES FOUND
 STRAND        2
 NO MATCHES FOUND
 NEW GEL READING DOES NOT OVERLAP. START A NEW CONTIG
 TRYING TO ENTER NEW GEL READING INTO DATABASE
 THIS GEL READING HAS BEEN GIVEN THE NUMBER        2
 ADDING NEW GEL READING TO CONSENSUS
 >>>>>>>>>>>>>>>>>>>>>>>>>>>>>>>>>>>>>>>>>>>>>>>>>>>>>>>>>>>>>>>>>>>>>>>>>>>>
 NAME OF NEW GEL READING=A1B7B.SEQ
 LENGTH OF NEW GEL READING=    179
 SEARCHING FOR OVERLAPS
 STRAND        1
 MATCH FOUND WITH CONTIG NUMBER =        2
            1          11         21         31         41         51
            CAGACCCCGA GTCAAAAGCG CCCAACCAAC AAACCCGCCC ACTTGACCCA GTCCCCCAAC
            ****** *** ******* ** * ******** ********** ********** *********
            CAGACCTCGA GTCAAAATCG CTCAACCAAC AAACCCGCCC ACTTGACCCA GTCCCCCAAC
            1          11         21         31         41         51
            61         71         81         91         101        111
            GACGCGACAA TTGAACATTG TGCCGATCAC AGGGGTCTGG GCCACGTTTT TTTCGGTTTT
            ********** ********** ********** ********** ********** ********
            GACGCGACAA TTGAACATTG TGCCGATCAC AGGGGTCTGG GCCACGTTTT TTTCGGTTTG
            61         71         81         91         101        111
 STRAND        2
 NO MATCHES FOUND
```

Figure 5.

(b)

```
TRYING TO ALIGN THE NEW GEL READING WITH CONTIG      2
TOTAL PADDING IN CONTIG=     0 AND IN GEL=     0
PERCENTAGE MISMATCH AFTER ALIGNMENT =  3.3
BEST ALIGNMENT FOUND
           1         11        21        31        41        51
        CAGACCCCGA GTCAAAAGCG CCCAACCAAC AAACCCGCCC ACTTGACCCA GTCCCCCAAC
        ****** *** ******* ** * ******* ********** ********** **********
        CAGACCTCGA GTCAAAATCG CTCAACCAAC AAACCCGCCC ACTTGACCCA GTCCCCCAAC
           1         11        21        31        41        51
          61        71        81        91       101       111
         GACGCGACAA TTGAACATTG TGCCGATCAC AGGGGTCTGG GCCACGTTTT TTTCGGTTTT
         ********** ********** ********** ********** ********** *********
         GACGCGACAA TTGAACATTG TGCCGATCAC AGGGGTCTGG GCCACGTTTT TTTCGGTTTG
          61        71        81        91       101       111
NEW GEL READING OVERLAPS CONTIG      2
TRYING TO ENTER NEW GEL READING INTO DATABASE
THIS GEL READING HAS BEEN GIVEN THE NUMBER       3
CALCULATING A CONSENSUS
>>>>>>>>>>>>>>>>>>>>>>>>>>>>>>>>>>>>>>>>>>>>>>>>>>>>>>>>>>>>>>>>>>>>>>>>>>>>
NAME OF NEW GEL READING=BBT.A1X
LENGTH OF NEW GEL READING=    242
SEARCHING FOR OVERLAPS
STRAND     1
NO MATCHES FOUND
STRAND     2
MATCH FOUND WITH CONTIG NUMBER =      1
           1         11        21        31        41        51
        CAACATCATC AAATCTGTCA TTCAAAACTC GTTGATTTCC GATGACTACA CTCGAAGCGT
        ********** ********** ********** ********** ********** **********
        CAACATCATC AAATCTGTCA TTCAAAACTC GTTGATTTCC GATGACTACA CTCGAAGCGT
         111       121       131       141       151       161
          61        71        81        91       101       111
         GTTACATCCT TTGAACGAGA AGAATGAAAA CAATCATGAA TTGAGTTGAG TAAAGACCAT
         ********** ********** ********** ********** ******      **    *
         GTTACATCCT TTGAACGAGA AGAATGAAAA CAATCATGAA TTGAGTGAGT AAAGACCATC
         171       181       191       201       211       221
         121       131
           CTCATCAGTG CA

           TCATCAGTGC AC
         231       241
MATCH FOUND WITH CONTIG NUMBER =      2
         136       146       156       166       176
         CATTTCCAAT TGAATGAGTT TAGAATTTTC AGTTGAATTT GAAT
         *  ******* ********** ********        *
         CCATTCCAAT TGAATGAGTT TAGAATTTCG TTGAATTTGA ATTA
           1         11        21        31        41
```

(c)

```
TRYING TO ALIGN THE NEW GEL READING WITH CONTIG      1
TOTAL PADDING IN CONTIG=     0 AND IN GEL=     1
PERCENTAGE MISMATCH AFTER ALIGNMENT =  0.0
BEST ALIGNMENT FOUND
           1         11        21        31        41        51
        CAACATCATC AAATCTGTCA TTCAAAACTC GTTGATTTCC GATGACTACA CTCGAAGCGT
        ********** ********** ********** ********** ********** **********
        CAACATCATC AAATCTGTCA TTCAAAACTC GTTGATTTCC GATGACTACA CTCGAAGCGT
         111       121       131       141       151       161
          61        71        81        91       101       111
         GTTACATCCT TTGAACGAGA AGAATGAAAA CAATCATGAA TTGAGTTGAG TAAAGACCAT
         ********** ********** ********** ********** ****** *** **********
         GTTACATCCT TTGAACGAGA AGAATGAAAA CAATCATGAA TTGAGT*GAG TAAAGACCAT
         171       181       191       201       211       221
```

```
121         131
CTCATCAGTG CAC
********* ***

CTCATCAGTG CAC
231         241
TRYING TO ALIGN THE NEW GEL READING WITH CONTIG      2
TOTAL PADDING IN CONTIG=      0 AND IN GEL=      3
PERCENTAGE MISMATCH AFTER ALIGNMENT =   0.0
BEST ALIGNMENT FOUND
        135         145         155         165         175
        CCATTTCCAA TTGAATGAGT TTAGAATTTT CAGTTGAATT TGAAT
        *** ****** ********** ********* * ******** *****
        CCA*TTCCAA TTGAATGAGT TTAGAATTT* C*GTTGAATT TGAAT
        1          11         21         31         41
OVERLAP BETWEEN CONTIGS      1 AND      2
LENGTH OF OVERLAP BETWEEN THE CONTIGS=     -67
CLASS OF JOIN=     14
ENTERING THE NEW GEL READING INTO CONTIG      1
TRYING TO ENTER NEW GEL READING INTO DATABASE
THIS GEL READING HAS BEEN GIVEN THE NUMBER      4
CALCULATING A CONSENSUS
COMPLEMENTING CONTIG      4
COMPLEMENTING CONTIG      2
TRYING TO ALIGN THE TWO CONTIGS
TOTAL PADDING IN CONTIG=      3 AND IN GEL=      0
PERCENTAGE MISMATCH AFTER ALIGNMENT =   0.0
BEST ALIGNMENT FOUND
        264         274         284         294         304
        ATTCAAATTC AAC*G*AAAT TCTAAACTCA TTCAATTGGA A*T
        ********** *** * **** ********** ********** * *
        ATTCAAATTC AACTGAAAAT TCTAAACTCA TTCAATTGGA AAT
        1          11         21         31         41
EDITING CONTIG      4
COMPLETING THE JOIN BETWEEN CONTIGS      1 AND      3
CALCULATING A CONSENSUS
```

Figure 5. (a) A typical DBAUTO run. Here we are starting on an empty database and so the first gel reading (A1E1.SEQ) does not overlap. The gel reading with file name A1B7B.SEQ is the first to match. The user input is underlined. (b) Here we see the alignment step for gel reading A1B7B.SEQ (which is no different to the original) and its entry into the database. Next gel reading BBT.A1X is compared and its complementary sequence is found to match two contigs (1 and 2). (c) A typical run of DBAUTO. Finally we see the alignment steps for BBT.A1X, its entry into contig 1, the alignment between the two contigs, and their subsequent joining. Note that the contig numbers keep changing because of the entry of the new gel reading and the complementing of the contigs.

7.1 **Running the automatic assembly program**

(i)   Start the program by typing DBAUTO and it will ask the first question.

(ii)   " ? PROJECT NAME="

       Type the name of the database for the sequencing project. Note that DBAUTO cannot start a project database; this can only be done by DBUTIL.

(iii) " ? COPY NUMBER(DEF=0)"

       Type the copy number of the database (the default is copy 0).

(iv)   " ? FILE OF GEL READING NAMES="

       Type the name of input file of file names.

(v)   " ? FILE FOR NAMES OF FAILED GELS="

       Type the name for the output file of file names for failed gel readings.

(vi)   " ? MINIMUM MATCH (>11)="

       Type a value for the minimum match length to define matches for the comparison algorithm. The default is 12 and any selected value must be at least this large. To avoid finding spurious matches it is best to use a value of about 15.

(vii)   " ? MINIMUM SLIDE (DEF=3)="

Type a value for the minimum slide length used by the alignment routine. The default value is 3 and this should be used except for special cases.

(viii) " ? MAX PADS ALLOWED IN GEL(DEF=5)="

Type a value for the maximum number of padding characters that the alignment routine is allowed to insert in any individual new gel reading. The routine may insert more than this value to achieve alignment, but if it does, although the alignment will be displayed, the gel reading will not be entered into the database.

(ix)   " ? MAX PADS ALLOWED IN CONTIG(DEF=5)="

Type a value for the maximum number of padding characters that the alignment routine is allowed to insert into any individual contig to try to align a single new gel reading.

(x)   " ? MAX PERCENT MISMATCH BETWEEN NEW GEL AND CONTIG(DEF=5)="

Type a percentage value to define the maximum permissible percentage mismatch allowed after alignment. Again, if this figure is exceeded, the alignment will be displayed but the gel reading will not be entered into the database. The values chosen for the maximum number of padding characters and percentage mismatch will depend on the quality of the gels being produced but typically we currently use about 6, 6 and 5.0 respectively.

(xi)   " TO PERMIT JOINING TYPE 1"

Occasionally a user may wish to stop joins being performed and so this is made selectable. To permit joins type 1, othewise type only carriage return and joins will not be performed, and all gel readings involved in joins will be written to the error file.

8. COMMAND PROCEDURES FOR SCREENING AND ASSEMBLING GEL READINGS

Usually the screening and assembly programs are used non-interactively as part of a procedure that would run several programs one after the other with no user intervention. Typically these would include:

(i)   SCREENV for any number of vector sequences arranged so that the output for each (a file of file names) is passed on to the next program.

(ii)   SCREENR for any number of restriction enzyme recognition sequences that should not be present.

(iii)   DBUTIL is run both to indicate the state of the database before the current batch of gel readings is processed and to make a copy that could be used in the event of a problem occurring during the DBAUTO run.

(iv)   DBAUTO is then run using as input the file of file names that is output from the last screening program.

(v)   DBUTIL is then run to show the final state of the database and to calculate a current consensus.

```
$SCREENV
LAMBDA.NAM
M13MP7.OUT
M13MP7.SEQ
15
$SCREENV
M13MP7.OUT
PBR322.OUT
PBR322.SEQ
15
$SCREENR
PBR322.OUT
BAMH1.OUT
BAMH1.SEQ
$DBUTIL
3
0
LAMBDA
0
13
A
0
2
$DBAUTO
LAMBDA
0
BAMH1.OUT
FAILED.OUT
15
3
6
6
5.
1
$DBUTIL
3
0
LAMBDA
0
8
LAMBDA.TMP
0
0
2
$DBCOMP
FAILED.OUT
LAMBDA.TMP
14
```

Figure 6. A command procedure for screening and assembling gel readings. This command procedure is for a VAX. It shows the screening of a batch of gel readings whose names are contained in a file called LAMBDA.NAM against, first M13, then all those that do not match are screened against pBR322. In both cases the programs look for a match of at least 15 consecutive nucleotides. Then all those that do not match are screened against the BAMH1 recognition sequence that is stored in a file called BAMH1.SEQ. All those that pass all these screening stages are passed on to DBAUTO but first DBUTIL is used to make a copy of the database to copy A. The database is called LAMBDA. DBAUTO writes the names of all the gels that fail to be entered into the database to a file called FAILED OUT. After DBAUTO is run DBUTIL is used to calculate a current consensus called CONSEN.TMP and then DBCOMP is used to compare all the failed sequences against the current consensus in CONSEN.TMP.

193

(vi)   DBCOMP is run finally to recompare all those gel readings that DBAUTO rejected. This is done because the positions of matches displayed by DBAUTO
could have been changed by the addition of further sequences.

An example of such a command file for a VAX is shown below in *Figure 6*. It shows
the screening of a batch of gel readings whose names are contained in a file called
LAMBDA.NAM against first M13, then all those that do not match are screened against
pBR322. In both cases the programs are looking for a match of at least 15 consecutive
nucleotides. Then all those that do not match are screened against the *Bam*H1 recognition sequence that is stored in a file called BAMH1.SEQ. All those that pass all these
screening stages are passed on to DBAUTO but first DBUTIL is used to make a copy
of the database to copy A. The database is called LAMBDA. DBAUTO writes the names
of all the gels that fail to be entered into the database to a file called FAILED.OUT.
After DBAUTO is run DBUTIL is used to calculate a current consensus called CON-
SEN.TMP and then DBCOMP is used to compare all the failed sequences against the
current consensus in CONSEN.TMP.

Similar procedures could be written for other operating systems. Note that the VAX
operating system VMS allows multiple copies of files with the same name (see Chapter
2, Section 4.5). On other machines extra steps to delete files would need to be introduced
into the procedure in order that it could be used repeatedly without changes.

9. INTERACTIVE OPERATIONS ON A PROJECT DATABASE

The majority of users computing time will be spent working interactively on their project databases. The database utility program DBUTIL is used for this purpose, and
because it is interactive, more detailed notes are required than for any of the other programs described. The main program menu has 18 options, some of which give access
to further menus. The main menu offers the following options:

        1 = HELP
        2 = STOP
        3 = OPEN A PROJECT DATABASE
        4 = EDIT CONTIG
        5 = DISPLAY A CONTIG
        6 = LIST A TEXT FILE
        7 = DIRECT OUTPUT TO DISK
        8 = CALCULATE A CONSENSUS
        9 = SHOW RELATIONSHIPS
       10 = ENTER NEW GEL READING
       11 = COMPLEMENT A CONTIG
       12 = JOIN CONTIGS
       13 = COPY THE DATABASE
       14 = CHECK THE DATABASE FOR LOGICAL CONSISTENCY
       15 = EXAMINE CONTIGS FOR QUALITY OF DATA
       16 = ALTER RELATIONSHIPS
       17 = SEARCH FOR A GEL READING BY NAME
       18 = SET DISPLAY PARAMETERS

The enter new gel reading menu offers:

    1 = HELP
    2 = GIVE UP
    3 = COMPLETE ENTRY
    4 = EDIT CONTIG
    5 = DISPLAY OVERLAP
    6 = EDIT NEW GEL READING

The join contig menu offers:

    1 = HELP
    2 = GIVE UP
    3 = COMPLETE JOIN
    4 = EDIT LEFT CONTIG
    5 = DISPLAY JOIN
    6 = EDIT RIGHT CONTIG
    7 = MOVE JOIN

The alter relationships menu offers:

    1 = HELP
    2 = QUIT
    3 = LINE CHANGE
    4 = EDIT SINGLE GEL READING
    5 = DELETE CONTIG
    6 = SHIFT
    7 = MOVE GEL READING
    8 = RENAME GEL READING

Help is available in the program from any point at which the user is asked a question. Similarly, escape to the previous menu is also possible from any such point. To start the program type DBUTIL and the main menu will appear.

9.1 Opening a project database

The first thing that needs to be done when using DBUTIL is to open a database. If any of the other options are selected before a database has been opened the program will give the message "YOU HAVE NOT OPENED A DATABASE!" and then rewrite the menu. Option number 3 is used to open databases. It allows a new database to be started (generally only done at the beginning of a project) or will open an existing one. Once the option is selected the following dialogue will occur.

(i)   " TO START A NEW DATABASE TYPE 1, TO OPEN AN OLD ONE TYPE 0"

    Type carriage return and the program will ask for the project name.

(ii)   " ? PROJECT NAME="

    The project name must be six characters in length and the database must have been created by 'starting a new database'.

    When the project name has been typed the program will ask for the copy number of the database to open.

(iii)   " ? COPY NUMBER(DEF=0)="
         This is a single letter or number. Normally 0 is the current copy and 1 a backup.
         Assuming the database exists (if not the message "ERROR OPENING PRO-
         JECT DATABASE" will be given) the main menu will reappear.

9.2 Displaying the aligned sequences in a contig

To see the aligned gel readings in a contig displayed as in *Figure 3* the display option
(option number 5 in all menus) should be selected.

It will display on the screen all of the gels covering any region of the sequence, lined
up in register and showing their gel numbers, their strand direction, their gel names,
and underneath the consensus. Note that this display can be directed to disk by prior
selection of the option 'disk output' and that this can be processed by program HIGH
to highlight problem areas in the sequence (see Section 10), or edited using a system
text editor (see Section 11).

The display contig option has two parameters that can be set by the user. The sequences can be displayed on the screen in lines of width 50 characters or 100 characters.
(For printing, the 100 character line width is most convenient.)

The second parameter concerns the score used to determine the consensus for the
display routine. (Note this is entirely separate from the option that calculates a consensus and writes it to disk). The threshold values used can either be 75% or 100%. If
100% is used, only those places in the contig where there is no disagreement at all
will give a definite assignment in the consensus. If 75% is chosen (the default), a majority decision, using the algorithm described for the consensus calculation, is employed.
Both of these parameters can be changed by using option 18, 'set display parameters'
in the main menu.

9.2.1 *Using the display contig option*

(i)   Select option 5.
(ii)   Define the contig you wish to display by the number of its left gel —
         " ? NUMBER OF LEFT GEL THIS CONTIG="
(iii)   Define the region of the contig you wish to have displayed. Carriage return only
         gives the defaults X and Y)
         " ? RELATIVE POSITION OF LEFT END(DEF=X)="
         " ? RELATIVE POSITION OF RIGHT END(DEF=Y)="
         Listing will then start and when it is finished the main menu will reappear.

9.3 Examining the relational information

Option 9 will show the relational information for contigs in the database. An example
of a display is given in *Figure 7* which shows a contig of length 689 nucleotides. The
left gel reading is number 6 and has archive name HINW.010, the rightmost gel reading
is number 2 and has archive name HINW.004. On each gel descriptor line is shown:
the name of the archive version, the gel number, the position of the left end of the
gel relative to the left end of the contig, the length of the gel (if this is negative it means
that the gel is in the opposite orientation to its archive), the number of the gel to the
left, and the number of the gel to the right.

196

```
CONTIG LINES
CONTIG          LINE   LENGTH                     ENDS
                                               LEFT    RIGHT
                 48     689                      6        2
GEL LINES
NAME          NUMBER POSITION LENGTH          NEIGHBOURS
                                               LEFT    RIGHT
HINW.010         6       1     -279             0        3
HINW.007         3      91     -265             6        5
HINW.009         5     137     -299             3       17
HINW.999        17     140      273             5       12
HINW.017        12     193      265            17       18
HINW.031        18     385     -245            12        2
HINW.004         2     401     -289            18        0
```

Figure 7. The relational information for a contig. Here we see a contig of length 689 nucleotides. The left gel reading is number 6 and has archive name HINW.010, the rightmost gel reading is number 2 and has archive name HINW.004. On each gel descriptor line is shown: the name of the archive version, the gel number, position of the left end of the gel relative to the left end of the contig, the length of the gel (if this is negative it means that the gel is in the opposite orientation to its archive), the number of the gel to the left, and the number of the gel to the right.

An alternative display in which all the contig descriptor lines and then all the gel lines are listed, is also available. When users select option 9 they are first asked if they want to display only selected contigs. If they do they are asked to name the contig. If they choose not to select contigs, all the relational information for the whole database will be listed. In this case, a sub-option of having the gel descriptor lines sorted into their left-right order for each separate contig, is also offered.

9.3.1 *Running the show relationships option*

(i)   Select option 9 and the first question is asked.
       " TO SELECT CONTIGS TYPE 1"

(ii)   Type 1 to obtain information for selected contigs, type carriage return only to see data for the whole database.

(iii) If you choose to see the whole database you will be asked if you want the data to be sorted into contigs.
       " TO SEE THE DATA SORTED INTO LEFT-RIGHT ORDER TYPE 1"

(iv)   If you choose to see only selected contigs you will be asked to identify the contig by its left gel number.
       "? NUMBER OF LEFT GEL THIS CONTIG="
       Type the number of any gel reading in the contig.

(v)   Define the section to be listed.
       " ? RELATIVE POSITION OF LEFT END(DEF=X)="
       " ? RELATIVE POSITION OF RIGHT END(DEF=Y)="
       Carrige return only, gives the defaults X and Y. After the data has been displayed you will be asked if you want to select another contig.
       " TO SELECT ANOTHER CONTIG TYPE 1"
       If you type 1 you will be asked to define the contig, etc. as above. Otherwise if you type only carriage return the main menu will reappear.

9.4 **Editing gel readings in the database**

Editing the sequences is obviously an essential part of managing a sequencing project. A basic part of the strategy used by the programs described in this article is that new gel readings should be correctly aligned throughout their whole length when they are entered into the database, and that when contigs are joined they are edited so that they are well aligned at the region of overlap. Alignment can be achieved by adding padding characters to the sequences, which is the way DBAUTO operates when entering new sequences into the database. Editing is required when new sequences are added, when contigs are joined, and when sequences are corrected.

The editor in DBUTIL operates in three modes:

(i)   Individual gel readings can be edited as they are being entered into the database.
(ii)   Gel readings in contigs can be edited with alignments maintained (the program always makes the same number of insertions or deletions in all sequences covering the edit position).
(iii) Gel readings in contigs can be edited without the maintenance of alignments. The only time that alignments are not maintained is when the editor is entered from the 'alter relationships' menu and the extra notes describing this menu should be read very carefully before using any of its options.

All three modes of editing look the same to the user and offer the following options.

" HELP=1,QUIT=2,INSERT=3,DELETE=4,CHANGE=5"

The help option simply lists the information given in this chapter; QUIT sends the user back to the menu from which the editor was entered; INSERT and DELETE add or take characters from the gel readings, and CHANGE replaces characters in sequences.

The editors are mostly used when new gel readings are entered into the database or when contigs are joined. In both these cases the display of aligned sequences shown to the user will consist of two sets of sequences: when a new gel is entered it will be shown aligned with the contig to which it is being added, and when two contigs are joined both will be shown aligned one above the other (see *Figures 8* and *9*). A mistake that is sometimes made when using the editor is to define the editing position using the positions from the wrong contig. For example users might define the position using the numbering for the contig when the edits were to the new gel reading. Referring to *Figure 8*, if the user wanted to edit the first mismatch by changing the T in the contig to an A, the edit position is 65, not 15. When the editor is entered from the 'enter' or 'join' options the program projects the user from this to a certain extent by only allowing edits in the region of overlap. If a point outside the permitted region is chosen the program will ask again for the position.

9.4.1 *Using the editor*

Select the editing option from the current menu. If the selection was made from the main menu or from 'alter relationships' you will be asked to identify the contig or gel reading to edit. If the editor is selected from anywhere else the program will known which sequences to edit.

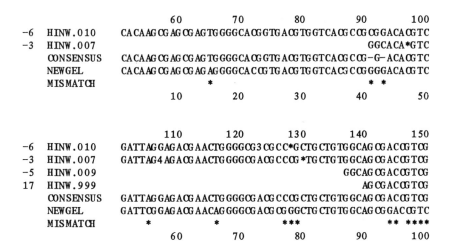

```
                       60         70         80         90        100
 -6   HINW.010   CA CA AG CG AG CG AG TG GGG CA CG GTG A CG TG GTCA CG C CG CG GACA CG TC
 -3   HINW.007                                               GGCACA*GTC
      CONSENSUS   CA CA AG CG AG CG AG CG AG TG GGG CA CG GTG A CG TG GTCA CG C CG –G– ACA CG TC
      NEWGEL      CA CA AG CG AG CG AG AG GGG CA CG TG A CG TG GTCA CG C CG GGGACA CG TC
      MISMATCH                        *                      *  *
                       10         20         30         40         50


                      110        120        130        140        150
 -6   HINW.010   GATTAG GAG A CG A ACTG GGG CG 3 CG CC*G CTG CTG TG GCAG CG ACG TCG
 -3   HINW.007   GATTAG 4 AG A CG A ACTG GGG CG A CG C CCG *TG CTG TG GCAG CG ACG TCG
 -5   HINW.009                                          GGCAG CG ACG TCG
 17   HINW.999                                          AG CG ACG TCG
      CONSENSUS   GATTAG GAG A CG A ACTG GGG CG A CG C CCG CTG CTG TG GCAG CG ACG TCG
      NEWGEL      GATTCG GAG A CG A ACAG GGG CG A CG CG GGCTG CTG TG GCAG CG GAC CG TC
      MISMATCH        *                  *          ***              **  ****
                       60         70         80         90        100
```

Figure 8. The display of the alignment of a new gel reading with a contig. The gel readings in the contig and their consensus are displayed with the new gel reading underneath. The mismatches are shown by *'s on the next line down.

```
                      1460       1470       1480       1490      1500
 56   HINW.100   TCT *G AG CAG TG TG GGCG CTG *C CGG
 33   HINW.300   TCT *G AG CAG TG TG GGCG CTG C* CG GGCT CG G AG GG
 -25  HINW.090   TCT *G AG CAG TG TG GGCG *T *G *CG GGCT CG G AG GG
 19   HINW.123   TCT CG AG CAG TG TG GGCG CTG * *CG GGCT CG G AG GGCATG AAG TAG AG CG
      CONSENSUS   TCT CG AG CAG TG TG GGCG CTG –CG GGCT CG G AG GGCATG AAG TAG AG CG
 -6   HINW.010   TCT CG AG CAG TG TG GGCG CTG C CCG GGCT CG G AG GGCATG AAG TTAG AG C
 -3   HINW.007                TG GGCG CTG C CCG GGCT CG G AG GGCATG AAG T*AG AG C
 -5   HINW.009                         GCT CG G AG GGCATG AAG T*AG AG C
      CONSENSUS   TCT CG AG CAG TG TG GGCG CTG C CCG GGCT CG G AG GGCATG AAG TTAG AG C
      MISMATCH                                 *                   ******
                       10         20         30         40         50
```

Figure 9. The display of a join using DBUTIL. This shows the right end of the left contig and the left end of the right contig and their mismatches (*'s). The left contig ends in gel readings 56, 33, 25 and 19, and the right contig starts with gel readings 6, 3 and 5. The mismatches marked are differences in the consensus.

Using the insert function

(i)   Selection option 3 and you will be asked for the position to edit.

(ii) " ? POSITION = "

Type the position. If the program asks the question again you have given a position outside of the permitted range (outside the contig, the gel or the region of overlap, depending on where the editor was entered from).

(iii) Type the number of characters to insert:

" ? NUMBER OF CHARACTERS = "

The maximum number allowed at any insertion point is 80.

(iv) Type the characters to insert at the edit position:

" ? CHARACTERS TO INSERT INTO GEL X = "

(Note that if you are editing a contig the program will ask for the characters to insert into each separate gel reading, hence allowing different changes to be made to each.)

If the user types only carriage return the program will automatically insert space characters. When all the gel readings have been done the menu reappears.

Using the delete function

(i)   Select option 4 and you will be asked for the position to edit.

(ii)   " ? POSITION="
        Type the position. If the program asks the question again you have given a region outside of the permitted range (outside the contig, the gel or the region of overlap, depending on where the editor was entered from).

(iii)   Type the number of characters to delete.
        " ? NUMBER OF CHARACTERS="
        Do not delete so many characters from a contig that some gel readings have zero length. To remove gel readings from contigs see Section 9.14.2. The program will then delete the given number of characters from the edit position and the main menu will reappear.

Using the change function

(i)   Select option 5. If you are not editing a new gel reading whilst entering it into the database you will be asked which gel reading you wish to edit.

(ii)   " ? GEL NUMBER="
        Type the number of the gel reading.

(iii)   Then you will be asked for the position to change.
        " ? POSITION="
        Type the position and you will be asked to type the number of characters to change.
        " ? NUMBER OF CHARACTERS="

(iv)   Type the new chracters:
        " ? NEW CHARACTERS="
        The program will then replace the characters at the edit position and the menu reappears.

9.4.2 *Further notes on editing*

Edits to the database are immediately carried out and the 'give up' options of 'enter' and 'join' do not undo them. Users must undo these themselves. *Do not kill the program during edit contig.*

When you are editing a single gel reading in a contig from 'alter relationships' (which you should not normally need to do) the program will correct the length of the individual gel reading, but it will not update the length of the contig if it has changed.

An alternative to this editor for making multiple edits to contigs is facilitated by PREDIT, POSTED and DBMERGE which make it possible to use system screen editors on the sequences. Their use is described in Section 11.

9.5 Complementing contigs

Before contigs are joined they must be in the same orientation. The 'complement a contig' function, option 11 in the main menu of DBUTIL will complement and reverse all of the gels in a contig. It automatically reverses and complements each gel sequence, reorders left and right neighbours, recalculates relative positions and changes each strandedness.

The only user input required is to define the contig to complement by the number of its leftmost gel. It will take a few moments for the program to complete this action. The main menu will then reappear. *Do not kill the program during this step.*

9.6 Entering new gel readings into the database

The enter option is used to enter new gel readings into the project database. (Note that normally DBAUTO performs this function and interactive entry is only required for those gel readings that cannot be sufficiently well aligned by the automatic alignment routines.) The new gel reading must have already been compared with the contents of the database by use of DBAUTO or DBCOMP in order to ascertain if it overlaps previously entered data. The user is expected to know if the gel reading overlaps, and if so, which contig it overlaps, and where it overlaps. The program takes the user through a series of questions to establish the nature of the overlap and then displays the overlapping sequences. The user is then offered a number of options, including editors for the new gel reading and the contig, to enable the correct alignment of the gel reading throughout its whole length.

9.6.1 *Running the enter option*

Select option 10 and the program will confirm that the enter option has been chosen and explains what the user needs to know.

"ENTER A NEW GEL READING INTO THE DATABASE
NOTE THAT BEFORE A GEL READING IS ENTERED IT MUST BE
COMPARED WITH THE DATABASE TO SEARCH FOR OVERLAPS USING
A PROGRAM SUCH AS DBAUTO OR DBCOMP. THE REPLIES TO ALL
THE QUESTIONS THAT FOLLOW ARE REPORTED BY THESE PROGRAMS"

(i)   The first question the program needs answering is the name of the file containing the new gel reading.
"FIRST WE NEED THE NAME OF THE FILE CONTAINING THE NEW
GEL READING
? FILE NAME OF GEL READING TO ENTER="
Type the name of the gel reading file.
If there is an error in the file name the program will report this and give the user further tries to give the correct file name.
If the gel has already been entered before the program will type "GEL ALREADY IN LINE X, ENTRY STOPPED" and return to the main menu of DBUTIL. The program gives the gel reading a unqiue number.
"THIS GEL READING HAS BEEN GIVEN THE NUMBER XXX"

(ii)   Next the program needs to know if the new gel reading overlaps any sequences that are already in the database.
"WE NEED TO KNOW IF THE GEL READING OVERLAPS DATA
ALREADY IN THE DATABASE

  IF THE GEL READING OVERLAPS DATA ALREADY IN THE DATABASE
  TYPE 1"
  Type 1 for yes, and 0 for no. If "no", entry is complete and the user is returned
  to the main options of DBUTIL. If "yes", the dialogue continues.

(iii) "WE NEED TO KNOW IF THE GEL READING OVERLAPS IN THE
  NORMAL OR COMPLEMENTARY SENSE
  IF IT IS THE COMPLEMENT OF THE NEW GEL READING THAT
  MATCHES TYPE 1"
  Type 1 if the comparison programs found a match with the complementary sense
  of the gel reading, 0 for the sense as read from the gel. If the user types 1 the
  program will reverse and complement the new gel reading.

(iv) "WE NEED TO KNOW WHICH CONTIG THE NEW GEL READING
  OVERLAPS
  ? NUMBER OF LEFT GEL THIS CONTIG="
  The user should type the number of the contig as reported by the comparison
  programs.

(v) "WE NEED INFORMATION ABOUT THE POSITION OF THE OVER-
  LAP. FIRST, THERE ARE TWO TYPES: THOSE THAT EXTEND
  THE CONTIG LEFTWARDS AND THOSE THAT START INTERNALLY.
  IF THIS NEW GEL EXTENDS THE CONTIG LEFTWARDS TYPE 1"
  Type 1 for a match where the left end of the gel reading protrudes from the
  left end of the contig, or 0 otherwise.

(vi) "SECONDLY, WE NEED TO KNOW THE EXACT POSITION OF THE
  OVERLAP."
  For gel readings that do not protrude from the left end of the contig the user
  will receive the following question.
  " ? POSITION IN CONTIG OF LEFT END OF GEL READING="
  To which the reply is the position in the contig (reported by DBAUTO or
  DBCOMP) at which the left base of the new gel reading overlaps.
  For gel readings that protrude from the left end of the contig the question is:
  " ? POSITION IN NEW GEL READING OF LEFT END OF CONTIG"
  Again this is as reported by the comparison programs.
  Once this is completed the program will display the first 100 bases of the overlap.
  The gel readings in the contig and their consensus are displayed with the new
  gel reading underneath. The mismatches are shown by *'s on the next line down
  (*Figure 8*).

(vii) The program then needs to know if the position of the left end of the overlap
  is correct.
  "IF JOINT CORRECT TYPE 1"
  If it is, the user should type 1, if not, 0, and the program will ask for the new
  position and display it.

 The program now offers a number of options to allow the user to align the new gel
correctly over its whole length with the data already in the contig. It is important that
sufficient edits are made to the new gel reading and the gel readings in the contig at
this stage to get the alignment correct. Once entry is completed the alignment is fixed

and cannot easily be changed (see 'alter relationships'). Alignment can be achieved by making insertions or deletions but deletion of data requires the original gels to be checked. For this reason at entry we usually make only insertions to achieve alignment. We use X or spaces as padding characters to achieve alignment and so can distinguish padding characters from characters assigned from reading the gels.

The menu that appears contains the following options.
"HELP=1,GIVE UP=2,COMPLETE ENTRY=3,EDIT CONTIG=4,DISPLAY=5,
EDIT NEW GEL READING=6"

The HELP function gives the information in this article; GIVE UP allows users to change their minds about entering the new gel reading and return to the main menu (the program will ask the user to confirm this choice); COMPLETE ENTRY is the command to add the new gel reading to the contig. The program updates the relationships accordingly (the user is asked to confirm this command); EDIT CONTIG gives the user access to a simple editor that allows insertions, deletions and changes to be made to the contig (the editor maintains alignments by making the same number of insertions or deletions in all sequences covering the edit position. It also protects the user by allowing edits only within the region of overlap. See Section 9.4); DISPLAY allows display of the region of overlap only. This is defined by the relative positions in the contig. The default is the whole of the region of overlap; EDIT NEW GEL is identical in use to EDIT CONTIG but the edits are performed on the new gel reading.

9.7 **Joining contigs**

Sometimes DBAUTO will find that a new gel reading overlaps two contigs but will be unable to obtain a sufficiently good alignment with both. In this case it will enter the new gel reading into one of the contigs but will not join them. This function allows contigs to be joined interactively. It allows the user to align the ends of the two contigs by editing each contig separately. It is important that the alignment achieved is correct because once the join is completed the alignment is fixed. The program needs to know which two contigs to join and where they overlap.

(i)   First which two contigs are to be joined.
        "?WHICH CONTIGS DO YOU WANT TO JOIN"
        Type the numbers of the left gel readings of the two contigs.
        "LEFT CONTIG
        ? NUMBER OF LEFT GEL THIS CONTIG=
        RIGHT CONTIG
        ? NUMBER OF LEFT GEL THIS CONTIG="
        The program checks that the two contig numbers are different (it will not allow circles to be formed!).

(ii)   Now we need to identify the exact position of overlap. This is defined as the position in the left contig that the leftmost character of the right contig overlaps.
        "WE NEED TO KNOW EXACTLY WHERE THE OVERLAP STARTS
        THE JOINT IS THE POSITION IN THE LEFT CONTIG THAT THE
        FIRST CHARACTER OF THE RIGHT CONTIG OVERLAPS
        ? POSITION IN LEFT CONTIG OF LEFT END OF RIGHT CONTIG="
        Type the position of the overlap. Normally this will have been established by

DBAUTO or DBCOMP comparing a single gel reading and finding an overlap with two contigs. The gel reading will have been entered into one contig and so the user must taken into account the corresponding change in the position of the overlap.

The overlap must be of at least one character. If this criteria is not fulfilled, the program prints "ILLEGAL JOIN. TO RETURN TYPE −9" and then asks the user to define the position of the overlap. If the user then types −9 the program returns to the main options in DBUTIL without making the join.

The program then displays the join showing all the gel readings overlapping the join from the left contig, their consensus, all the gel readings from the right contig that overlap the join, their consensus and then asterisks to denote mismatches between the two consensuses (*Figure 9*).

It is essential that the user aligns the two contigs throughout the whole region of overlap before completing the join because it is only at this stage that the two contigs can be edited independently. Once the join is completed the alignment can only be altered using the routines supplied by 'alter relationships'. The program offers the user options to facilitate the alignment of the two contigs. These options are:

"HELP=1,GIVE UP=2,COMPLETE JOIN=3,EDIT LEFT CONTIG=4,
DISPLAY JOIN=5,EDIT RIGHT CONTIG=6,MOVE JOIN=7"

The HELP function gives the information in this article; GIVE UP allows users to change their minds about making the join and return to the main menu (the program will ask the user to confirm this choice); COMPLETE JOIN is the command to alter the relationships so that the two contigs become one; EDIT LEFT CONTIG and EDIT RIGHT CONTIG give the user access to a simple editor that allows insertions, deletions and changes to be made to either contig (the editor maintains alignments by making the same number of insertions or deletions in all sequences covering the edit position. It also protects the user by allowing edits only within the region of overlap, see Section 9.4); DISPLAY allows display of the region of overlap only. This is defined by the relative positions in the left contig. The default is the whole of the region of overlap; MOVE JOIN allows the position of the joint to be changed.

9.8 Calculating a consensus

The object of the project is to produce a consensus sequence from all the aligned gel readings. To analyse the consensus with further programs it must exist in a file with a suitable format. The consensus used by DBAUTO is not stored as a disk file and exists only while the program is running. The consensus shown at the base of the aligned sequences using the display function is also calculated only while the program is running (for each screen width). The option called 'calculate a consensus' is used to calculate a consensus and store it in a disk file. The format of the file produced is described in Section 4.1, and it is suitable for programs such as DBCOMP or sequence analysis programs (6).

Users can calculate a consensus for selected contigs or for the whole database. If contigs are selected it is also possible to calculate consensuses for restricted sections of them.

9.8.1 *The consensus algorithm*

Both the 'calculate a consensus' function and the display routine use the rules outlined here to calculate a consensus from aligned gel readings.

Individual characters are assigned the following scores:

definite assignments, i.e. A,C,G,T,B,D,H,V,K,L,M,N = 1
probable assignments, i.e. 1,2,3,4 = 0.75
any other character = 0

For each position in a contig the routine calculates 4 base totals: one each for A,C,G,T. These totals are calculated by adding up all of the individual scores for each of the characters that contributes to each base total at each position. For example only A, B, M and 3 contribute to the total for base A (see Section 1.4). When all of the gel readings covering a position have been added to the four base totals the routine calculates the sum of these four base totals for each position. The routine then looks at each position to see if any of the four base totals is ≥ 75% of the sum for that position. If a base total is sufficiently high its corresponding character is put into the consensus, otherwise a dash (-) is assigned.

Note that these rules differ from those used by the 'EXAMINE QUALITY OF A CONTIG' function: only defined uncertainty codes are counted by the consensus calculation whereas the quality examiner counts all characters. By ignoring undefined characters the consensus calculation is more likely to produce a definite assignment in the consensus, but by counting all characters, the quality routine can indicate places where padding has been used, and hence show possible problems in the sequence.

9.8.2 *Running the consensus calculation option*

(i) Select option 8 and you will be asked for a file name for the consensus sequence.
" ? NAME FOR CONSENSUS FILE = "
Type a legal file name and you will be asked if you wish to select contigs.

(ii) " TO SELECT CONTIGS TYPE 1"
If you wish to select contigs type 1, otherwise type carriage return only.

(iii) If selecting contigs you will be asked to type the number of the left gel reading of the contig.
" ? NUMBER OF LEFT GEL THIS CONTIG = "

(iv) If selecting contigs define region of the contig:
" ? RELATIVE POSITION OF LEFT END(DEF = X) = "
" ? RELATIVE POSITION OF RIGHT END(DEF = Y) = "
Type the positions [carriage return only gives the whole contig (X to Y)].
The program will give the message "CALCULATING A CONSENSUS" until it has finished, and then will give the user the opportunity to cycle round steps (iii) and (iv) by typing 1 in response to the next question.
"TO SELECT ANOTHER CONTIG TYPE 1"
If the user types only carriage return the program will ask if the user wishes to write a new consensus file. Also if contigs were not selected, the program will ask the same question when it has finished calculating the consensus for the whole database.

"TO CALCULATE ANOTHER CONSENSUS TYPE 1"

(v) If the user elects to calculate a new consensus the routine loops back to step (i), otherwise the main menu will reappear.

9.9 Copying the database

As is described in Section 2.5.1 every database has a copy number. The copy that users normally operate on is copy 0, and it is advisable to regularly make a backup to say, copy 1. This is particuarly important if any of the functions in the 'alter relationships' menu are about to be used, or if a machine that is likely to crash unexpectedly is being used. If the relational information gets corrupted by either of these causes it can be very hard, or even impossible, to get the database into a useable state. It does not take long to make a copy and we routinely do it as part of the command procedure that runs DBAUTO. A further use of the copy routine is in expanding a full database. Users can choose a new database size each time a copy is made. The program prompts for a copy number of the new copy and then for the size of the new database. If the new size is too small or >1000 the current size will be used. (This means carriage return will leave the size unchanged.) It will take a few seconds to run depending on the computer and the amount of data.

9.9.1 *Using the database copying function*

(i) Select option 13 from the main menu of DBUTIL and the program will confirm that the copy option has been selected and then ask the first question.
" MAKE A COPY OF THE DATABASE"
" ? NEW COPY NUMBER (DEFAULT=1)="
(ii) Type a legal file name symbol for the copy number. Typing only carriage return will create copy 1. The next question will then be asked.
" ? SIZE FOR NEW DATABASE (DEFAULT=CURRENT SIZE)="
(iii) To change the size type the new value, otherwise type only carrige return. The copy will be made and the main menu reappears.

9.10 Finding gel readings by name

To identify gel readings it is usually easiest to use the number given them by the assembly programs, but sometimes it is necessary to find the number when only the archive name is known. This function allows a search to be made for a gel by its archive name. If the gel is found its gel descriptor line is typed, if the gel is not found the program prints "NOT IN DB"

9.10.1 *Running the gel reading search option*

(i) Select option 17 and you will be asked for the archive name.
" ? ARCHIVE NAME="
(ii) Type the name of the archive, and if the gel reading is in the database its gel descriptor line will be typed. The program will then ask if you want to search for another gel reading.
"TO SEARCH FOR A GEL TYPE 1"

(iii)   To search for another type 1, otherwise type only carriage return and the main menu will reappear.

9.11 **Printing aligned sequences or relational information**

It is sometimes useful, especially if working at a visual display terminal to be able to make printed copies of displays produced by the progams. It is also necessary to be able to save the display of aligned gel readings of a contig in a disk file for processing by screen editors (see Section 11) or by the program HIGH (see Section 10). Option 7, 'direct output to disk', is used to direct output from 'display' or 'show relationships' to a disk file. After it has been selected the very next time either of these two options is used output will go to the named disk file and will not appear on the screen. Subsequent use of these two options will produce output on the screen until 'direct output to disk' is selected again. The user is asked to type a file name.

9.11.1 *Running the direct output to disk option*

(i)   Select option 7
        "DIRECT OUTPUT TO DISK"
        " ? NAME FOR DISK FILE="
(ii)   Type a legal file name and the main menu will reappear. Select the 'display contig' or 'show relationships' option.

9.12 **Examining the quality of a contig**

The quality of a consensus depends on the number of times it has been sequenced and the particular uncertainty codes used in each gel reading. Although the display of aligned gel readings shown in *Figure 3* contains all the information needed to assess the accuracy of a consensus sequence, it still needs to be examined very carefully to find any doubtful regions. The program HIGH can be used to simplify this process but as a final check, option 15, 'examine contigs for quality' should be used. This function can tell us if there are any doubtful nucleotides or if any have only been adequately determined on one strand of the DNA. It reads through a contig and divides every position in the consensus sequence into one of five numbered categories:

1.   Well determined on both strands and they agree. code=0
2.   Well determined on the plus strand only. code=1
3.   Well determined on the minus strand only. code=2
4.   Not well determined on either strand. code=3
5.   Well determined on both strands but they disagree. code =4.

It gives a summary of the quality of the data in a contig, in terms of the percentage of positions that have been placed in each category (see *Figure 10*), and can produce a listing of the codes for the whole contig consensus (see *Figure 11*).

CODE	0	1	2	3	4
PERCENTAGE	80.0	8.0	9.0	2.0	1.0

Figure 10. The summary produced by the quality checking routine.

	10	20	30	40	50	60
	0000001000	0002000000	3000000000	1111000000	0000000000	2222222220

	70	80	90	100	110	120
	1000000000	0000033000	4100000000	0000111000	0000120000	0000000000

Figure 11. The listed quality codes for a contig.

9.12.1 *The contig quality algorithm*

The five categories are defined using an algorithm that is similar to that used to calculate the consensus but which has two important differences: each strand is treated independently, and every symbol in the gel readings contributes to the calculation. Each of the possible characters in a sequence is given an individual score to be added to the appropriate base total (e.g. A, B and M contribute 1 to the total for A, but 0 to C, G and T totals).

A,C,G,T,D,B,H,V,K,L,M,N=1
1,2,3,4=0.75

anything else = 0, but all characters contribute a value of 1 to the number of times the sequence has been determined on each strand. For each position of each strand the program calculates five numbers: using the individual values for each uncertainty codes it calculates the sum for each of the four nucleotides A, C, G, T and the number of times the sequence has been determined for each position. The score for each base, in each position, on each strand, is divided by the number of times the sequence has been determined on that strand, and mutiplied by 100. The user of the program is asked to supply a cutoff score as a percentage before the scan is done. If the score for any of the four bases at a position equals or exceeds this percentage then that position is called 'well determined' for that strand. For example, a single code of 2 on one strand will be called 'well determined' if the percentage is ≤75, but not if the percentage is >75. Or two A's aligned with one G will be called 'well determined' if the percentage is <66.6.

9.12.2 *Running the contig quality examiner*

(i)   Select option 15 and the program will confirm this by typing
    "EXAMINE A CONTIG FOR THE QUALITY OF DATA"

(ii)   It will then request the user to type a percentage cutoff score.
    " ? PERCENTAGE="
    Type a value such as 80.

(iii)   The program will then ask you to identify the contig to examine
    " ? NUMBER OF LEFT GEL THIS CONTIG="
    Type the gel number, and the program will ask which section you want examined.

(iv)   " ? RELATIVE POSITION OF LEFT END(DEF=1)="
    " ? RELATIVE POSITION OF RIGHT END(DEF=Y)="
    Y is the total length of the contig so if you want to examine the whole contig

type only carriage return for both positions, otherwise define a restricted region. When the calculation is complete the program will list a summary of its findings showing the proportions of the consensus that fall into each of the five categories (see *Figure 10*).

(v)   The program now asks if you want to see the sequence of codes for the contig listed on the screen.

        " TO GET CODES LISTED TYPE 1"

        The listing is simply a representation of the contig consensus in which the consensus is replaced by the quality codes (see *Figure 11*).

        If you want to see such a listing type 1, otherwise carriage return only and the next question will be asked.

        " TO WRITE THE QUALITY CODES TO A DISK FILE TYPE 1"

(vi)   If you want to send the listing to a disk file where it can be processed by other programs type 1 and you will be asked for a file name; otherwise type only carriage return and you will be asked if you want to examine another contig.

(vii)   " TO EXAMINE ANOTHER CONTIG TYPE 1"

        To examine another contig type 1, otherwise type only carriage return and the main menu will reappear.

9.13 Checking the logical consistency of a database

It is possible for the relational information in a project database to get corrupted and so it is advisable to check to see if it is consistent by using option 14. The sources of inconsistencies include incorrect use of the options in the 'alter relationships menu' (see Section 9.14), and killing the program when it is changing the relational information. DBAUTO always performs a check on the logical consistency of a database when it is opened, and will only tolerate a "GEL N IS NOT USED" error (see below); if any other error is found the program will stop.

The function peforms a check on the logical consistency of the database. (Note in the messages below A and B refer to gel numbers and N and M are numeric values.)

(i)   If A is the left neighbour of B, is B the right neighbour of A? If not, the error message is "HAND HOLDING PROBLEM FOR GEL A" followed by the gel descriptor lines for gels A and B.

(ii)   Are there any contig lines with no left gel? The error message is "BAD CONTIG LINE NUMBER A"

(iii)   Do the gels that are described as left ends on contig lines agree that they are left ends? The error message is "THIS CONTIG LEFT END HAS A LEFT NEIGHBOUR A"

(iv)   Are there gels that are in more than one contig? The error message is "GEL NUMBER A IS USED N TIMES"

(v)   Are there gels that are not in any contig? The error message is "GEL A IS NOT USED"

(vi)   Do the relative positions of gels agree with their position as defined by left and right neighbourliness? The error message is "GEL NUMBER A REL POSITION N IS LNBR OF GEL NUMBER B REL POSITION M"

(vii)   Are there are loops in contigs? If so no further checking is done. The error message is "LOOP IN CONTIG A NO FURTHER CHECKING DONE".
The program then prints the gel numbers in the looped contig up to the start of the loop.

(viii) Are there any contigs of length <1? The error message is "THE CONTIG IN LINE NUMBER N HAS ZERO LENGTH"

(ix)   Are there any gels used in only one contig that have zero length? The error message is "GEL NUMBER N HAS ZERO LENGTH"

9.13.1 *Using the relationships checker*

Select option 14 and the program will type the message "CHECKING DATABASE FOR LOGICAL CONSISTENCY". If there are any problems the program will list them as just described. If not the program will type "DATABASE IS OK". Either way the main menu will then reappear. If the inconsistencies are minor fix them with functions in the 'alter relationships menu'.

9.14 **Altering the relationships**

This function exists to cope with any problems that occur with the database. Its use requires a good understanding of the data structure. A skilled user can correct any mistakes and misalignments in the data but, be warned, it is also very easy to make a horrible mess if you are inexperienced.

Using the options here you can edit individual gel readings in contigs, move one section of a contig relative to another, break contigs, remove contigs, remove gel readings, etc. To give flexibility most of the commands do only one thing. This means that several commands may have to be executed to complete any change. At the end of this section are notes on breaking contigs and removing gel readings from the database.

The following options are offered:

"HELP=1,QUIT=2,LINE CHANGE=3,EDIT GEL=4,DELETE CONTIG=5,
SHIFT=6,MOVE GEL=7,RENAME GEL=8"

(i)   HELP gives this information.

(ii)   QUIT returns to the main menu of DBUTIL.

(iii)   LINE CHANGE allows the user to change the contents of any line in the file of relationships. The line is selected by number, the program prints the current line and prompts for the new line. The user should type the new values, each followed by a comma or in I6 format (i.e. evenly spaced, each number occupying exactly 6 column positions). It then asks the users to type "1" if they are sure of the changes made. If the user types anything else the line is left unchanged.

(iv)   EDIT allows the user to edit an individual gel independently of any others it may be related to. The commands are as for 'edit contig' (see Section 9.4). The edit positions are relative to the contig. The effect of this editing on the length of this gel is taken care of, but if it changes the length of a contig, or its relationship to others, this must be accounted for by use of the line change function.

(v)   DELETE CONTIG is a function that deletes a contig line by moving down all the contig lines above by one position. It prompts only for the line to delete.

It does not delete any of the gels or gel lines for the deleted contig but it does reduce the number of contigs on line IDBSIZ by 1.

(vi)   SHIFT allows the user to change all the relative positions of a set of neighbouring gels by some fixed value, i.e. it will shift related gels either left or right. It can therefore be used to change the alignment of the gels in a contig or as part of the process of breaking a contig into two parts (see below). It prompts for the number of the first gel to shift and then for the distance to move them (note a negative value will move the gels left and a positive value right). It then chains rightwards (i.e. follows right neighbours) and shifts each gel, in turn, up to the end of the contig. (This means that only those gels from the first to shift to the rightmost are moved). It updates the length of the contig accordingly.

(vii)   MOVE GEL is a function to renumber a gel. It moves all the information about a gel on to another line. The user must specify the number of the gel to move and the number of the line to place it. It takes care of all the relationships. Of course gels must not be moved to lines occupied by other gels! It can be used as part of the process of removing a gel from the database (see below).

(viii)   RENAME GEL is a function that is used to rename the archive names of gels in the database; it only changes the name in the .ARN file of the database (where N is the copy number).

9.14.1 *Breaking contigs*

Occasionally it is necessary to break a contig into two parts and this can be achieved using the options in 'alter relationships'. *First make a copy.* You have to do several things including creating two new contigs. Let us call the new contigs L and R, and the contig you wish to break O; let the rightmost gel of the new contig L be called x, and the leftmost gel of the new contig R be called y (so initially x and y will point at one another). We will need to create a new contig line so we must increase the number of contigs in the database by 1; this number is stored on the last line of your database so change this line accordingly using LINE CHANGE; next change the lines describing x and y using LINE CHANGE: x must be given a zero right neighbour, y a zero left neighbour; next, again using LINE CHANGE, create a new contig line to describe contig R (make this a copy of the line describing contig O but with x as the leftmost gel). We will use the line describing contig O to describe contig L, so using LINE CHANGE, change this line by updating its rightmost gel and its length; now using SHIFT, shift all the gels in R leftwards so that the position of the left end of gel y will become 1 (y is the first gel to shift and the distance will be negative). SHIFT will update the length of the contig accordingly. This should have completed the break.

9.14.2 *Removing gels from contigs*

Gels can be removed from contigs if they are not essential for holding the contig together (i.e. are not the only gel covering a particular region). Suppose the gel to remove is gel b with left neighbour a and right neighbour c. Using LINE CHANGE change the right neighbour of a to c, and the left neighbour of c to a. To tidy things up: suppose there are x gels in the database, using MOVE GEL, move gel x to line b; then using LINE CHANGE decrease the number of gels in the database (stored in the last line) by 1.

9.15 **Safeguarding databases**

It is advisable to copy regularly (using the copy function of DBUTIL) from say copy 0 to copy 1 in case of errors.

The give-up options in DBUTIL allow the user to change his mind about entering a new gel reading or joining two contigs without affecting the file of relationships. But if the edit contig option from either of these two functions has been used the edits will remain even though the user has 'given up'. To leave the files completely unchanged the user could, if required, undo any edits before 'giving up'

There are various checks within the programs to protect users from themselves.

(i)   All user input is checked for errors − e.g. reference to non-existent gel readings or contigs, incorrect positions in the contig or gel readings. If an error is detected the programs usually reprompt for the input or ask the user to try again.

(ii)   Before entering a gel reading the system checks to see if a gel of the same name has already been entered.

(iii)   Join will not allow the circularizing of a contig.

(iv)   Both enter and join functions restrict the region that the user is allowed to edit (using edit contig) to the region of overlap.

(v)   Users may escape from any point in the program by typing a negative number in response to a question.

(vi)   Help is available from all points in the program either explicitly as an option or if the user types −99 in response to a question.

It is essential that users do not kill the program while it is doing anything that involves changing the contents of the database, i.e. during complete entry, complete join, complement contig, edit contig. This could corrupt the database so badly that it is impossible to fix. The program should always be left using the STOP option.

10. HIGHLIGHTING DISAGREEMENTS IN CONTIGS

To simplify the task of finding disagreements in contigs the program called HIGH is used. This is a program for highlighting differences between individual working versions of gel readings and their consensus. The program expects an input file written to disk by the display function of DBUTIL. It writes an output file that looks like the display output but which only marks those bases that are different to the consensus; all identical bases are set to . characters.

For example if the contig shown in *Figure 3* were passed through HIGH the output would look like that shown in *Figure 12*.

10.1 **Preparing for the highlighting program**

This is described in Section 9.2, but to summarize:

(i)   Run DBUTIL by typing DBUTIL

(ii)   Open the project database using option 3.

(iii)   Direct output to disk using option 7.

(iv)   Display the contig using option 5.

(v)   Stop DBUTIL using option 2.

212

```
                     10        20        30        40        50
 -6  HINW.010  ...................................................
               GCGACGGTCTCGGCACAAAGCCGCTGCGGCGCACCTACCCTTCTCTTATA

                     60        70        80        90       100
 -6  HINW.010  .........................................C.G.......
 -3  HINW.007                                          G.C...*....
               CACAAGCGAGCGAGTGGGGCACGGTGACGTGGTCACGCCG-G-ACACGTC

                    110       120       130       140       150
 -6  HINW.010  ......................3....*.....................
 -3  HINW.007  ......4...................*.....................
 -5  HINW.009                              ...........
 17  HINW.999                              ...........
               GATTAGGAGACGAACTGGGGCGACGCCCGCTGCTGTGGCAGCGACCGTCG

                    160       170       180       190       200
 -6  HINW.010  ...*................*....................*
 -3  HINW.007  ...*................C*...................*
 -5  HINW.009  ...*...............*.*G*.................*
 17  HINW.999  ..................**....................G
 12  HINW.017                             .......*
               TCTCGAGCAGTGTGGGCGCTG-CCGGGCTCGGAGGGCATGAAGTAGAGCG
```

Figure 12. An example of the results from program HIGH.

10.2 Using the highlighting program

(i)   Start the program by typing HIGH and the first question will be asked.

(ii) " ? NAME OF DISPLAY FILE CREATED BY DBUTIL = "
      Type the name of the file created when you directed output to disk from DBUTIL
      and the highlight display will be shown.

11. SCREEN EDITING OF CONTIGS

The amount of editing that is required during a sequencing project depends on the quality
of the gel readings produced. As is described above most of the entry and alignment
of gel readings is performed by the program DBAUTO, but the interactive program
DBUTIL contains three simple editors for operating on the sequences in a project
database. Although these editors are adequate for producing alignments during entry
of new gel readings and contig joining, when the project is nearing completion it is
necessary to clean up whole contigs, and a faster editor is required.

Generally the contigs are checked by displaying them and using the program HIGH
(Section 10) and the routine in DBUTIL to examine the quality of the data in a contig.
It is convenient to make a printed copy of the display and to check the films whose
sequences cover problem areas. The required changes can be noted on the printed
display, which can then be used as a guide when making the changes to the contents
of the database. At this stage many edits may be required and it is convenient to have
a faster editor than the ones so far described. Here I describe an alternative method
of editing contigs that makes use of the system editor resident on the computer being
used to process the sequences.

As the data in the database is not stored as it appears on the screen when using the 'display' function it would be difficult to write a screen editor as part of DBUTIL that would work sufficiently fast. Also such an editor would be specific to the particular type of terminal used. The new method is therefore not part of DBUTIL but consists of three extra programs: PREDIT, POSTED and DBMERGE and the use of the system editor (say EDT on a VAX, or WORDSTAR on an IBM PC). The editing is done using the system editor and the programs are used first to prepare the data for this and then to put it back into the database.

The 50 character line 'display' output of DBUTIL is written to a disk file; this is edited with the system editor, and then put back into the database. In order to simplify the problem of putting the data back into the database the program PREDIT reads the 'display' file and adds < characters to the left ends of lines and > characters to the right ends of lines. It also adds the numbers 1000, 1001 and 1002 to lines containing respectively, numbers, consensus sequences and blank lines. These numbers appear as gel numbers on the corresponding lines and can be used later by POSTED to check that lines are in the correct order. Users can then edit the resulting file with the system editor. Their edited file is then interpreted by the post editing program (POSTED) and put into a temporary database for checking. Checking may include running DBUTIL and looking at the results of the 'display contig' function, and perhaps running the 'check logical relationships' function. Once the data has been verified in this way it can be put back into the original database using the program DBMERGE.

The program POSTED can check that lines still contain the < and > symbols in allowed range of positions and that lines are in the correct order, but users must check the actual changes by comparing with the original database. Although it may sound complicated, in practice it is very simple and quick.

11.1 Rules for screen editing

There are some limitations on the changes that can be made to the contigs when using the screen editor. Users are unlikely to want to break the rules in order to achieve changes to contigs, but nevertheless the constraints need to be defined and they are given below.

The display output is formed of rows and columns of symbols and to aid explanation I define two special column positions. L is the column of the leftmost < symbol, and R is the column of the rightmost > immediately after the display output has been passed through PREDIT.

(i)   The display output must be passed through PREDIT before editing with the screen editor.

(ii)   Alignments must be maintained during editing.

(iii)   Whole lines of sequence should not be deleted or added unless the order of the gel readings in the contig is preserved.

(iv)   Editing can move the < symbol no further left than L−1, and right no further than R+10.

(v)   Editing can move the > symbol right no further than R+10, and left no further than L−1.

(vi)   If you extend a gel reading so far leftwards that its left end becomes left of that of its left neighbour you will need to do the following. After making the temporary database containing your edited contig complement it twice using the complement function of DBUTIL before merging it with the original database; this will sort out the order of the neighbours.

(vii)   Only one contig can be worked on at a time.

11.2 Using the screen editing procedure

This is a summary of the steps involved in using the screen editing procedure. The details of using DBUTIL are given in Section 10, and users are expected to be familiar with their system editor. In the steps below suggested file names are given to show how the data passes from one program to the next, but of course any legal file names could be used.

(i)   Run DBUTIL.

(ii)   Select the 'direct output to disk' option (create file DISPLAY.OUT).

(iii)   Select the 'display contig' function and display the contig you want to edit.

(iv)   Run PREDIT and it will ask for the name of the display file from DBUTIL.
         " ? NAME OF DISPLAY FILE FROM DBUTIL = "
         Type the file name (DISPLAY.OUT) and the program will ask for the name of an output file.
         " ? NAME FOR TRANSFORMED DISPLAY FILE = "

(v)   Type a name for the transformed file (say PREDIT.OUT). This is the file containing the special symbols < and > etc.

(vi)   Edit the transformed file (PREDIT.OUT) using the system editor, but following the rules given in Section 11.1.

(vii)   Run POSTED to create a temporary database containing the edited contig. It will ask first for the name of the edited file.
         "? NAME OF EDITED FILE = "

(viii)   Type the name of the edited file (PREDIT.OUT) and the program will ask for the name of a temporary database.
         " ? NAME OF TEMPORARY PROJECT DATABASE = "

(ix)   Type the name for the temporary database (say POSTED). The program will give the message "DATABASE POSTED, COPY T SUCCESSFULLY STARTED". Note that the copy number is T. If the editing has not broken the rules given in Section 11.1 the message "FINISHED CHECKING DATA, NO ERRORS FOUND" will be given. If errors are found the message "LINES OUT OF ORDER" will be given and you should use the editor to check for the cause of the problem. Assuming no errors are found we proceed as follows.

(x)   Run DBUTIL on the temporary database (say POSTED) copy T. Use the 'display contig' function to examine the edits made, and perhaps the 'check logical consistency' function (the only allowed error is "GEL N IS NOT USED"). When satisfied that the edits made are correct proceed as follows.

(xi)   Run DBMERGE to merge the temporary database with the original. Once started it will ask the first question.

(xii)   " GIVE NAME OF ORIGINAL PROJECT DATABASE"
         " ? PROJECT NAME = "
         " COPY NUMBER(DEF = 0) = "

Type the project name and copy number and the next question will be asked.
(xiii) " ? NAME OF TEMPORARY DATABASE=″

Type the name of the temporary database (say POSTED). The program will replace the contig in the original database with the edited one. It will list the relationships for old and new versions of gel readings as it proceeds. When finished, if successful it will report "FINISHED TRANSFER WITH NO ERRORS FOUND", otherwise it will report errors.

(xiv) To tidy up delete the temporary files (DISPLAY.OUT, PREDIT.OUT, POSTED.RLT, POSTED.SQT, POSTED.ART) created.

12. SEARCHING FOR MISSED OVERLAPS

Towards the end of a project, if it is suspected that some overlaps may have been missed, the following procedure can be used to check the contents of the database. We need to compare the ends of contigs to look for matches, but, as has been said, comparisons can only be made between new gel readings and the consensus for the database. It is therefore necessary to take copies of the gel readings at the ends of all the contigs in the database, and write them to separate files. These can then be compared with the consensus. The program ENDSOUT will remove the gel readings at the ends of contigs, write each of them to a separate file, and also write a file of file names. This file of file names can then be used by a program called DBCOMP to compare all of the gel readings with the current consensus for the database. Each gel reading should only match one contig: a double match indicates a possible join. DBUTIL can be used to make joins interactively. Missed matches can be found during the latter stages of a project by looking for shorter overlaps and by checking if the gel patterns confirm the joins. Sometimes the gel patterns can be seen to overlap even though the sequence cannot be read with absolute certainty.

12.1 Steps used to search for possible missing overlaps

12.1.1 *Running the program ENDSOUT*

(i)   Type ENDSOUT to start the program and it will ask for the project name.
       " ? PROJECT NAME=″
(ii)   Type the project name and you will be asked for the copy number.
       " ? COPY NUMBER(DEF=0)=″
(iii)   Type the copy number and you will be asked for the name for a file of file names.
       " ? FILE OF FILE NAMES=″
(iv)   Type the file of file names and the program will start to copy out the end gel readings for each contig. It will give them their original names, but prefixed by the letter E.

12.1.2 *Calculating a database consensus*

Calculate a consensus for the database using DBUTIL (see Section 9.8).

12.1.3 *Comparing end gel readings*

Compare the batch of end gel readings using the program DBCOMP,

(i)   Start the program by typing DBCOMP, and it will ask for the file of file names created by ENDSOUT.

      " ? FILE OF FILE NAMES = "

(ii)   Type the file name and you will be asked for the name of the consensus sequence.

      " ? NAME OF CONSENSUS SEQUENCE FILE = "

(iii)   Type the name of the file created by the consensus function of DBUTIL and you wil be asked for the minimum match length.

      " ? MINIMUM MATCH(DEF = 12) = "

(iv)   Type 12 to find the shortest allowable match and the program will start. Any matches found will be displayed and the numbers of the contigs involved will be given.

13. REFERENCES

1.   Sanger,F., Coulson,A.R., Barrell,B.G. and Roe,B.A. (1980) *J. Mol. Biol.*, **143**, 161.
2.   Bankier,A.T. and Barrell,B.G. (1984) In *Techniques in Nucleic Acids Biochemistry*, Flavell,R.A. (ed.), B508, p.1, Elsevier Scientific Publishers, Ireland.
3.   Grymes,R.A., Travers,P. and Engelberg,A. (1986) *Nucleic Acids Res.*, **14**, 87.
4.   Staden,R. (1982) *Nucleic Acids Res.*, **10**, 4731.
5.   Baer,R., Bankier,A.T., Biggin,M.D., Deininger,P.L., Farrell,P.J., Gibson,T.J., Hatfull,G., Hudson,G.S., Satchwell,S.C., Seguin,C., Tuffnell,P.S. and Barrell,B.G. (1984) *Nature*, **310**, 207.
6.   Staden,R. (1986) *Nucleic Acids Res.*, **14**, 217.

CHAPTER 9

Automatic reading of DNA sequencing gel autoradiographs

J.K.ELDER and E.M.SOUTHERN

1. INTRODUCTION

The increasing size of DNA sequencing projects has meant that the speed and accuracy with which sequences can be read from gel autoradiographs has become a limiting factor in the process. Reading by eye is time-consuming, errors arise due to mistakes in transcription and even experts are prone to fatigue. A partial solution to this problem is to record band positions using a digitizing tablet (1,2) or a sonic digitizer (3) linked to a computer (Chapter 8). Human interpretative skills are retained and transcription errors are eliminated, but the problem of fatigue, although less severe, remains. Another approach, described in this chapter, is to automate the reading process completely by using a digital scanner to scan the autoradiograph, followed by analysis using computer programs which locate tracks and bands and read a sequence. This method eliminates fatigue as well as transcription errors. Human skills are replaced by the ability of the automatic system to quantify inter-band distance and band intensity, and the use of these measurements in determining a sequence. The system we shall describe is that developed in our laboratory.

There are several steps in the automatic sequencing process. First, the autoradiograph is scanned and a digital image is stored in computer memory or on disc. For an ideal sequencing gel in which all tracks were straight and parallel to each other, and all bands were straight and perpendicular to the direction of the tracks, the gel could be scanned in the same way as a one-dimensional restriction fragment gel, by summing pixel values across the width of each track during the scan, and recording only a one-dimensional density profile for each track (Chapter 7). In practice these conditions are rarely satisfied, and a full two-dimensional scan is necessary so that no loss of information results from distortions in the shapes of tracks and bands. The following steps detect and correct these distortions. The positions of the track boundaries are found and the tracks are straightened. The characteristic band shapes in local sectors of each track are found and the track images are reduced to one-dimensional density profiles by summing pixel values across the band shapes. The track profiles are then registered with respect to each other so that the peaks in the profiles occur in the correct order when the profiles are superimposed. Finally a sequence is read from the four profiles by looking for an ordered set of bands using band intensity and inter-band spacing information.

2. SCANNING AND DIGITIZATION

The scanner must satisfy requirements concerning field size, spatial resolution and data storage, and optical density (OD) range and resolution. We discuss these requirements and give examples of two suitable scanners.

2.1 Scan dimensions

The dimensions of a sequencing gel autoradiograph are typically $200-300 \times 400$ mm, and the autoradiograph may contain more than 40 tracks. Since it is more efficient to scan all sets of tracks simultaneously, the scanner should be capable of digitizing a field of this size. For most gels, a two-dimensional scan is necessary, so that the effects of distorted tracks and bands can be corrected.

2.2 Spatial resolution, amount of data and storage

The required spatial resolution of the scanner is determined by the most closely spaced bands in the gel. A pixel size of 100 μm is usually adequate. Digitizing a 200×400 mm autoradiograph at this resolution would produce 8 Mbytes of data. However, since the bands are thin lines, lying more or less perpendicular to the tracks, there is no need for high resolution across the width of the track. The amount of data can therefore be reduced by taking the mean of successive groups of adjacent pixels across the width of the gel.

This filtering operation serves two purposes: the amount of scan data is substantially reduced and the effects of noise are suppressed. The entire scan can be performed at full resolution, and the filtering operation carried out afterwards, but a better method is to perform the filtering during the scan, hence reducing the amount of storage required for the scan data. A 200×400 mm scan using a pixel size of 100 μm and averaging pixel values four at a time in each row, will generate 2 Mbytes of scan data, assuming that each pixel value occupies one byte. The scan data may be stored either on disc or in computer memory. The rapidly falling cost of memory makes it possible to perform the scan and all subsequent operations without any disc access for data, greatly reducing the time needed for the scan and the analysis.

2.3 Optical density range and resolution

The scanner must be able to detect faint bands, and to resolve strong bands which overlap. The charge-coupled device (CCD) array scanner used in our studies (4) has a linear density range of $0-2$ OD and can distinguish density differences of 0.01 OD. This has proved adequate.

Pixel OD readings are usually stored as 8-bit digital values. It is useful to be able to match the scanner's OD range to that of the autoradiograph being scanned so that the digital range is used effectively. This can be done automatically by scanners which use a software lookup table to convert intensity to OD. The OD range of the autoradiograph is found by a fast coarse resolution scan, a lookup table is generated which maps this range to the digital range and the full scan is then performed.

2.4 Scanners

As detailed in Sections 2.1 $-$2.3, the scanner should be able to scan a large field area in a reasonable time at 100 μm resolution and with a moderate OD performance. Television cameras cannot digitize the full length of a sequencing gel at adequate resolution and suffer from low dynamic range. Two-dimensional CCD array cameras have adequate dynamic range, but again do not have sufficient spatial resolution (5). Two types of scanner which satisfy the requirements listed are those which scan mechanically in one axis and use a laser beam or a CCD linear array to scan in the other axis. We give brief descriptions of two such scanners which are suitable for scanning sequencing gels.

The European Molecular Biology Laboratory laser scanner (6) consists of a flat-bed stage moved along its length by a motor-driven screw-spindle, and a laser beam which scans the width of the stage by means of a galvanometer-driven mirror. The maximum field size is 250 × 400 mm and the pixel size can be varied from 64 to 256 μm. The data acquisition time is 10 μsec/pixel. The scanner is linked to a Motorola MC 68000 microprocessor (Chapter 1).

The large format scanner built in our laboratory (4) consists of a linear bearing stage driven in one axis by a stepping motor and a 2048-element CCD linear array camera which scans the width of the stage. The maximum field size is 350 × 450 mm and the pixel size can be varied from 25 to 175 μm by changing the position of the camera. Illumination across the width of the stage is provided by a white fluorescent lamp driven by a 20 kHz AC power supply. A software shading correction factor is applied to each pixel element in the CCD array, to compensate for variation in diode sensitivity and uneven field illumination. Pixel intensities are digitized as 12-bit values and are converted to 8-bit OD values by a software lookup table. A typical scanline integration time is 5 msec, corresponding to a maximum data rate of 400 kpixels/sec. Since data cannot be stored at this rate, a lineskip facility is used which only accepts data from every nth scanline, where n is chosen to match the rate at which data can be processed and stored. The stage speed is reduced accordingly so that no part of the field is missed in the scan. The scanner is controlled by a Motorola MC 68000 microprocessor with a 20 Mbyte disc. The data reduction operation described in Section 2.2 is performed in real time during the scan and the rate at which the filtered data is stored in memory is 15 kpixels/sec, corresponding to a scan time of 135 sec for a typical sequencing gel image of 2 Mbyte.

3. ANALYSIS

In a large sequencing project all autoradiographs will usually be produced in a standard format. The total number of tracks in a gel, the number of blank tracks (if any) between each set of four sequencing tracks, and the bases to which the tracks correspond can therefore be entered once by an operator, for use in the analysis of all gels in the project.

3.1 Track boundary detection

Before tracks can be straightened, their boundaries must be found. We start with the knowledge that each track runs roughly parallel to its neighbours, that the tracks are of approximately equal width and are approximately equispaced, and that changes take place in a continuous manner (*Figure 1*). These constraints are valuable in determining the track boundaries.

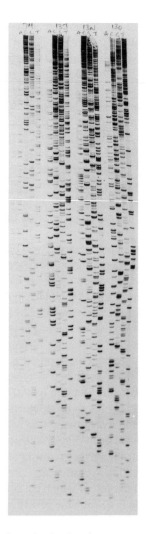

Figure 1. A sequencing gel autoradiograph, showing the curvature of the tracks.

Since the track boundaries do not change rapidly, it is not necessary to use the full resolution image to detect them. Instead, the image is compressed along its length by summing pixel values n at a time in each column. With a pixel size of 100 μm, a value of n of 20 is suitable. Each row in the compressed image then represents 2 mm down the gel, which is a small enough distance to detect any change in the track boundaries (*Figure 2*). As well as reducing the time taken to detect boundaries, compressing the image also suppresses noise and artefacts.

The boundaries are found by examining the first derivative of the rows of the compressed image. Well defined left and right boundaries of a track are indicated by significantly positive and negative derivative values respectively. Although the boundaries of each track will not always be well defined along the entire length of the gel, their position can still be estimated from the positions of neighbouring track

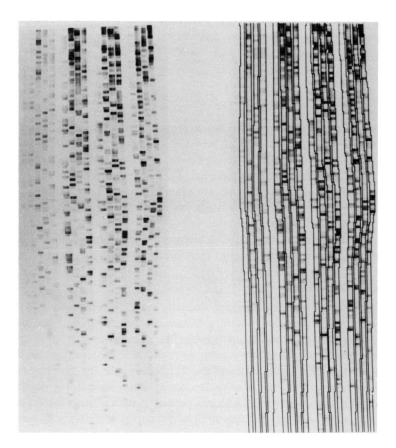

Figure 2. A compressed digital image (**left**) of the gel in *Figure 1* and (**right**) with track boundaries superimposed. The gel was digitized using a pixel size of 50 μm and pixel values were summed 8 at a time across the gel. The compressed image used for finding boundaries was then made by summing pixel values 40 at a time along the length of the gel. For display a non-square aspect ratio has been used.

boundaries, which act as constraints. Derivatives are estimated by a quadratic convoluting function (7) (Section 3.6).

An initial set of boundaries in a single row of the image is used as a starting point. This set is found by searching each row of the compressed image for an interleaved sequence of positive and negative derivative values and, by using the known number of tracks and the spacing between the derivative values, evaluating each sequence for its suitability as a starting set of boundary positions.

The chosen set of initial boundary positions is then extended row by row along the length of the gel. Boundary positions are constrained by allowing each boundary and the distance between adjacent boundaries to change by at most one pixel column from row to row. Within a row, it is not necessary to estimate the boundaries of each track simultaneously, since the position of a track's boundaries are influenced only by the few closest tracks on each side. Instead, the new boundary positions for each track are estimated separately, taking into consideration the feasible boundary positions of only the two closest tracks on each side.

For each group of five adjacent tracks within a row, the application of the constraints on boundary position still leaves many feasible sets of boundaries to be evaluated. For a track t, a candidate set of left and right boundary positions l_k, r_k ($k = t-2, \ldots, t+2$) and a row of pixel values $z[j]$ ($j = 1, \ldots, m$), a simple objective function is

$$\sum_{k=t-2}^{t+2} z'[l_k] - z'[r_k]$$

For a good set of boundary positions this expression will be large and positive, since the derivative will be positive at the left boundaries and negative at the right boundaries. From the set of boundary positions giving the maximum value of the objective function the new boundaries for track t are recorded. This process is repeated for every track in the current row, and the boundaries are extended row by row in this manner until the ends of the gel are reached. Finally, the boundaries are smoothed to remove local random fluctuations (*Figure 2*).

Once the boundaries have been found, each set of four tracks is treated separately in the analysis. The following sections therefore refer to a single set of four tracks.

3.2 Track straightening

Straightened versions of the tracks are produced by moving an imaginary scan line down the full resolution gel image perpendicular to the mean direction of the track boundaries, and reading off pixel values lying between the boundaries. The four tracks are straightened simultaneously to help in retaining registration. Pixels at the edges of tracks are ignored so that the straightened tracks are of uniform width. The straightening process substantially reduces the amount of data, since all other scan data is discarded.

In autoradiographs where the tracks run parallel to each other and track curvature is modest, it is not necessary to make straightened copies of the track images; instead the later stages of the analysis can use the positions of the boundaries to work directly from the appropriate parts of the original gel image.

3.3 Background subtraction

The background pixel value varies from track to track and along the length of a track. To aid in the estimation of band shapes and the registration of track profiles, it is helpful to subtract this background. Each track is divided into 15 sectors of equal length, and for each sector a frequency histogram of pixel values is calculated. A smooth background is then constructed by interpolation from the 10% cumulative frequency pixel values in the histograms.

3.4 Estimation of band shapes and generation of track density profiles

Each track is to be reduced to a one-dimensional density profile, so that bands can be detected by finding the peaks in the profile. If all the bands are known to be straight and perpendicular to the track, a density profile can be produced by simply summing all the pixels in each row between the track boundaries. In practice however, the bands are often sloping and curved, and generating a profile in this way can smear the peaks to such an extent that individual bands are difficult to resolve (*Figure 3*). The problem of curved and sloping bands can be avoided by using only the central pixel value across the width of a track, but the resulting profile will then be sensitive to noise and artefacts.

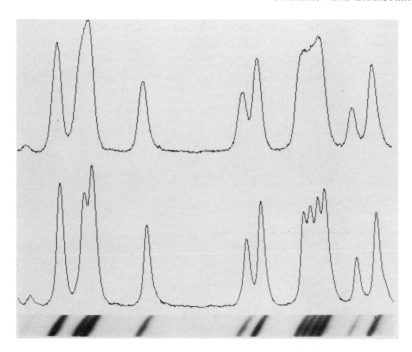

Figure 3. A track sector with sloping bands. In the upper profile, pixels were summed across the track perpendicular to its length; in the lower profile, pixels were summed along the band shape.

To obtain a high resolution profile it is therefore necessary to determine the shapes of bands and to sum pixel values along these shapes. Since band shapes are locally similar within a track, the track is divided into the same sectors as those used for finding the background (Section 3.3), and a band shape is found for each sector. Within a sector the characteristic band shape is found by choosing a path across the width of the track which maximizes the cross correlation between pairs of pixel columns, where one column is displaced with respect to the other in accordance with the shape of the path. It is not sufficient to examine cross correlations between adjacent columns only, since for a section of the gel containing band shapes of small but non-zero gradient, maximum cross correlation would be obtained for zero displacement of every pixel column, and the band shape would be taken to be straight and perpendicular. We therefore take into consideration the cross correlation between all pairs of pixel columns.

The track is assumed to have been straightened so that it runs parallel to the pixel columns. For a rectangular sector of a track with rows m_1 to m_2 and columns 1 to n, consisting of pixel values z_{ij} ($i = m_1, ..., m_2; j = 1, ..., n$), a cross correlation function for columns s and t is

$$R_{st}(d) = \sum_{i=m_2}^{m_2} z_{is}\, z_{i-d,t}$$

where d is the displacement of column t with respect to column s. We want to choose a set of column displacements $\{d_1, ..., d_n\}$ which maximizes

$$C(d_1, ..., d_n) = \sum_{s=1}^{n-1} \sum_{t=s+1}^{n} R_{st}\, (d_t - d_s)$$

225

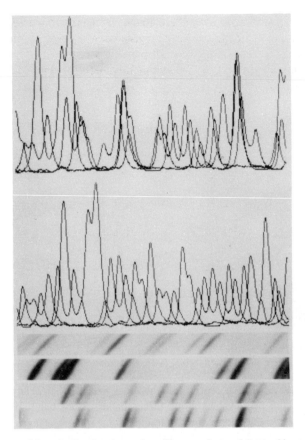

Figure 4. A sector of 4 tracks showing the track profiles superimposed (**top**) without and (**bottom**) with registration.

The range of displacements over which this expression is evaluated must be sufficiently large to accommodate the most severely sloping band shape likely to be encountered. However, since the number of possible band shapes increases rapidly as the displacement range increases, it is not practical to evaluate C for every possible set of displacements. The band shape is therefore found in two stages. An approximate shape is first found by maximizing C over a wide range of displacements, varying each displacement in coarse increments. This band shape is then refined by maximizing C over a narrow range of displacements centred about the set of displacements chosen in the first stage, varying each displacement in fine increments.

This process is repeated for every sector in the track and the track image is converted into a density profile by summing pixel values along the shapes (*Figure 3*).

3.5 Profile registration

Before the sequence can be read, the four profiles must be registered with respect to each other so that when they are superimposed, the bands from all four tracks lie in the correct order. If the profiles are superimposed without registration, loss of band order can occur (*Figure 4*).

In the same way that band shapes are found by maximizing inter-column cross correlation across a track, track profiles are registered by choosing displacements which minimize the cross correlation between all pairs of profiles.

The four profiles are divided into the sectors used before (Section 3.3). For a sector consisting of rows m_1 to m_2, a cross correlation function for profiles p^u and p^v is

$$R_{uv}(d) = \sum_{i=m_1}^{m_2} p_i^u p_{i-d}^v$$

where d is the displacement of profile p^v with respect to profile p^u. We want to choose a set of profile displacements $\{d_1, d_2, d_3, d_4\}$ which minimizes

$$C(d_1, d_2, d_3, d_4) = \sum_{u=1}^{3} \sum_{v=u+1}^{4} R_{uv}(d_v - d_u)$$

As with the determination of band shapes, the displacements are found in two stages. An initial set of displacements is calculated from the slopes of the characteristic band shapes in the sector. The registration of the profiles is then refined by minimizing C over a small displacement range centred about the initial set.

For each profile a smooth displacement function for all positions along the profile is calculated by interpolation from the sets of chosen displacements for the profile sectors. In practice, the displacements need not be applied to the profiles themselves, but only to the band positions (Section 3.6).

3.6 Band detection

Bands in each track are found by applying a peakfinder which uses the first and second derivatives of the density profile. A profile consists of a sequence of equispaced density values $\{p_i\}$, and 5-point estimates of the first and second derivatives at p_i are

$$p_i' = (-2p_{i-2} - p_{i-1} + p_i + 2p_{i+2}) / 10$$
$$p_i'' = (2p_{i-2} - p_{i-1} - 2p_i - p_i + 2p_{i+2}) / 7$$

These estimates are the first and second derivatives of the least squares best quadratic functions for the set of points $\{p_{i-2}, p_{i-1}, p_i, p_{i+1}, p_{i+2}\}$. For an extensive list of sets of convoluting integers for first and second derivatives of polynomials of various degrees and numbers of points, see (7).

Isolated peaks are found by looking for zero crossings of p'. Positions in the profile where p' changes sign from positive to negative and p'' is negative are recorded as peaks. In practice a small negative threshold is set which p'' must not exceed. The threshold is chosen to be small to err in favour of recording spurious peaks rather than missing genuine ones, since the spurious peaks will be discarded during the reading of the sequence.

This procedure successfully finds peaks with well-defined summits, but fails to find secondary peaks whose presence is indicated only by an inflection on the side of a major peak. The sides of peaks already found are therefore examined for shoulders. The presence of a shoulder is indicated by a series of changes in sign of p'' from positive to negative to positive again and the position of the peak causing the shoulder is recorded as the position where p'' is minimum and negative.

There are many more sophisticated peak finding methods (8), using methods such as moment analysis and fitting sums of Gaussian curves, but we have found the method described adequate for sequence reading.

3.7 Reading the sequence

Before a sequence is read, the bands in each track are registered by applying the displacements described (Section 3.5), and the registered bands are merged and sorted into ascending order up the gel. Associated with each band is its track, its position along the track and its intensity, which is the height of the corresponding peak in the track profile.

An initial inter-base distance is estimated by selecting all the bands near the bottom of the gel whose intensity is at least 5% of the maximum band intensity, and calculating the median inter-band distance. Starting from the bottom of the gel, a sequence is then read by weighting all bands within a window ahead of the current position and choosing the band with maximum weight as the next base in the sequence. Other bands with high weights are recorded as possible alternative bases. The weighting function is

$$w(h, x) = (0.5 + 0.67x)h \quad (0.3 \le x \le 0.75)$$
$$w(h, x) = (1.75 - x)h \quad (0.75 \le x \le 1.75)$$

where h is the band intensity and x is the distance between the band and the current position, expressed as a fraction of the standard inter-band distance. As the window moves up the gel, the standard inter-band distance is updated according to the distances between the last few bases.

The output from the sequence analysis is therefore a list of bases and their weights. This information can either be used directly, or can form the input to a software package which obtains a consensus sequence from overlapping sequences.

This method of reading a sequence is a simple one which works surprisingly well (4), but since only one base is read at a time, it does not use all the available information from surrounding bands. We plan to replace the method by one which examines larger windows extending over several inter-base distances. Within a window, all subsets of bands will be assigned weights using inter-band distance and band intensity information, and the subset with the greatest weight will be chosen as the sequence within the window. A suitable weighting function will be found by measuring inter-band distances and relative band intensities in gels where the sequence is known.

3.8 Time and cost

At present the analysis is performed on a Motorola MC 68000 microprocessor with floating point operations done in software; the analysis of a single set of four tracks takes 15 min. The processor will shortly be upgraded to a 32-bit MC 68020 with a floating point co-processor and the analysis time should then be substantially reduced.

The total cost of the scanner and computer is approximately £15 000. This is considerably more expensive than a semi-automatic system using a hand-held digitizer, but a system such as ours can also be used for scanning and analyzing one-dimensional restriction fragment gels and two-dimensional protein gels.

4. ACKNOWLEDGEMENTS

We thank Daryll Green for assistance with the scanner, Donald McLeod and Robert Hill for providing autoradiographs and Douglas Stuart for photographic assistance.

5. REFERENCES

1. Gingeras,T.R., Rice,P. and Roberts,R.J. (1982) *Nucleic Acids Res.,* **10**, 103.
2. Komaromy,M. and Govan,H. (1984) *Nucleic Acids Res.,* **12**, 675.
3. Staden,R. (1984) *Nucleic Acids Res.,* **12**, 499.
4. Elder,J.K., Green,D.K. and Southern,E.M. (1986) *Nucleic Acids Res.,* **14**, 417.
5. Toda,T., Fujita,T. and Ohashi,M. (1984) *Electrophoresis,* **5**, 42.
6. Marsman,H. and Wijnaendts van Resandt,R. (1985) *Electrophoresis,* **6**, 242.
7. Savitzky,A. and Golay,M.J.E. (1964) *Anal. Chem.,* **40**, 1627.
8. Grushka,E. (1975) in *Methods of Protein Separation*, Vol. **1**, Catsimpoolas,N. (ed.), Plenum Press, New York, p. 161.

CHAPTER 10

Identifying coding sequences

G.D.STORMO

1. INTRODUCTION

This chapter describes methods for identifying coding regions in nucleic acid sequences. It is divided into two main classes of searches, by signal and by content. Search by signal is the identification of translation initiation sites by the sequence features that distinguish them from other sites. This is presumed to be related to the mechanism used by the ribosome to find the sites. Search by signal also includes finding splice sites based on their sequences, and identifying the introns and exons of the gene. Search by content identifies coding regions by the constraints placed upon them by codon usage. All known species make unequal use of all the codons. There are biases both in the proportions of each amino acid and in the choice of codons for any particular amino acid. This codon usage bias can aid in the identification of coding sequences. Neither search method is totally reliable, and using combinations of them gives improved results.

Prokaryotes and eukaryotes present different problems when trying to identify their coding sequences. The mechanisms of translation initiation are quite different (1,2). Prokaryotic mRNAs may contain several genes, while eukaryotic mRNAs are nearly always mono-cistronic. Eukaryotic genes are often interrupted by introns in the DNA, sequences that are removed from the mRNA before it is translated (3). The two kingdoms are, therefore, treated as separate problems. Examples from each are given and the strengths and limitations of the various methods are shown.

The most reliable method of identifying the coding region of a particular protein is to know the amino acid sequence and find the appropriate region by translation of the DNA sequence. This is not practical in general because the rate of determining DNA sequences is much faster than the methods for purifying and sequencing proteins. Therefore computer methods for finding coding sequences in DNA are valuable, but their value depends on their reliability. For instance, it would be easy to make a rule that would find 100% of all known eukaryotic and >90% of all prokaryotic initiation sites: look for ATGs (4,5). This would, however, find additional thousands of sites that do not function as initiation sites (6). At the other extreme, it would be easy to make a rule that only found initiation sites, but at the cost of missing many real sites. For instance, one could make a list of the 50 bases around each known site. Searching the entire library of sequences with such a list would only find those sites again, but would also fail to identify the next real coding sequence to be discovered. The objective is to simultaneously maximize the number of correct sites found while minimizing the number of incorrect sites found. Stated somewhat differently, the goal is to extract, from some examples, what is the essence of being an initiation site or a coding sequence so that other sequences can be searched for containment of that essence.

Each method has strengths and weaknesses for reliability. Some are prone to miss certain kinds of coding regions while others miss different kinds. In general, the greater the variety of methods used in any one search the better the reliability. Each method is described individually, for both prokaryotes and eukaryotes, pointing out its usefulness and its limitations. Then combinations of methods are employed to improve the reliability of the search. The fact that uncertainty remains after using the best approaches indicates that finding the coding sequences by computer has much progress behind it, and much more ahead.

2. SEARCH BY SIGNAL

2.1 **Consensus sequences**

A common approach to locating sites of all kinds is to search for similarities to 'consensus sequences'. While this method is sufficient for applications such as finding restrictions sites, it is inadequate for most sites of biological interest. The method suffers from the fact that individual sites are not usually identical to the consensus, and different positions vary in their importance within the consensus. For example, the consensus sequence for *Escherichia coli* promoters in the -10 region is TATAAT (7).

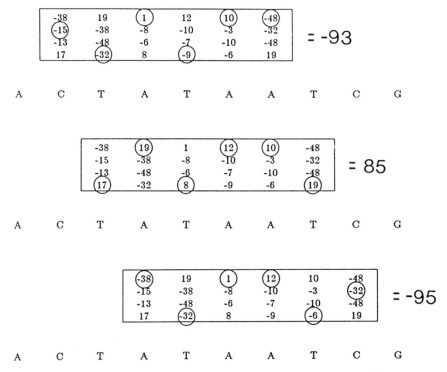

Figure 1. Matrix evaluations of a sequence. The matrix contains an element for each possible base at 6 positions. At each alignment of the matrix above the sequence a score is calculated based on the matrix elements for the sequence. The matrix rows are in the order A, C, G, T from top to bottom. The sequence TATAAT scores the highest by this matrix, with a value of 85.

However, very few promoters actually have that sequence, most matching at only three or four positions. Furthermore, the divergence from the consensus is not uniform across the site. In fact, the 3' T is conserved in nearly all promoters, the 5' TA is quite highly conserved and the other positions are more highly variable. Such descriptions are not easily accommodated into consensus-type searches.

2.2 Matrix evaluation of sequences

A superior method is to search using a matrix. *Figure 1* demonstrates the use of a matrix in evaluating sequences. The matrix contains an element for each possible base at every position within a site. The evaluation of each potential site involves summing the elements that correspond to the sequence at that site. If one only has information about the consensus sequence, a matrix can be used to find all sites within some range of similarity. For instance, the matrix of *Figure 2a* would give values of 60 to all -10 consensus sequences, 50 to all sites with one mismatch and so on down to zero for a site with no matches to the consensus. As more information is gained about what is necessary to be a -10 promoter site, that information can be added to the matrix. The matrix of *Figure 2b* is an example of including more information about a site. The consensus sequence is still evaluated to 60, but similar sequences with mismatches in different positions are evaluated differently; mismatches at the most conserved positions give the largest reductions in evaluation. *Figure 2c* is a matrix whose elements are the percentages of occurrences of each base in 112 examples of *E. coli* -10 promoter regions (7).

a

pos:	1	2	3	4	5	6
A	0	10	0	10	10	0
C	0	0	0	0	0	0
G	0	0	0	0	0	0
T	10	0	10	0	0	10

b

pos:	1	2	3	4	5	6
A	0	10	8	10	10	0
C	1	0	1	1	3	0
G	1	0	1	1	1	0
T	10	0	10	1	1	10

c

pos:	1	2	3	4	5	6
A	2	95	26	59	51	1
C	9	2	14	13	20	3
G	10	1	16	15	13	0
T	79	3	44	13	17	96

Figure 2. Different matrices for -10 promoter region. **(a)** Matrix for consensus sequence. **(b)** Matrix with different penalties for different mismatches to consensus. **(c)** Matrix with elements proportional to the frequency of each base at each position in a collection of promoters (7).

Searching with this matrix gives a score based on the similarity of any site to the collection of known functional sites.

Searching for sites using matrices is a simple and effective way that allows one to include all available information about what is involved in being a site. In searching for promoters one would also have to include a matrix for the -35 region and information about the spacing between the two elements. Such searches have been done and are quite effective $(8-10)$. It is even possible to find matrices that give good correlations between site evaluations and functional activities (10). In fact, given enough data one can solve for the matrix that gives the best fit to such data (11).

2.3 Alternative matrices

The matrices of *Figures 1* and *2* use mono-nucleotides at particular positions as the informational elements that are summed to evaluate any sequence. Higher units of information than mono-nucleotides can also be used as the matrix elements. For example, the matrix of *Figure 2a* could have used di-nucleotides instead. If overlapping di-nucleotides were used, the matrix would have five columns and 16 rows, one for each possible di-nucleotide. Matrices of this kind have been used in searches for transcriptional termination sites (12) and for quantitative evaluations of other sites (11). Matrices can also be constructed that are composites of several pieces, each with its own formulation. The columns need not correspond to single positions, either. They might include the sequence composition over several positions. Such extensions to the matrix evaluation method have not been used much yet, but as more information becomes known about some common recognition proteins they will replace the simpler ones. I mention them here to point out that matrices are very general ways of encoding the information about sequences that determine their functionality. The difficult part is finding a matrix that describes well the function under consideration.

This chapter uses search by signal to find coding sequences, but the general method is applicable to any DNA or RNA interaction involving sequence recognition. Although matrices have been used effectively in finding *E. coli* promoters, signal searches usually use consensus sequences. It is possible to take a few examples of an interaction site and construct a matrix that is consistent with the information contained in those sites (13,14). This approach should prove more valuable as a general method for finding new occurrences of sites.

3. SEARCH BY CONTENT

Search by content methods rely on the fact that coding for a protein puts constraints on the sequence (15). The most obvious constraint is for a long open reading frame (ORF), since proteins are generally over 100 amino acids in length (16). This constraint on the mRNA sequence may not be evident in eukaryotic DNA, however, due to the presence of introns (3). In addition, proteins usually have a typical amino acid composition, regardless of function or organism (17). This causes biased base position preferences in coding sequences. Species also use a biased codon choice for their amino acids (18). Usually one codon for an amino acid will be used much more frequently than any of the other synonymous ones. The codons of choice vary between different taxanomic groups (19). The combination of the constraints makes coding sequences

Table 1. Occurrence/1000 amino acids[a]

Ala	86	Arg	49	Asn	43	Asp	55
Cys	29	Gln	39	Glu	60	Gly	84
His	20	Ile	45	Leu	74	Lys	66
Met	17	Phe	36	Pro	52	Ser	70
Thr	61	Trp	13	Tyr	34	Val	66

[a]Data from Dayhoff (17).

statistically very different fron non-coding sequences, and that difference can be used to locate them.

3.1 Open reading frames

In a random sequence, the probability of any codon being followed by an ORF of over 50 codons is less than 10%. The probability of the ORF being longer than 100 codons is less than 1%. Since most proteins are more than 100 amino acids in length, searching for long ORFs is an effective way of finding genes, at least in prokaryotes. The existence of introns in eukaryotic genes diminishes the usefulness of merely finding long ORFs in DNA sequences. However, most exons are at least 20 codons long (20) and are, therefore, found in ORFs of over 50 codons most of the time. In genomes with high G-C content the reliability decreases because the probability of encountering a stop codon, in any frame, is decreased. Nonetheless, ORF searches can be a useful first step in finding gene sequences, even in eukaryotes.

An important point about using ORFs as screens for potential coding regions is to realize that the coding constraints affect both strands of the DNA. One result is that coding regions tend to have long ORFs even on the non-coding strand (14,15). This is due to the fact that many of the preferred codons are complementary to each other. Therefore, even methods that look at codon bias can identify the wrong strand from a coding region. Using the signal search methods already described can help in determining the proper strand.

3.2 Base/position preferences

Fortunately, other cues besides the length of the ORF can be used. Proteins with different functions and from all species studied share a typical amino acid composition (*Table 1*, 17). This fact can be used to rate ORFs on their likelihood of being translated. This is particularly useful in genomes with highly biased base compositions. For instance, bacteriophage T4 is about two-thirds A-T, yet the most abundant amino acids in its proteins are alanine and glycine, with codons GCN and GGN, respectively. The result is that 75% of the third position bases are A and T, and 80% of all G's are in the first two positions. In fact, G is the most common base in the first position of codons. By merely counting the proportion of each base at each codon position for each possible reading frame one can reliably identify the coding regions and the reading frame. T4 has the additional advantage of having the non-coding reading frames open for only short lengths because of the numerous out of frame stop codons in an A-T rich genome. In genomes that have a high G-C content long ORFs abound, but positional differences in base compositions can still be used reliably to find coding frames.

a

	U	C	A	G	
	26	59	22	12	U
U	69	67	63	12	C
	5	1	0	0	A
	8	9	0	13	G
	10	11	5	152	U
C	23	0	38	64	C
	0	21	16	1	A
	267	104	112	0	G
	44	58	6	3	U
A	163	95	113	34	C
	0	5	173	0	A
	95	13	57	0	G
	129	107	72	179	U
G	23	27	133	87	C
	91	86	213	0	A
	48	97	66	4	G

b

pos:	1	2	3
U	366	1001	895
C	859	1089	612
A	824	760	1011
G	1362	561	893

Figure 3. Codon usage and base position preference in *E. coli*. Data is from the genes *rplK, rplA, rplJ, rplL, rpoB* and *rpoC* (32). (**a**) The number of times each codon is used. (**b**) The number of times each base occurs in each codon position.

Even in species with nearly equivalent base frequencies this method can be used to identify ORFs that are coding. One approach is to take a collection of known genes and count the frequency with which each base occurs in each frame. For instance, in a collection of *E. coli* genes 48% of the G's are in the first position, 20% are in the second position and 32% are in the third position (*Figure 3*). These numbers can have subtracted from them the frequency expected in a random sequence, 33% in each position, to give a G-preference vector of (15, -13, -1). The same can be done for each other base, the result being a standard base/position preference vector with 12 numbers. The same type of vector can be calculated for each possible reading frame of a test sequence. The sum of the correlation coefficients between the standard vector and each test vector is zero. Usually the correct reading frame gives a high correlation coefficient, and the other two frames give negative ones. In non-coding regions there is usually no frame that is particularly high. In tests on ORFs of known function I have found that those with correlations above 0.7 were nearly always coding, those below 0.3 were nearly always non-coding, and those in between could be of both classes.

It is also possible to test the likelihood of a sequence being coding when one does not have any information about the coding bias for that species. Shepard (21) first noticed that in coding sequences there is a triplet-associated autocorrelation for each nucleotide. That is, if one counts the distances between each pair of A's in a sequence, there is a distinct bias for them to be multiples of three apart. The same is true for each other base. Every base has preferred positions in codons and over a large collection of codons this preference is significant enough to distinguish coding from non-coding regions.

Fickett (22) has used a simple set of measurements to devise a test of coding sequences. He determines eight parameters for a sequence: the four base compositions and four base/position preferences. The latter is determined by counting the number of occurrences of each base in each codon position and dividing the maximum of those by the minimum ($+1$). He then weights the parameters according to their predictive ability. For instance, the T position preference parameter is more important in distinguishing coding from non-coding sequences than are any of the base content parameters. On a test of sequences this method mis-identified 6% of the non-coding and 3% of the coding sequences, while offering 'no opinion' on 18% of the sequences. If the sequences were less than about 200 bases in length the reliability decreased.

The method of Fickett is designed to ask whether sequences of length 200 or more are within coding regions. Since in prokaryotes there are very few non-coding regions that long (excluding rRNA and tRNA genes), it is not particularly useful. However, eukaryotes have long non-coding regions, including within introns, so it is a useful screen of genomic DNA. It may still miss short exons, and does not indicate the reading frame.

3.3 Codon bias

All species deviate from a random choice of synonymous codons. The exact nature of the deviation is characteristic of the taxonomic group (19). The codon usage for a given species can be used to identify coding regions in that species. This is because the alternative reading frames, while they may have similar amino acid compositions and base compositions, have very different codon compositions. Some data indicate that highly expressed genes are the most biased in their use of codons. Lowly expressed genes use a greater fraction of the rare codons, but still predominantly use the common ones (18).

By using the known codon bias for a species one can evaluate the likelihood of a particular reading frame being coding, even for fairly short segments. Staden and McLachlan (23) have used Bayesian statistics to devise a probability measure for which of three reading frame is most likely to be coding. The idea is as follows:

(i)   Codon C_i occurs in coding sequences with frequency F_i;

(ii)   A collection of codons C_1 to C_n is expected to occur in a coding sequence with probability $= \prod_{i=1}^{n} F_i$;

(iii)   If the probabilities from (ii) for each of the three reading frame sequences are p_1, p_2 and p_3, then the relative proabilities that each of the three reading frames is the correct one is $P_i = \dfrac{p_i}{(p_1 + p_2 + p_3)}$ for $i = 1$ to 3.

Note that in this treatment $\sum_{i=1}^{3} P_i = 1$. That is, the P_i are the probabilities that frame i is coding, assuming that one of the three frames is. In practice this method also distinguishes well between coding and non-coding sequences because in non-coding sequences no particular frame is strongly favoured over the other two, whereas in coding sequences one frame is usually very much favoured. This is true even when the unknown segments are fairly short, such as 20 codons. Gribskov *et al.* (24) have used a similar method except that theirs normalizes for amino acid composition. That is, instead of the F_i values being the frequency of a particular codon, it is the fraction of that codon being used for its amino acid. This may be important in identifying genes with unusual amino acid compositions.

4. PROKARYOTIC CODING REGIONS

4.1 Search by signal

4.1.1 *Initiation site features*

Over 300 prokaryotic translation initiation sites have now been sequenced (2,5). They have also been studied extensively, both genetically and biochemically. Two features are well established. The initiation codon is usually an ATG, although about 10% of the time it is a GTG or, occasionally, a TTG. One natural initiation codon is ATT. ATA, CTG and ACG have all been shown to function *in vivo*. There is usually a sequence 5′ to the initiation codon that is complementary to the 3′ end of the 16S rRNA. Shine and Dalgarno proposed that the mRNA and the 16S rRNA base pair to one another as part of the initiation site selection (25). The complementary region of the mRNA is often called the Shine − Dalgarno sequence for its gene. Substantial genetic and biochemical work has verified the importance of this interaction to the selection of the initiation site.

4.1.2 *Consensus sequence*

An initiation codon and an appropriately spaced Shine − Dalgarno sequence are not, however, sufficient to define an initiation site. Many different search rules based on these features have been tried (6). On the criterion that a rule must find at least 80% of the known sites, the best rule was 'ATG preceded at six to nine bases by AGG, GAG, or GGA'. In a test on a large collection of real sequences, this rule found 83% of the known sites and about 60% more non-sites. The number of false positives could be reduced to half the number of sites found by increasing the length of the Shine − Dalgarno sequence to four bases, but at the expense of missing 33% of the known sites.

Statistical analyses of initiation sites showed that other positions around the initiation codons displayed biased base compositions. These biases could be included in searches to give improved discrimination between real sites and other similar sites. The rules also became more complex and the searches more complicated (6). Some data indicated that there could be a trade-off between features involved in determining what is a site. For instance, initiation sites with poor Shine − Dalgarno sequences were more likely to have consensus bases at the other biased positions, as if interactions of those bases with the ribosome could compensate for poor base-pairing to the 16S rRNA. It was

pos:	-60	-59	-58	-57	-56	-55	-54	-53	-52	-51	-50	-49	-48	-47	-46	-45	-44	-43	-42	-41	-40	-39	-38
A:	7	-2	13	-2	-8	-13	-18	5	0	-5	13	8	-15	9	-4	-7	9	0	-8	-11	-10	-6	-7
C:	-21	-6	-11	-21	0	8	-7	-12	-1	1	0	-19	12	-3	-1	10	2	-8	-5	-11	8	1	23
G:	-6	-9	-7	0	8	-16	-4	-2	-16	1	-4	8	-14	5	11	-13	-24	3	7	22	-11	-9	-15
T:	5	1	-3	9	-14	7	15	-5	3	-16	-17	4	18	5	-3	-1	2	4	5	-5	7	8	-5

-37	-36	-35	-34	-33	-32	-31	-30	-29	-28	-27	-26	-25	-24	-23	-22	-21	-20	-19	-18	-17	-16	-15	-14	-13	-12
-5	-6	-12	-1	-27	-3	-6	0	-12	-3	-4	-7	14	-2	-4	-6	0	12	5	-9	0	-11	-11	10	8	2
6	-5	2	-14	-3	-8	-10	-21	2	0	-2	-1	-11	-3	-1	5	-11	-4	7	0	-14	6	-8	-20	-7	-36
10	-4	4	-5	-6	-3	-1	-4	-1	-4	-15	0	-14	3	10	-19	-3	-10	-7	-7	7	1	-8	-6	15	21
-15	6	3	4	16	-4	7	11	-4	-1	12	8	10	-1	1	8	2	-10	-16	11	1	-3	16	-3	-36	-8

-11	-10	-9	-8	-7	-6	-5	-4	-3	-2	-1	0	1	2	3	4	5	6	7	8	9	10	11	12	13	14
8	0	-3	-5	4	-20	-11	5	6	-2	-15	66	-69	-52	-5	-4	6	8	-24	-7	-10	-7	13	14	-9	-18
-44	-15	-50	-43	-35	-38	-29	-29	1	-9	1	-87	-55	-64	-45	11	-22	-14	-20	-15	-15	-10	-22	-5	2	6
42	35	22	16	-6	-5	-15	-25	-33	-28	-53	-36	-50	107	-5	-37	-44	-27	-15	-23	-16	-29	-47	-17	-29	-15
-27	-53	-27	-26	-23	2	-7	-14	-40	-28	0	-53	75	-62	-20	-40	-10	-35	-5	-12	-1	4	14	-23	7	-2

15	16	17	18	19	20	21	22	23	24	25	26	27	28	29	30	31	32	33	34	35	36	37	38	39	40
14	-12	-42	1	-5	-4	-32	12	-10	20	-6	-1	3	-4	4	-10	-1	-2	-14	11	14	-3	2	-13	5	5
6	-8	19	-7	9	-3	17	-2	3	-9	5	22	22	8	-1	1	18	6	11	-10	-8	7	10	0	7	14
-23	-7	-1	-6	-17	-4	0	-15	-14	-4	-17	-10	-5	-13	-8	10	-13	-13	9	-4	-3	10	2	4	-8	-21
-26	1	4	-7	3	-4	0	-10	8	-18	7	-22	-21	8	4	-3	-6	7	-8	1	-5	-16	-16	7	-6	0

Figure 4. Matrix W101 for *E. coli* translation initiation sites. This matrix evaluated all initiation sites higher than any non-initiation sites in a library of almost 80 000 bases of mRNA (26). The initiation codon is located at positions $0-2$.

then decided to look for a matrix that would weight each possible base at each position such that the sum of the weights would be higher for real sites than other similar sequences.

4.1.3 *E. coli initiation site matrices*

Figure 4 is a matrix used to find initiation sites in *E. coli*. It was found using a pattern learning algorithm (the 'perceptron' algorithm, 26) on a collection of known sites. The program was given the sequences, from -60 to $+40$ of the initiation codon, of 124 known start sites. It was also given a collection of other sites and the algorithm produced a matrix that distinguished the two sets. After a series of trainings the matrix of *Figure 4*, called W101, was obtained. W101 gave each of the 124 known sites a higher evaluation than any of the other sites in a library of sequences containing about 80 000 bases of mRNA.

Ribosomes protect from nucleases about 35 bases of mRNA around initiation sites. W101 presumably uses the information from more bases to find the sites than does the ribosome. Shorter matrices can also be found that distinguish the sites in that collection (26), but a matrix of only 35 positions was unable completely to separate the two classes. This is presumably because more than just the linear sequence information is relevant to ribosome binding. In tests on sequences not included in the training set W101 was the best at predicting initiation sites, so it will be used in the example to follow. Some improved results can be obtained by using several matrices on the same sequence and combining the results (26).

While these matrices are good qualitative predictors, this is only an approximation to ribosome binding site function. Sites vary greatly in their efficiencies of translation

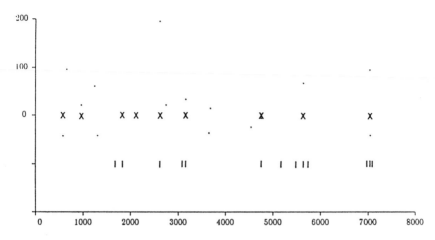

Figure 5. Initiation sites and sites identified as initiation sites in the *unc* operon. The horizontal axis represents the positions within the sequence. The X's are the 9 actual initiation sites. The vertical bars along the bottom are the sites identified by the search rule. The dots are the sites found by W101, with the value as given by the scale on the left of the plot.

initiation *in vivo*. The matrix evaluations are actually somewhat correlated with initiation site efficiency, the better sites often getting higher evaluations. Some current experimentation is devoted to finding matrices that are good predictors of quantitative activity (11,27).

4.1.4 *Other prokaryotes*

Section 4 has focused on initiation sites in *E. coli* because most of the prokaryotic research has been done there. However, some other organisms have been studied to a lesser extent. The 16S rRNAs that have been sequenced share considerable homology to that of *E. coli*, particularly at the 3' end. In some cases messages from one organism have been translated *in vitro* or *in vivo* in another organism. In general the same proteins are made, though their relative abundances will vary. These facts together strongly suggest that the ribosomes from different prokaryotes initiate via the same general mechanism. The same identification methods should then work, and in fact they do to a large extent. As more information is gained for any particular organism, alterations can be made to the matrix used. For instance, *Bacillus subtilis* initiation sites require longer Shine − Dalgarno sequences, but use more non-ATG initiation codons (2,28). An alternative to W101 could be made for *B. subtilis* that reflected these differences.

4.1.5 *Finding initiation sites in the unc operon*

The *unc* operon sequence (29) was not included in the training set that generated W101, so it is a valid test of the matrix. There are nine genes in this operon (including one putative gene). *Figure 5* and *Table 2* shows the positions of the genes and the site identified by W101 and by the search rule. The nine gene starts are marked by X's on the centre line of *Figure 5*. The vertical bars at the bottom are those sites identified by the search rule 'ATG preceded at six to nine bases by AGG, GAG or GGA'. The points

Table 2. Gene start identifications.

Sequence position	Gene start	W101 value	Search rule site	Alternate start[a]	Reading frame
565	+				1
577		−43		+	1
650		96			2
966	+	21			0
1249		61			1
1308		−42		+	0
1678			+		1
1828	+		+		1
2129	+				2
2614	+	196	+		1
2743		22		+	1
3085			+		1
3160	+	34	+		1
3650		−36			2
3678		15			0
4540		−24			0
4751	+	−6	+		2
5168			+	+	2
5480			+		2
5641	+	69	+		1
5741			+		2
6976			+		1
7044	+	98	+		0
7050		−41		+	0
7089			+	+	0

[a]In frame with overlapping gene and leading to a peptide of at least 50 amino acids.

above the centre line are the nine sites evaluated above zero, and are therefore identified as gene starts, by W101. The height of the point above the line indicates the value given to the site by the matrix. The points just below the centre line are the next highest sites evaluated by W101.

The search rule finds 13 sites. Six of these are real gene sites. The other three real sites are not found. Of the seven false positives, the two at positions 5168 and 7089 are in frame with the gene beginnings at 4751 and 7044, respectively. These could possibly be internal start positions leading to alternate proteins from a single coding region. Even if they are not, the rule has identified a coding region although not the correct start site.

Five of the nine W101-identified sites correspond to the known sites. The next highest evaluated site, with a score of −6, is the known gene start at 4751. Six of the ten highest scores correctly identify six of the nine known start sites. The other four are false positives. Three of those four are certainly false positives, indicating start sites that do not lead to long ORFs. The fourth false positive, at position 2743, is in frame with the start at 2614 (see *Table 2*). This could possibly function as an internal start site, leading to two proteins with the same carboxy terminus.

Three of the nine known sites are not identified within the highest 10 scores. Two, in fact, are not found by W101 until many false positives have been identified. The putative gene, thought to start at position 565, is also not found by W101. However, position 577, four codons downstream and in the same reading frame, is identified by W101 with the 15th highest score. The results from W101 suggest that position 577 is more likely than 565 to be an initiation site, but both would lead to nearly identical proteins. Another of the top 15 sites by W101 is a potential start site. The gene that starts at position 7044 is found with a high score (98) by W101. In addition, position 7050, two codons 3' and in frame, is found with a score of −41. It is possible that both ATG's function as initiation codons, although most translation should come from position 7044.

Of the nine known genes, only the one at 2129 is not found by either the search rule or W101. The search rule finds six of them, with five false positives and two possible internal starts. W101 finds seven of the starts (counting 577 as correct) in the highest 15 scores, with five false positives and three potential internal starts. The five sites identified by both the search rule and W101 are all correct sites. All the false positives were found by only one of the two signal searches.

4.1.6 *Strengths and limitations*

The major advantage of the matrix signal search method is that only the information around the codon is important. Presumably the ribosome does not 'know' how long the ORF following an initiation codon is. Some important peptides are, in fact, quite short, such as the leader peptides involved in regulation of many of the biosynthetic operons (30). The method will also find genes with unusual codon usage provided the start signal is reasonably typical. More reliable matrices than W101 could probably now be found. Some work is in progress to find a matrix that is related to activity (11,27) that will probably require an alternative form, such as with di-nucleotides in the Shine−Dalgarno region.

However, the matrix search method is never likely to succeed absolutely in identifying real initiation sites. One problem is that mRNA is capable of forming secondary structures that can hide sites from ribosomes. That information is lost in this search, and it is unlikely that good enough predictions of structure are likely to exist in the near future. This means that there will be sites that are identified as starts that, in fact, do not function as such.

There will probably also remain functional sites that are not found by the signal search method. It is known that some sites are activated by a termination event nearby (31). This method is used by *E. coli* to couple the synthesis of many genes; the downstream gene being translated only if the upstream gene is. While most of these coupled genes have fairly normal start signals that are somehow non-functional in the absence of upstream translation, some activation events work at otherwise silent sites. They may be considered equivalent to promoter sites that need to be activated. The range of possible activation events is unknown, but some such sites will undoubtedly escape identification by signal searches.

4.2 **Search by content**

4.2.1 *Open reading frames*

Escherichia coli proteins average over 30 000 daltons in molecular weight (16). Very

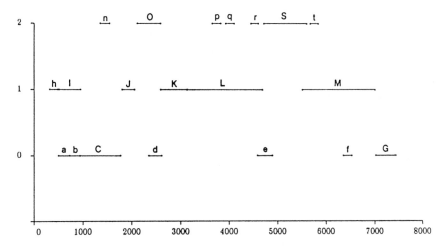

Figure 6. ORFs in the *unc* operon. All 20 of the ORFs of at least 50 codons are shown. The reading frame is indicated on the left axis. Those labelled with capital letters include the known genes.

few are under 10 000, so most have ORFs of over 100 amino acids. *Figure 6* shows the 20 ORFs in the *unc* operon that are over 50 codons in length. Those labelled in capital letters contain the known gene sequences. The eight longest are each coding sequences. The ninth coding sequence, marked J, is the eleventh longest reading frame. Eight of the nine ORFs over 100 codons are coding sequences. If, for each pair of overlapping ORFs, the shorter were eliminated, all the known genes would be identified and, in addition, non-coding ORF h would be included. This has proven to be a reliable method of locating known genes. Limitations include not knowing where within an ORF is the start point and whether or not to include the short ORFs.

4.2.2 *Base/position preference*

Figure 3 shows both the codon usage and base/position preference for a collection of *E. coli* genes. These numbers come from four ribosomal genes and two RNA polymerase genes (32). These are all highly expressed genes and indicate the strong bias in codon usage. Genes that are expressed at lower levels use some rare codons, but still primarily use these same ones.

Figure 7 shows the correlation of each of the 20 ORFs shown in *Figure 6* with the base/position preferences of *Figure 3*. The nine highest correlations are the proper reading frames. Seven of the ORFs have correlations above 0.83, and the two lowest are still above 0.53. The highest non-coding ORF has a correlation less than 0.35. Most of them have negative correlations. While this has given a clean separation between the coding and non-coding ORFs, in practice it is uncertain in which group to place those with correlations between 0.3 and 0.6.

4.2.3 *Codon bias*

For prokaryotes, the above analyses are reliable for identifying ORFs that function as coding sequences. However, they do not indicate the start sites, since they are correlations to the whole ORF. To locate the coding initiation site one needs to evaluate small

ORF	corr. coef.	ORF	corr. coef.
L	0.99	b	0.38
G	0.98	h	0.34
M	0.98	n	0.33
K	0.97	a	0.21
O	0.96	q	0.01
S	0.91	f	-0.16
J	0.84	t	-0.19
C	0.59	d	-0.31
I	0.54	r	-0.36
		p	-0.48
		e	-0.73

Figure 7. Correlations coefficients for each ORF in *Figure 6.* The correlation coefficients are based on base/position preferences as compared to the genes in *Figure 3,* according to the method described in the text.

regions of the sequence at a time. Using the codon preference, shown in *Figure 3,* and the method of Staden and McLachlan (23) it is possible to evaluate short regions for the relative potential of each frame to be the coding one. *Figure 8* shows the evaluation of the *unc* operon, taken 20 codons at a time. This analysis was done with shifts of 10 codons to the next evaluated position. Shifts of one codon would give higher resolution, but even here it is usually clear where coding sequences begin. In comparison with *Figure 6,* notice that ORF M begins about 5530. In *Figure 8* the high correlation does not begin until about 5640. In fact, the gene begins at 5641. Most of the other genes are preceded by short non-coding ORFs, and so the maps of *Figure 6* and *8* are nearly equivalent. Only the gene of ORF I is not well resolved in *Figure 8.* This is the putative gene of unknown function, and may not, in fact, be made.

4.2.4 *Strengths and limitations*

In *E. coli* nearly every ORF over 100 codons is a gene. And nearly every gene has an ORF of at least 50 codons. These two criteria can limit the regions and frames of potential genes. The main limitation is the slight overlap between short genes and the long non-coding ORFs that makes it less than 100% reliable. Using a simple base/position preference correlation can help eliminate the long non-coding ORFs from consideration. Finally, using codon usage frequencies can not only confirm the coding function of some ORFs, but can also help determine where, within the ORF, is the actual gene start position. In combination these search by content methods can give highly reliable information about the positions of genes.

4.3 **Combined methods**

In looking for functional proteins, rather than short regulatory peptides, it is best to

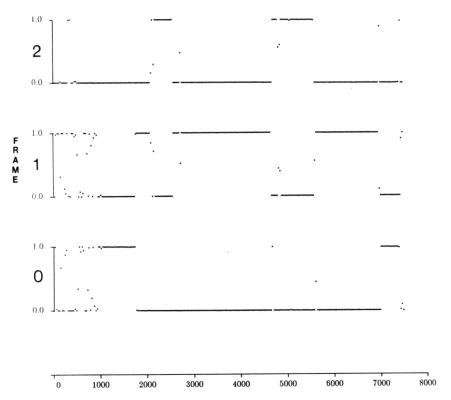

Figure 8. Codon usage predictions. The horizontal axis is the sequence position. The vertical axis shows the probability that each of the 3 frames is the coding one. Twenty codons were included for each point, with a shift of 10 codons to the next point. The probabilities are calculated by the method of Staden and McLachlan (23) and described in the text.

start with the ORFs. *Table 3* shows the 20 ORFs of over 50 codons and their evaluations by each of the methods independently. Frames G, K, L, M and S are identified by each method. The two signal search methods give the proper start sites. ORF J is missed by W101 but found by each other method. ORF C is found by codon usage and W101, but not by the search rule and is indeterminate by the base/preference correlation. ORF O is found by both content methods, and missed by both signal methods. It may be translated poorly or under some control. ORF I is missed by the search rule and only weakly found by each other method. This is the putative gene and all this evidence leaves its existence speculative.

Of the non-coding ORFs, t is found by the search rule but nothing else. ORFs n, h and b are indeterminate by base/position correlation but not found anywhere else, except that h is weakly identified by codon usage. ORF h is not overlapping any other gene and might actually be used, although it may not even be transcribed (29). By using the signal and content methods together all the genes are found to be stronger than any of the non-coding ORFs. There is some indecision at the boundary, of whether I and h are genes, but all others are quite clear.

Table 3. Combined ORF identifications.

ORF	Corr. coef.[a]	Codon bias	W101 site	W101 value	Search rule site
a	0.21	−			
b	0.38	−			
C	0.59	+	966	21	
d	−0.31	−			
e	−0.73	−			
f	−0.16	−			
G	0.98	+	7044	98	7044
h	0.34	+/−			
I	0.54	+/−	(577)	−43	
J	0.84	+	−		1828
K	0.98	+	2614	196	2614
L	0.99	+	3160	34	3160
M	0.98	+	5461	69	5641
n	0.33	−			
O	0.96	+	−		
p	−0.48	−			
q	0.01	−			
r	−0.36	−			
S	0.91	+	(4571)	−6	4751
t	−0.19	−			5741

[a]Correlation coefficients.

5. EUKARYOTIC CODING REGIONS

Eukaryotic sequences that come from mRNA (cDNA sequences) can have their coding sequences identified as easily as in prokaryotic sequences. It is actually easier because one can assume only one gene per mRNA and almost always the first AUG is the initiator. Even in genomic DNA, if the 5′ end of the transcript is known then it is fairly easy to find the initiating ATG, but the rest of the gene is still unknown. Any knowledge about the gene can be used in its identification. For instance, knowing its size might tell you if it needs to be made of more than one exon and information about its amino acid composition limits the possibilities considerably. In the following example I assume that nothing is known about the region of sequence under investigation except that it contains a (complete) coding sequence for some protein.

5.1 Search by signal

5.1.1 *Initiation site features*

Initiation sites seem to be found in a fundamentally different way by eukaryotic ribosomes than by prokaryotic ones (1,2). The 3′ end of 18S rRNA is missing the Shine−Dalgarno complementary sequence from prokaryotic rRNA. Likewise, there is not the equivalent conserved sequence 5′ to the initiation codon. Initiation codons are always ATG, although some evidence with mutants and *in vitro* experiments show that other codons can work inefficiently (33), and may yet be found. The most likely mechanism for initiation site location by eukaryotic ribosomes is called the 'scanning model' (1). This model states that ribosomes (actually the small subunit) first bind to the 5′ end of the

pos:	-4	-3	-2	-1	0	1	2	3	4	5
A	-1	12	3	0	14	-40	-40	-1	1	-5
C	8	-18	3	7	-40	-40	-40	-18	5	-5
G	-8	-7	-11	-6	-40	-40	14	9	-4	1
T	-8	-32	0	-7	-40	14	-40	-8	-6	5

Figure 9. Matrix for eukaryotic translation initiation sites. The initiation codon is at positions $0-2$. The elements are calculated as described from the data in ref. 35.

mRNA, which usually has an unusual 'cap' structure. The ribosome then migrates in the 3' direction until it encounters an ATG, at which point translation is initiated. 95% of all known initiation sites begin at the 5'-most ATG of the mRNA.

If one knows the 5' end of an mRNA it is easy to locate its translation initiation site with high reliability. However, from a genomic DNA sequence it is not currently known how to find mRNA 5' ends. Fortunately, other nucleotides around initiating ATG's are biased, and that information can help in finding real initiation sites (4). The consenesus sequence for initiation sites is 'AccATGg', where the lower case letters are less highly conserved. In the 5% of the cases when the 5'-most ATG is not the initiation codon used, it occurs in a context that is quite different from the consensus. It has been shown that the surrounding bases can affect the efficiency of utilizing a particular initiation site (34).

5.1.2 *Search matrix*

One can design a search matrix that contains the extra information around the ATG. *Figure 9* displays a matrix based on nucleotide frequencies around real initiation sites (35). The matrix elements are 10 times the logarithms of the frequencies of the bases divided by 0.25, the expected frequency in a random sequence. This makes them proportional to the log of the probability of finding that base at an initiation site, relative to finding it in a randomly chosen position. The evaluation of a site would then be proportional to the log of the probability of finding that sequence at an initiation site relative to any other site, if neighbouring bases were uncorrelated. This is clearly an approximation since neighbouring bases are not independent in real sequences, but the matrix is, nonetheless, useful in narrowing down the possible start sites. The highest value given by this matrix is 91, and the lowest is -207. Sites with an ATG range in value from -45 to 91. Sequences without an ATG can have values between -207 and 37. Since all known start sites use an ATG initiation codon, choosing a cutoff of 38 will require every site found to have an ATG. This cutoff also helps discriminate between used ATG's and non-initiation sites. Over 200 initiation sites are known (4), and over 85% of them score above 37 with this matrix.

5.1.3 *Finding the human β-tubulin initiation site*

The human β-tubulin gene is contained in a 5117 base pair region of sequenced DNA (36). *Figure 10* shows the position of the 71 ATG's in this sequence, and of the 24 sites that score above 37 using the matrix of *Figure 9*. The 5'-most of those found with

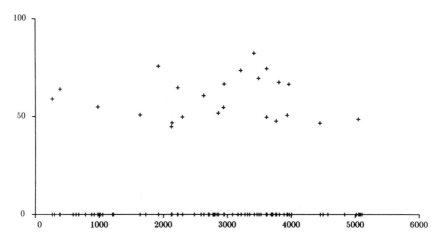

Figure 10. Potential initiation sites in human β-tubulin gene. There are 71 AUGs, marked on the zero line. Above that are the 24 sites identified by the matrix of *Figure 9* with values above 37.

the matrix, at position 267, is the real initiation site. This is the second ATG within the sequence, the first being at position one. In this case, the real initiation site is the 5′-most ATG on the mRNA, which starts at position 107. If the 5′ end of the mRNA is unknown, the matrix would only be of marginal help in finding the site. It reduces the number of potential sites about three-fold over looking for ATG's, but with about a 15% risk of not identifying the real site.

5.1.4 *Strengths and limitations*

The major advantage of using the search matrix is that most ATG's are eliminated as potential initiation sites. When the mRNA 5′ end is known, the matrix will also indicate those rare instances where the 5′-most ATG is not used. However, an experienced person can do as well searching 'by eye', so that the value of translation initiation site signal searching is quite marginal. As more information becomes known it may be possible to find the signals responsible for mRNA initiation. That could be coupled with initiation site searches to greatly increase their value. In the meantime, eukaryotic initiation site searches will remain poor predictors of coding regions.

5.1.5 *Splice junction site features*

Knowing the initiation codon for a eukaryotic gene does not guarantee that one will know the coding region because of the possible existence of introns in the DNA. That is, there are sequences that are transcribed which are removed before the mRNA is translated. The removed sequences are called introns, and the remaining regions are called exons. The coding sequence will be a long ORF in the processed mRNA but not necessarily in the DNA sequence.

Introns come in very different lengths, from under 100 bases to many kilobases (20). There are conserved sequences around the intron/exon boundaries (3,37), though these are not as informative as one would like. In some species, like yeasts, there are cona

pos:	-3	-2	-1	0	1	2	3	4	5
A	5	9	-11	-35	-35	9	10	-11	-4
C	5	-8	-15	-35	-35	-24	-10	-35	-7
G	-10	-8	11	14	-35	2	-8	12	-11
T	-12	-7	-7	-35	14	-14	-8	-16	9

b

pos:	-11	-10	-9	-8	-7	-6	-5	-4	-3	-2	-1	0	1
A	-14	-5	-8	-3	-7	-21	-9	0	-19	14	-35	-1	-4
C	0	2	3	1	4	4	1	-1	9	-35	-35	-3	-1
G	-9	-15	-13	-11	-13	-17	-17	0	-35	-35	14	7	0
T	9	7	7	6	6	8	8	2	2	-35	-35	-11	4

Figure 11. Matrices for splice junction sites. (a) Matrix to find donor sites. Intron begins at position 0. (b) Matrix to find acceptor sites. Exon begins at position 0. The elements are calculated as described from the data in ref. 37.

served sequences within the intron (38), but this does not seem to be a general feature.

5.1.6 *Matrices*

Figure 11 shows two matrices for finding splice junction sites. *Figure 11a* has the matrix for the donor site, at the 5' end of the intron. There is a highly conserved GT sequence there, at position zero of the matrix. The elements for the matrix are determined in the same way as those of *Figure 9,* using the data from Mount (37). Sequences can have values between 93 and −185 by this matrix. Those with GT can range from 93 to −87, and those without a GT from 44 to −185. *Figure 11b* shows the matrix for acceptor sites, at the 3' end of introns. These have a conserved AG site at position −2 of this matrix. Sequences can have values between 101 and −225. Those with the conserved AG will vary from 101 to −127, and those without from 52 to −225. Unfortunately there is quite a lot of overlap between the values of real sites and unused sites. Probably other information besides the primary sequence is used, such as the secondary structure of the RNA.

5.1.7 *Finding introns in β-tubulin*

Figure 12 shows the positions of potential donor and acceptor sites in the human β-tubulin gene. Only those sites are shown that are in the upper one-third of the allowed range for sites with the conserved bases. For instance, sites with GT can be evaluated by the donor matrix between 93 and −87. The upper third of that range is those sites above 36, which are shown in the figure. Likewise, only those sites above 25 with the acceptor matrix are shown. Notice that there are many more potential acceptor sites found than donor sites. In each case only two of the three sites are located. This points to the poor predictability of locating these sites. Nonetheless, this information can be useful in conjunction with the search by content methods.

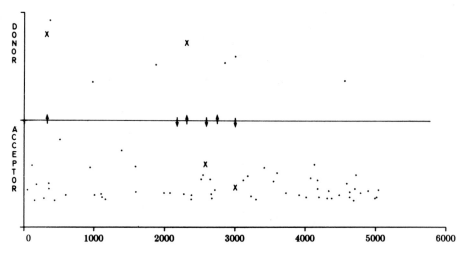

Figure 12. Potential splice junction sites in the human β-tubulin gene. Above the line are the sites evaluated above 36 by the matrix of *Figure 11a*. Below the line are the sites evaluated above 25 by the matrix of *Figure 11b*. The vertical height of the point is proportional to the value from the matrix. The up and down arrows are the actual donor and acceptor sites, respectively. The X's mark the sites correctly found by the matrices.

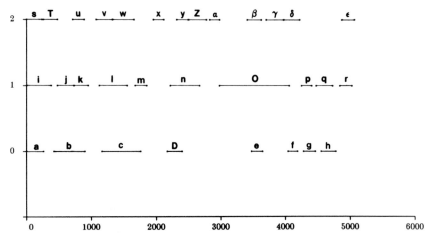

Figure 13. ORFs in the human β-tubulin gene. All the ORFs of at least 50 codons are listed. The 4 labelled with capital letters contain the known coding regions.

5.1.8 *Limitations*

Signal searching for splice junctions suffers from a lack of reliability. There are too many possible sites found by the matrices. As with the initiation site signal search, where the 5' end of the mRNA is crucial, important information about what determines whether a site functions as a donor or acceptor site is missing from the matrices. The two signal searches together are not much help, although they can be used with the search by content methods to limit the number of possibilities.

a

	U	C	A	G	
	6	4	10	3	U
U	25	25	28	8	C
	1	1	0	0	A
	4	3	0	13	G
	5	11	8	5	U
C	31	25	18	20	C
	5	9	6	2	A
	58	5	40	12	G
	8	11	4	2	U
A	36	31	24	23	C
	1	6	17	4	A
	30	7	47	9	G
	1	19	14	3	U
G	15	48	46	42	C
	5	6	15	6	A
	41	10	71	12	G

b

pos:	1	2	3
U	131	272	114
C	260	221	445
A	260	348	84
G	354	164	362

Figure 14. Codon usage and base/position preference in human genes. These numbers are from the genes for beta-actin, lipoprotein apoAI and adenosine deaminase (HUMACTB, HUMAPOA1 and HUMADA, respectively, from GenBank release 40.0). **(a)** The number of times each codon occurs in the collection of genes. **(b)** The number of times each base occurs in each codon position.

5.2 Search by content

5.2.1 *Open reading frames*

Figure 13 shows the 31 ORFs of at least 50 codons in the sequence of a human β-tubulin gene. The four with capital letters contain the four exons of the protein.

ORF	corr. coef.	ORF	corr. coef.
O	0.96	g	-0.02
T	0.84	f	-0.10
D	0.81	m	-0.13
s	0.54	r	-0.15
w	0.48	c	-0.16
p	0.43	a	-0.17
j	0.31	u	-0.18
h	0.21	k	-0.18
α	0.16	q	-0.18
x	0.14	b	-0.22
ϵ	0.13	n	-0.22
1	0.12	v	-0.23
Z	0.12	β	-0.25
δ	0.11	i	-0.58
y	-0.01	e	-0.58
		γ	-0.63

Figure 15. Correlation coefficients for each ORF. The correlation coefficient for the base/position preference between each ORF of *Figure 13* and the standard from *Figure 14*. The method is described in the text.

Even though the first three exons are between 19 and 37 codons long, they each occur in an ORF of at least 50 codons. Only the fourth exon, labelled O, is of substantial enough size to be distinguishable from the non-coding ORFs. Many of the non-coding ORFs are longer than the exon-containing ones. By itself, this method can find some ORFs that are almost certainly exons or entire genes, but it cannot reliably find short exons because of their overlap in size with non-coding ORFs.

5.2.2 Base/position preference

Figure 14 has the base/position occurrences and codon usages for a few human genes (39). These are the standards used in the comparisons to follow. *Figure 15* has the correlations for base/position preferences between each of the ORFs of *Figure 13* and those of the standards. Three of the four exon containing ORFs, O, T and D, show high enough correlations to be reliably called coding. The fourth, labelled Z, is slightly positive but in the range where it would be considered non-coding. None of the non-coding ORFs are high enough to be called coding, although four of them are in the 0.3 to 0.6 range of indeterminacy.

Figure 16 shows the method of Fickett (22) applied to the same sequence. Two hundred long windows of sequence were analysed, each shifted 50 bases from the previous. The horizontal lines divide the sequence into three zones, predicted coding above, predicted non-coding below and not predicted in the middle. The bars at the bottom

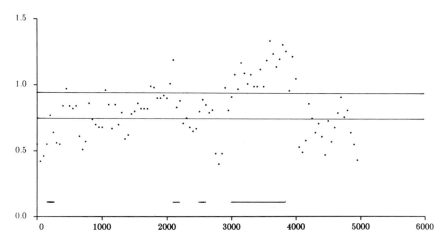

Figure 16. Coding potential by the method of Fickett. 200 base sections of sequence were analysed by the method of Fickett (22). 50 bases were shifted to the next analysis position. Scores above 0.95 are usually coding, scores below 0.75 are usually non-coding, and scores in between are indeterminant. Near the bottom are horizontal bars indicating the collection of points that include exon sequences.

show the points that overlap the coding sequences with at least half the region. The fourth exon, ORF O, shows up clearly as a coding region. Sequences 3' to it are consistently non-coding or indeterminate. 5' to ORF D most regions are non-coding or indeterminate, with a couple of peaks into the coding zone. One of those peaks corresponds to the exon in ORF D. The other isolated peaks into the coding zone do not correspond to coding regions. The region of the third exon, for ORF Z, is entirely in the indeterminate region. The region of the first exon, in ORF T, is mostly predicted to be non-coding. This exon is only 19 codons long, so all the 200 base windows that include it also contain predominantly non-coding sequence.

5.2.3 *Codon bias*

Figure 17 shows the codon usage bias prediction for each reading frame. The sequence was put into 20 codon windows, and each window was shifted 10 codons to the next. The vertical scale is the probability for each reading frame being coding, assuming that one of them is. The coding region of the fourth exon, in ORF O, is evident in frame one. No long coding regions are evident in frame two, even though there are two short exons in that frame. In frame zero there could be several coding regions. One region of five consecutive high points, near 2200, corresponds to the exon in ORF D. Most of the other high points occur in regions of indeterminate coding potential, from *Figure 16*. The stretch at the 3' end looks to be non-coding in *Figure 16*. Moreover, none of the ORFs in that region score well, or even positive, in the correlation analysis of *Figure 15*.

5.2.4 *Strengths and limitations*

Using the various search by content methods together can help find coding regions in eukaryotic DNA. The fourth exon, in ORF O, is evident in all of these methods. Its

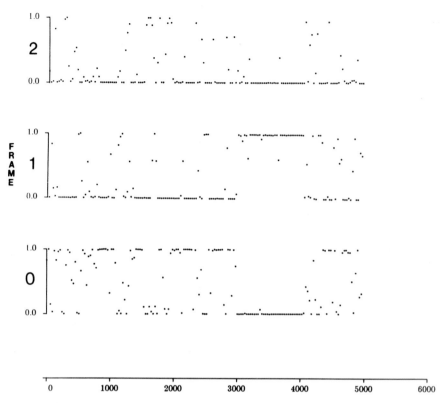

Figure 17. Codon usage predictions. The horizontal axis is the sequence position. The vertical axis shows the probability that each of the 3 frames is the coding one. 20 codons were included for each point, with a shift of 10 codons to the next point. The probabilities are calcualted by the method of Staden and McLachlan (23) and described in the text.

size allows it to be found as easily as prokaryotic genes. The other three exons are small and much more difficult to identify. They are included in the ORFs of over 50 codons, but are not special within that collection of 30 (besides ORF O). The simple analysis of correlation to base/position preferences indicates that ORFs T and D should also contain coding regions, which they do. This did not, however, find the exon in ORF Z and left four other non-coding ORFs within the range of normal indeterminacy. The method of Fickett largely eliminates the region 3' to ORF O as being coding, and a short region 5' of it. Also the very 5' end of the sequence. Most of the rest of it is scattered between non-coding and indeterminate. The ORF D region shows up, though its small size makes the identification less reliable. ORF Z is still not found, and no evidence is found to support ORF T. The codon usage method also points out ORFs O and D, but misses T and Z and has some false positives. However, those false positives are negated by the analyses of *Figure 15* and *16*. That is, the ORFs that correspond to the high points in *Figure 17* are shown to be uncorrelated in *Figure 15* and sometimes indicated non-coding in *Figure 16*.

The main limitation to the effectiveness of this approach is the resolution. It is usually required that one take an average codon usage over a region several codons long.

Even abundant proteins occasionally use a rare codon, just not very often. That means that to be reliable one has to look at several codons at a time. This can be a problem when trying to pinpoint the initiation site amongst several possibilities, or if one needs to know the splice junctions precisely. All of the small exons are hard to identify and their end points, even when they can be seen in the analyses, are not precisely identified.

5.3 **Combined methods**

The data from all these analyses can be combined to eliminate many of the possible gene arrangements from consideration, and to rank the remaining possibilities. It is safest to start with the most reliable predictions and add to them the more speculative alternatives. ORF O is found by all the content criteria quite strongly and is almost certainly a coding region. On the 3′ side there is nothing strong detected by Fickett's method (*Figure 16*) or by the correlation of *Figure 15*. The Staden analysis shows a potential for coding in frame zero (*Figure 17*), but ORF O has no strong donor sites (*Figure 12*) so the most likely possibility is that it terminates the coding region.

One possibility is that the entire gene is within ORF O. There are several strong initiation sites (*Figure 10*), all of them in the correct reading frame (data not shown). It also has several good acceptor sites (*Figure 12*). If we look upstream there are possible coding regions that could be connected as exons. ORF D is found by all the analyses of *Figures 15 − 17*. It also has a strong donor site and so could be connected to ORF O, or some other exon between. Of all the possible acceptor sites in ORFs O, only the two at positions 3003 and 3417 are in the proper reading frame to be translated with ORF D.

Since nothing between ORF D and ORF O is particularly strong by any of the content analyses, the most likely possibility would be that they are spliced together at one of those acceptor sites. This answer would be partially correct. Position 3003 is the acceptor site for ORF O, and 2312 is the donor site for ORF D, but they are connected to each other with the exon from ORF Z in between. None of the search by content methods evaluates ORF Z high enough to expect it to contain an exon. Only the strong acceptor site found in *Figure 12*, in which the frame is correct for connecting ORF Z to ORF D, might make one choose this correct alignment. Even if it were chosen, there is no good donor site with ORF Z with which to connect it to ORF D. The real arrangement of exons in this region is not among the most likely possibilities from any of the analyses.

None of the potential initiation sites (*Figure 10*) in ORF D is in the proper frame. There is only one acceptor site found 5′ to the strong donor site, but it is only 12 codons upstream. This is shorter than most exons (20). In fact, the real acceptor site is not found by the signal search matrix.

Of the ORFs 5′ to D, c shows up strongly by codon bias (*Figure 17*), but is negatively correlated with base/position preference (*Figure 15*). ORF T has a strong base/position correlation and two high points in the codon bias analysis. In addition, ORF T has two strong initiation signal sites, both in the correct frame, and a strong donor site. None of the other ORFs in this region have strong indicators from any of the analyses. In this case the most likely prediction of ORF T being spliced to ORF D would be correct, although the acceptor site in D is not identified.

5.4 Evaluation

The tubulin gene example has highlighted the strengths and weaknesses currently inherent in locating eukaryotic genes. Long exons are usually easily identified by a number of search by content methods. Genes that are made entirely of long exons can be found nearly as reliably as in prokaryotes. Short exons present more problems. They can sometimes be identified by content methods, but are also often missed. Genes that contain short exons are not reliably distinguished from the background of non-coding sequences. Signal searches, for both initiation sites and splice junction sites, often identify the correct sites, but the number of false positives is usually substantial. More information is required to find the sites than is contained in the matrix, such as the 5' end of the mRNA and the secondary structures of the introns.

While the combined analysis has not predicted the structure of the gene exactly, it has greatly reduced the number of possibilities. Given the number of compatible ways in which the 31 ORFs could have been joined, and the number of potential start sites and splice junction sites, the remaining likely candidates are an enormous reduction. Much of the final stages of this analysis has been done 'by hand', that is, by sorting through the computer analyses and finding likely, compatible options. The enumeration of alternatives can be automated, but there is currently no substitute for experience in choosing amongst the alternatives. This is an area ripe for expert systems. As more is learned about splice sites and promoters, and as the statistics of codon usage and variation get better, improved predictions will be made. In the meantime, computer analyses can usually get one close to the correct answer in locating eukaryotic genes, but solutions are not completely reliable.

6. AVAILABILITY OF THE PROGRAMS USED

With the description of each method used in this chapter, I have indicated the original source. In all cases, programs can be obtained by writing to that source. In addition, there are some packages that have incorporated all, or nearly all, of the methods described. I have used the Delila package (40,41) of programs for manipulating and analysing DNA sequences. It now contains all of the methods used in this chapter. Information about obtaining a copy of the entire system, or selected programs, can be had by writing to me. The package of programs by Rodger Staden (14,42) contains nearly all of the methods described. He introduced many of the search by content algorithms and has also added new ideas to the matrix method of searching for sites. His package also includes some convenient graphics routines for displaying the results of particular searches. The package of programs available from the University of Wisconsin (24) contains many of the methods described, and also has some excellent graphics routines. Some other academic sources, and now many commercial sources, provide some or all of these methods in available packages (see Chapter 3).

7. CONCLUSIONS

Search by signal and search by content methods of finding coding regions each have their advantages. We know enough about *E. coli* translation initiation sites to find them fairly reliably. The fact that functional proteins are long, and therefore have long ORFs, is alone quite indicative of coding regions. Coupled to analyses based on base/position preferences and codon usage, the search by content method is nearly infallible. In total, the methods for finding coding regions in prokaryotes, at least *E. coli*, are very reliable.

In eukaryotes the problem is much harder. The initiation site signal is coupled to the generation of 5′ ends, the mechanism of which is not known in general. Furthermore, long ORFs are not a reliable indicator because the gene may be broken into short exons, and some long ORFs may not even be present on the mRNA. Signal searching for the splice junction sites can help, but is not yet very reliable. As shown in the example, there are cases that are clearly coding regions, and there are known coding regions that are not found by any of these methods. Nonetheless, current predictive methods contribute significantly to the identification of coding regions. Many long ORFs can be eliminated by statistical analyses. The number of potential start and splice signals can be limited by matrix searches. In the end, the number of compatible alternatives is enormously reduced from the number seen prior to the analyses.

8. REFERENCES

1. Kozak,M. (1983) *Microbiol. Rev.*, **47**, 1.
2. Stormo,G.D. (1986) In *Maximizing Gene Expression*. Reznikoff,W. and Gold,L. (eds), Benjamin-Cummings Publishing Company, Inc., p. 195.
3. Breathnach,R. and Chambon,P. (1981) *Annu. Rev. Biochem.*, **50**, 349.
4. Kozak,M. (1984) *Nucleic Acids Res.*, **12**, 857.
5. Gren,E.J. (1984) *Biochimie*, **66**, 1.
6. Stormo,G.D., Schneider,T.D. and Gold,L. (1982) *Nucleic Acids Res.*, **10**, 2971.
7. Hawley,D.K. and McClure,W.R. (1983) *Nucleic Acids Res.*, **11**, 2237.
8. Harr,R., Haggstrom,M. and Gustafsson,P. (1983) *Nucleic Acids Res.*, **11**, 2643.
9. Staden,R. (1984) *Nucleic Acids Res.*, **12**, 505.
10. Mulligan,M.E., Hawley,D.K. and McClure,W.R. (1984) *Nucleic Acids Res.*, **12**, 789.
11. Stormo,G.D., Schneider,T.D. and Gold,L. (1986) *Nucleic Acids Res.*, **14**, 6661.
12. Brendel,V. and Trifonov,E.N. (1984) *Nucleic Acids Res.*, **12**, 4411.
13. Schneider,T.D., Stormo,G.D., Gold,L. and Ehrenfeucht,A. (1986) *J. Mol. Biol.*, **188**, 415.
14. Staden,R. (1985) In *Genetic Engineering: Principles and Methods*, Setlow,J.K. and Hollaender,A. (eds), Vol. **7**, p. 67.
15. Staden,R. (1984) *Nucleic Acids Res.*, **12**, 551.
16. Savegeau,M.A. (1986) *Proc. Natl. Acad. Sci. USA*, **83**, 1198.
17. Dayhoff,M.O. (1978) *Atlas of Protein Sequence and Structure*, Vol. **5**, supplement 3.
18. Ikemura,T. (1985) *Mol. Biol. Evol.*, **2**, 13.
19. Grantham,R., Gautier,C., Gouy,M., Mercier,R. and Pave,A. (1980) *Nucleic Acids Res.*, **8**, r49.
20. Naora,H. and Deacon,N.J. (1982) *Proc. Natl. Acad. Sci. USA*, **79**, 6196.
21. Shepard,J.C.W. (1981) *Proc. Natl. Acad. Sci. USA*, **78**, 1596.
22. Fickett,J.W. (1982) *Nucleic Acids Res.*, **10**, 5303.
23. Staden,R. and McLachlan,A.D. (1982) *Nucleic Acids Res.*, **10**, 141.
24. Gribskov,M., Devereux,J. and Burgess,R.R. (1984) *Nucleic Acids Res.*, **12**, 539.
25. Shine,J. and Dalgarno,L. (1974) *Proc. Natl. Acad. Sci. USA*, **71**, 1342.
26. Stormo,G.D., Schneider,T.D., Gold,L. and Ehrenfeucht,A. (1982) *Nucleic Acids Res.*, **10**, 2997.
27. Childs,J., Villanueba,K., Barrick,D., Schneider,T.D., Stormo,G.D., Gold,L., Leitner,M. and Caruthers, M. (1985) In *Sequence Specificity in Transcription and Translation*, Liss,A.R. (ed.), p. 341.
28. Hager,P.W. and Rabinowitz,J.C. (1985) In *The Molecular Biology of the Bacilli*, Vol. **2**, Dubnau,D. (ed.), p. 1.
29. Locus ECOUNC in GenBank release 40.0, and references therein.
30. Kolter,R. and Yanofsky,C. (1982) *Annu. Rev. Genet.*, **16**, 113.
31. Nomura,M., Gourse,R. and Baughman,G. (1984) *Annu. Rev. Biochem.*, **53**, 75.
32. Locus ECORPLRPO in GenBank release 40.0, and references therein.
33. Zitomer,R.S., Walthall,D.A., Rymond,B.C. and Hollenberg,C.P. (1984) *Mol. and Cell. Biol.*, **4**, 1191.
34. Kozak,M. (1986) *Cell*, **44**, 283.
35. Kozak,M. (1981) *Curr. Top. Microbiol. Immunol.*, **93**, 81.
36. Lee,M.G.-S., Lewis,S.A., Wilde,C.D. and Cowan,N.J. (1983) *Cell*, **33**, 477.

37. Mount,S.M. (1982) *Nucleic Acids Res.*, **10**, 459.
38. Langford,C.J., Klinz,F.-J., Donath,C. and Gallwitz,D. (1984) *Cell*, **36**, 645.
39. Loci HUMACTB, HUMADA and HUMAPOA1 in GenBank release 40.0, and references therein.
40. Schneider,T.D., Stormo,G.D., Haemer,J.S. and Gold,L. (1982) *Nucleic Acids Res.*, **10**, 3013.
41. Schneider,T.D., Stormo,G.D., Yarus,M.A. and Gold,L. (1984) *Nucleic Acids Res.*, **12**, 129.
42. Staden,R. (1986) *Nucleic Acids Res.*, **14**, 217.

CHAPTER 11

Secondary structure prediction of RNA

MANOLO GOUY

1. INTRODUCTION

Under natural conditions an RNA molecule will twist and bend and the bases will interact, folding the chain into local helices separated by single-stranded regions. The precise two-dimensional folding pattern the molecule adopts is called its *secondary structure*.

Secondary structure is involved in a variety of important biological phenomena. Several hundred sequenced tRNA molecules have a common folding potential: the well known cloverleaf. This folding is one of the fundamental structural motifs that characterize tRNA molecules. A common three-dimensional L-shaped conformation is superimposed over the cloverleaf. Secondary structure of rRNA molecules is known to be important in both self-assembly and functioning of the ribosome. Messenger RNA (or pre-mRNA) secondary structure is involved in translation initiation, intron splicing, transcriptional pauses and regulation by transcriptional attenuation of bacterial operons. Therefore secondary structure is an essential element of structure – function relationships in RNA molecules.

Numerous authors have devised computer tools and algorithms that provide help in various aspects of secondary structure analysis. We describe in this chapter the use of these methods: stability computation (Section 2); secondary structure prediction algorithms (Sections 4 and 5); stable local foldings (Section 6); computer-guided model building (Section 7); model drawings (Section 8).

Not all the programs mentioned here were accessible to the author for testing because some are device-dependent, or written in rarer programming languages or not widely distributed. Therefore not all methods could be described in equal detail.

Secondary structure prediction methods are far from perfect. Different algorithms often yield radically divergent predictions when applied to the same molecule. For this reason the algorithms are described in detail so that readers may locate the origin of differences in program behaviour (Section 3).

2. THE BASICS OF SECONDARY STRUCTURE MODELLING AND STABILITY COMPUTATION

2.1 Elementary motifs of secondary structure models

Any secondary structure that an RNA molecule may adopt is a combination of several substructures each matching one of a small number of elementary motifs. *Figure 1* shows an imaginary molecule folded in a structure containing each of the elementary motifs that are known to occur in natural molecules.

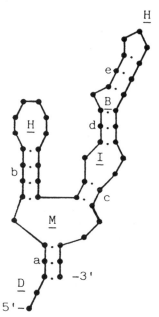

Figure 1. Nomenclature of elementary secondary structure components. Large dots indicate nucleotides, small dots base pairs. H, hairpin loop; B, bulge loop; I, internal loop; M, mutliple loop; D, dangling end. Helices are labelled a−e. A multiple loop can be closed as shown here or open as it would be if helix 'a' were absent.

A folded RNA molecule contains single-stranded (ss) and double-stranded (ds) parts. In ds regions, hereafter called *helices*, the molecule folds into a double helix in which nucleotides pair and adjacent pairs stack one upon the other. Helices can be as short as 2 bp. Most base pairs follow the classical Watson−Crick G.C or A.U rule, but other kinds of pairing may occur. In particular, the so-called 'wobble' G.U pair is frequently found in helices, although not all authors allow this pair to form at helix ends (see Section 2.2). Moreover, in Ninio's view of secondary structure, any base pair may occur but at the expense of varying degrees of helix destabilization and at a maximum of one per helix. Non-standard base pairing (*odd pairs*) has been found in a number of artificially synthesized RNA molecules (reviewed in ref. 1) and in several natural molecules (see for instance the consensus 5S rRNA folding in ref. 2).

Topologically speaking, ss regions fall into five classes, depicted graphically in *Figure 1. Hairpin loops* are believed to have a minimal length of three nucleotides. The other classes are called *bulge, internal* or *multiple loops* and *dangling ends*. Other motifs (*Figure 2*) have nevertheless been observed in some rare cases (3). *Knotted structures* are not allowed by most methods of secondary structure prediction or representation. Indeed, all methods presented here do not allow them.

The above terminology is not universal. Other authors would call a helix a stem, stacking region, paired region or segment. Likewise, multiple loops are sometimes qualified as bifurcated or multi-branched.

2.2 **Energy models for the computation of folding stabilities**

In a secondary structure each part of the molecule in one of the previously defined

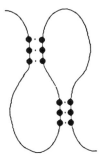

Figure 2. A knotted structure, never allowed in the secondary structure prediction methods described here.

elementary motifs contributes to the total structural stability. Helices contribute to increased stability, while ss regions tend to destabilize neighbouring helices. Thermo-dynamically, structural stability is measured by the loss in free energy (ΔG) in kcal/mol when going from a totally linear molecule to a given folding with all its helices and ss regions. Note that, because we are dealing with free energy losses, stabilizing contributions have negative values. Short RNA molecules, in agreement with the laws of equilibrium thermodynamics, naturally fold into the structure of minimum free energy, the optimal folding. Long sequences have such a large number of potential foldings that it is not clear whether they adopt the absolute optimal structure or some local minimum. A fundamental feature of folding thermodynamics is that, according to the Tinoco—Uhlenbeck postulate, *the overall free energy loss equals the sum of independent contributions of each elementary motif in a structure.* A formal presentation of the additive nature of substructure contributions and its consequences for secondary structure prediction algorithms has been given by Sankoff *et al.* (4). They show that when a structure is optimal for a molecule, substructures induced on both sides of each helix are optimal as well. Intuitively this is very simple: if a better substructure could exist on one side of a helix, it would confer a better overall stability to the complete molecule since free energies are additive.

Therefore, a set of free energy values describing the contribution of each elementary motif will allow one to compute the stability of any secondary structure. Such a data set will be called an *energy model*. A large number of experiments with chemically synthesized short RNA polynucleotides have been carried out (mainly by I.Tinoco, O.C.Uhlenbeck, D.M.Crothers and co-workers) in order to obtain energy values for this idealized model (reviewed in refs 1 and 5). However it has not been possible to obtain experimental values for each elementary motif of any molecule. The contribution of a ss region depends not only on its length, but also on its sequence and the nature of base pairs at its ends. This has two consequences. Firstly, no data set can be considered perfect: each relies on a number of assumptions and simplifications. Secondly not all workers use the same data set: two are commonly used and each will be fully presented here. Note also that none of these energy sets contains any data about knotted structure contributions.

2.2.1 *Salser's energy model*

The above-mentioned series of experiments analyzing the energy parameters of RNA

```
5' -  G -- C -- G -- A -- C -- U -- G
     -4.3 -1.3 -0.3 -2.1 -2.1 -2.1
      C -- G -- U -- U -- G -- A -- C  - 5'
```

Figure 3. Computation of a helix free energy contribution. Values for stacked base pairs are taken from *Table 1*. Calculating from either end yields the same result.

Table 1. Base pairing and stacking free energies.

| | | SALSER's energy model (I) (kcal/mol) 3'-side base pair | | | | |
		C.G	G.C	U.A	A.U	G.U
5'-side base pair	C.G	−4.8	−3.0	−2.1	−2.1	−1.3
	G.C	−4.3	−4.8	−2.1	−2.1	−1.3
	U.A	−2.1	−2.1	−1.2	−1.8	−0.3
	A.U	−2.1	−2.1	−1.8	−1.2	−0.3
	G.U	−1.3	−1.3	−0.3	−0.3	−0.3

G.U and U.G. pairs are not distinguished.
Odd pairs and helix-terminal G.U pairs are not allowed.

chain folding has been clearly compiled by Salser (5). The resultant data set, hereafter called Salser's energy model, has then been used for all secondary structure prediction methods except that of Ninio (1). Experimental work has continued since Salser's compilation (6,7) but no new synthesis of free energy values has been published, so that Salser's data are still widely employed.

(i) *Helix contribution.* The total free energy contribution of a helix is obtained by adding contributions of neighbouring stacked base pairs as shown in *Figure 3*. *Table 1* lists all necessary values taking into account both the pairing and stacking energies involved (5). This model is totally additive with respect to helices: the stability of a helix can be computed recursively by adding the contribution of a new base pair to that of the remaining helix fraction. This is an absolute requirement of recursive prediction methods (see Section 3.2.1).

(ii) *Single-stranded regions.* *Table 2* summarizes values used to compute the free energy contributions of ss regions according to their topology, length and closing base pairs (5). Since no experimental data exists for the computation of multiple loop contributions, the model assimilates them as internal loops. As stated above, the overall free energy of a complete secondary structure is computed by adding the contributions of all elementary motifs derived from *Tables 1* and *2*.

2.2.2 Ninio's energy model

Because experimental results do not cover all molecular substructures necessary for computing folding stabilities, Ninio and his co-workers wanted to estimate the missing items for an empirical energy model. This work has been done by a 'trial and error' process. They attempted to modify Salser's data in reasonable ways in order to find a data set able to predict two classes of known secondary structures, that of tRNA and 5S rRNA molecules (1,8) as these two families of molecules each have well established secondary structures (2,9,10). The resulting model is more sophisticated than Salser's

Table 2. Free energy values for the computation of contributions of ss regions.

Type of loop	SALSER's energy model (II) (kcal/mol) Loop length (in nucleotides)																
	1	2	3	4	5	6	7	8	9	10	12	14	16	18	20	25	30
H(G.C)	–	–	8.40	5.90	4.10	4.30	4.50	4.60	4.80	4.89	5.03	5.16	5.27	5.37	5.46	5.65	5.88
H(A.U)	–	–	8.00	7.50	6.90	6.40	6.60	6.80	6.90	7.00	7.13	7.25	7.36	7.46	7.55	7.74	7.90
I(G + G)	–	0.10	0.90	1.60	2.10	2.50	2.62	2.72	2.82	2.90	3.05	3.18	3.29	3.38	3.47	3.66	3.85
I(G + A)	–	0.95	1.75	2.45	2.95	3.35	3.47	3.57	3.67	3.75	3.90	4.03	4.14	4.23	4.32	4.51	4.70
I(A + A)	–	1.80	2.60	3.30	3.80	4.20	4.32	4.42	4.51	4.60	4.75	4.88	4.99	5.08	5.17	5.36	5.55
Bulge	2.80	3.90	4.45	5.00	5.15	5.30	5.45	5.60	5.69	5.78	5.93	6.05	6.16	6.26	6.35	6.54	6.70

H(G.C): hairpin loop closed by a G.C pair; H(A.U): hairpin loop closed by an A.U pair; I(G + G): internal loop closed by two G.C pairs; I(G + A): internal loop closed by a G.C and an A.U pair; I(A + A): internal loop closed by two A.U pairs.

Linear interpolation is to be used for unspecified loop lengths between 11 and 29.

In Zuker's program these values are rounded to two significant digits.

Bulge loops: add to the value given above the stacking energy of the two base pairs encompassing the bulge taken from *Table 1*.

Multiple loops; treat as internal loops using one of the three above sets of values according to whether the loop is closed by G.Cs only, A.Us only, or both.

Dangling ends: have a null contribution.

Table 3. Base pairing and stacking free energies.

	NINIO's energy model (I)												(kcal/mol)
	<internal>						<terminal>						
	C.G	G.C	U.A	A.U	U.G	G.U	U.G	G.U	G.G	U.U	C.C / C.A	A.A / A.G	U.C
C.G	−4.1	−3.1	−3.1	−3.1	−1.3	−0.8	−1.4	−0.9	−1.1	−0.9	−0.5	−0.1	+0.9
G.C	−3.6	−4.1	−3.1	−3.1	−1.8	−1.8	−2.1	−1.6	−1.1	−0.9	−0.5	−0.1	+0.9
U.A	−3.1	−3.1	−2.0	−2.0	−0.3	−0.3	−1.3	−0.3	−0.5	+0.5	+0.5	+0.9	+0.7
A.U	−3.1	−3.1	−2.0	−2.0	−1.3	−0.8	−1.3	−0.6	−0.5	+0.5	+0.5	+0.9	+0.7
U.G	−1.8	−0.8	−0.8	−0.3	+0.5	−0.3	+0.7	+1.2	+1.2	+0.7	+1.2	+1.2	+0.7
G.U	−1.8	−1.3	−1.3	−0.3	+0.1	+0.5	−0.3	+0.2	+1.2	+0.7	+1.2	+1.2	+0.7
U.G	−1.6	−0.9	−0.6	−0.3	+0.2	+1.2							+0.7
G.U	−2.1	−1.4	−1.3	−1.3	−0.3	+0.7							
G.G	−1.1	−1.1	−0.5	−0.5	+1.2	+1.2							
U.U	−0.9	−0.9	+0.5	+0.5	+0.7	+1.2							
C.C,C.A	−0.5	−0.5	+0.5	+0.5	+1.2	+0.7							
A.A,A.G	−0.1	−0.1	+0.9	+0.9	+1.2								
U.C	+0.9	+0.9	+0.7	+0.7	+0.7								

The first base pair is taken in the row, the second in the column. For odd pairs, and in the sequel X.Y and Y.X are not distinguished.
Add 1.0 kcal for each helix; a helix less stable than one of the two helices obtained by suppressing either terminal base pair is discarded.
Helices of length two: must contain either two G.Cs or one G.C and one A.U; subtract 0.4 kcal for every G.C.
Helices of length three or more: subtract 0.8 kcal for every terminal G.C stacked over a G.C, A.U or G.U pair, add 0.7 kcal for every terminal A.U stacked over a G.C.
Odd pairs: only one per helix, cannot be at helix ends, must not adjoin a terminal G.U nor be sandwiched between two G.Us; if there are four or more pairs on both sides of the odd pair, consider rather two helices separated by an internal loop; if not, count the numbers I and J of G.Cs or A.Us on either sides, let $K = \min(I,J)$, add $(K − 2) \times 0.5$ kcal.
Subtract 1.0 kcal for each of these two cases (Y = pyrimidine):

5′-UAG or 5′-UUG
  AGY-5′   GUC-5′

Table 4. Free energy values for the contribution of ss regions.

Type of loop	NINIO's energy model (II) (kcal/mol) Loop length (in nucleotides)											
	0	*1*	*2*	*3*	*4*	*5*	*6*	*7*	*8*	*9*	*10*	*11*
Hairpin	–	–	–	8.0	5.0	4.4	4.1	4.0	4.0	4.1	4.2	4.5
Internal	–	–	3.2	3.2	3.2	3.3	3.4	3.5	3.7			
Bulge	–	2.2	2.9	3.6	4.4	5.2	6.0	6.9	7.8			
Multiple	1.6	1.1	1.1	1.2	1.5.	2.0	2.5	3.0	3.5	4.0		
Dangling				0.2 kcal/mol per nucleotide								

Each side of a multiple loop contributes separately according to its length, whereas for internal loops, the accumulated lengths of their two sides is used.

Hairpin loops of length three: subtract 2.5 kcal for one U or 3.5 kcal for two or three Us in loop.

Long hairpin loops: beyond the length $N = 11$, compute the loop contribution as: $10 \times \log(N) - 6.0$ kcal (use logarithms to the base ten).

Long internal loops: beyond $N = 8$ residues, apply: $10 \times \log(N) - 5.5$ kcal.

Asymmetrical internal loops: let $N1$ and $N2$ be the lengths of the two loop branches, let $N = N1 - N2$, $M = \min(5, N1, N2)$, $f(1) = 0.7$, $f(2) = 0.6$, $f(3) = 0.4$, $f(4) = 0.2$, $f(5) = 0.1$, add $\min[6, N \times f(M)]$.

G.A. pairs in internal loops: subtract 1.5 kcal for each terminal G.A. pair in loop (do not count the same base twice).

Long bulge loops: beyond $N = 8$ residues, apply: $10 \times \log(N) - 1.2$ kcal.

Pyrimidine-facing bulge loops: if the two helix-terminal nucleotides facing the bulge are both pyrimidines, and if the loop length N is 8 or less, subtract $0.2 \times (9 - N)$ kcal.

Multiple loops: add 1.5 kcal for closing the loop, when it is closed.

Long multiple loops: beyond $N = 9$ residues, apply: $10 \times \log(N) - 5.5$ kcal.

Very short multiple loops: if the cumulated length N of all branches is two or less, add $1.5 \times (3 - N)$ kcal.

because it not only incorporates experimentally determined ΔG values but also includes a number of features found necessary to obtain good structural predictability for known structures.

(i) *Helix contribution. Table 3* shows the pairing and stacking energies of classical, G.U and odd base pairs used to compute helix contributions. Where helices containing more than two odd pairs occur, the model preferentially considers several helices separated by internal loops. Therefore the limitation of one odd pair per helix is not restrictive in practice. As before, the energies of adjacent base pairs are accumulated taking care to distinguish internal from helix-terminal G.U pairs. However, non-additive rules are also used. These non-additive features do not permit helix stabilities to be computed base pair by base pair and hence prevent the use of this model with recursive prediction algorithms.

(ii) *Single-stranded regions. Table 4* indicates the rules for computation of single-stranded contributions. While in Salser's model single-stranded contributions depend on the nature of loop-closing base pairs, no such dependence appears in *Table 4*. In Ninio's model a similar dependence is allowed for by the rules in *Table 3* through helix stability corrections according to the identity of terminal base pairs.

2.2.3 *An example of overall free energy computation*

Table 5 details the usage of Salser's and Ninio's energy models on the same short molecule.

Table 5. Usage of Salser's and Ninio's models for the same molecular folding.

Motif	Ninio's model	Salser's model
GCG		
CGC (a)	−7.3	−7.3
GUGA		
CGCU (b)	−4.8	−4.7
GC		
CG (c)	−3.3	−4.3
ACG		
UGC (d)	−5.3	−5.1
GCC		
CGG (e)	−8.3	−9.1
Internal loop	+2.3	+1.75
6-base hairpin	+4.1	+6.4
3-base hairpin	+4.5	+8.4
Bulge loop	+0.6	−2.0
Multiple loop	+5.3	+0.9
Dangling end	+0.4	0
Overall free energy	−11.8	−15.05

Both models are applied to the sequence:
5'-UUGCGAGUGACAUAGUUCGCGCGACGAGCCUAUGGCCGUUAGCAUCGC-3' folded in as *Figure 1*.
All values are in kcal/mol. The folding used here is not necessarily optimal for this molecule.

3. ALGORITHMS FOR THE PREDICTION OF SECONDARY STRUCTURE IN RNA MOLECULES

Secondary structure prediction relies on a fundamental assumption: two-dimensional folding is primarily determined by the sequence itself and only marginally by the molecular environment. In particular, the three-dimensional structure is built from the 2-D structure. The difficulty of secondary structure prediction is therefore 2-fold: the choice of the energy model (Section 2) and the design of an efficient algorithm. A classification of published algorithms reveals three main classes.

(i) *Combinatorial algorithms.* These work in two steps. First they compute the list of all potential helices in the molecule. Second, the algorithm evaluates all possible helix combinations and retains one or several that yield molecular foldings of minimal free energy. This and the following classes of algorithms are guaranteed to find the optimal folding associated with the chosen energy model.

(ii) *Recursive algorithms.* These derive the best structure of a given fragment of an RNA molecule from the best structures of smaller fragments included in it. Starting with all pentanucleotide fragments, and extending them one nucleotide at a time in both directions, these methods ultimately furnish an optimal folding of the whole molecule. (There may be more than one structure of minimum energy.)

(iii) *Heuristic algorithms.* These have been created in order to overcome the limits in sequence size and computation duration from which other methods suffer. Heuristic algorithms make several *a priori* assumptions on the probable nature of the optimal folding; consequently, they do not find with certainty the absolute free energy minimum. Moreover, they have not been extensively tested on the prediction of known structures, nor are they widely used. Therefore these algorithms are only briefly alluded to here.

Combinatorial and recursive algorithms applied to the same molecule often turn out to predict widely different foldings. In order to identify the origin of such behavioural differences, it is necessary to give further details of the algorithms.

3.1 Combinatorial algorithms

3.1.1 *Precursors*

One of the first methods for secondary structure prediction was published by Pipas and McMahon (11) and applied to a set of tRNA sequences. It was a crude combinatorial method: all helix combinations were evaluated and the best foldings retained. Out of the 62 studied tRNAs, only 32 were found with a cloverleaf as the best structure.

Since evaluation of all helix combinations is much too time-consuming for molecules longer than tRNAs, a new approach was taken by Studnicka *et al.* (12). Using an elegant mathematical formalization of the problem, the number of evaluated helix combinations was reduced. The algorithm has a computation time depending on the fifth power of sequence length. Salser's energy model with a null contribution for multiple loops was used. The program has been successfully applied to prediction of *Anacystis nidulans* 5S rRNA secondary structure. With the appearance of recursive algorithms (Section 3.2), that are very efficient when used with Salser's model, this algorithm has found little further use.

3.1.2 *Ninio's algorithm for secondary structure prediction*

The only combinatorial method now in use was developed by Ninio, Dumas, Papanicolaou and Gouy (1,8,13,14). As a first step, the program establishes the list of all potential helices in the molecule. The list is not restricted to helices of maximum length (those which cannot be extended because they are bounded by non-pairing nucleotides). Indeed a shorter helix may yield a better local structure because it allows the formation of another helix in its vicinity. Thus, potential helices include those of maximal length plus all sub-helices down to length two. In a sequence N nucleotides long, the number, L, of potential helices grows approximately as N^2, yet the exact growth is highly sequence dependent. A molecular folding can be seen as a combination of a small number of potential helices, ss regions being formed by the nucleotides that do not belong to any helix in the combination. Therefore, the search for the optimal secondary structure requires evaluation of all helix combinations. There are 2^L combinations of potential helices, although not all of these correspond to feasible structures (structures cannot be knotted, a nucleotide cannot belong to two helices at a time). Even so, the number of potential combinations is very large and grows extremely rapidly with sequence length.

Helix combinations are generated in Ninio's program by a *tree search* procedure. A first helix, taken from the list of potential helices, initiates (depth one) the combination, then another helix, compatible with the first one, is added (depth two) to the combination. This process is continued with increasing depth. Suppose now that helix *h* has been chosen at depth i and that all further combination extensions have been generated, helix h is then removed, replaced by helix h' compatible with all helices chosen up to depth $i - 1$, and the process is reiterated. When all candidate helices at depth i have been used, the helix chosen at depth $i - 1$ is removed and replaced. Thus, at any time the program is considering one helix combination, called the *current combination*, and scans all potential combinations.

A number of methods are used to make the tree search efficient. These methods have been described in detail (8,13,14) and are summarized here. First, the program attributes to each potential helix:

(i)   a helix stability;
(ii)   a helix compatibility set (the list of all other helices which may occur with it).

The program also maintains three quantities associated with the current combination:

(i)   its exact free energy, that is the energy of the corresponding secondary structure taking into account all ss regions;
(ii)   its approximate energy, that is the accumulated stabilities of all its helices;
(iii)   its compatibility set as follows. Every time a new helix is added to the current combination, the program updates the compatibility set for the new combination by suppressing from it all helices incompatible with that newly added. In this way the compatibility set of the current combination contains exactly the candidate helices for further combinations.

Next, helices are placed into groups of mutually incompatible helices called *incompatibility islets*. A structure may contain at most one helix per islet. In each islet there is at most one helix of maximal stability that pertains to the compatibility set of the current combination. Accumulating the maximal stabilities over islets one obtains an upper bound of the best approximate stability that can result from extending the current combination. Two cases can occur. If the upper bound is less than the approximate stability of the least stable current best structure minus a *free energy interval*, the current combination is not extended further so that branches of the tree do not require evaluation to full depth. The free energy interval allows for the distortions between differences in approximate stabilities and exact free energies. In the opposite case, the exact free energy of the combination is computed and retained if it is better than the least stable current best structure, and the combination is further extended.

Finally, any helix divides the molecule into two parts: the loop between the two helix strands on one side, and the two regions extending from the helix to the ends of the molecule on the other (in *Figure 1*, helix c divides the molecule into the part that covers the internal loop, the bulge and the hairpin on one side, and the rest of the molecule on the other). The additivity principle (Section 2.1) states that the best structures containing a given helix will be formed with the best structures in each of the two parts of the molecule. Therefore each time the current helix combination is initiated with one helix, the whole process explained above, islet grouping and combination extension, is independently repeated on the two resulting parts of the molecule. Finally, the best structures for each part are combined.

This algorithm can find the best structure corresponding to any energy model. But it can also, with virtually no loss in computation time, give any number of the next best structures.

In practical terms, Ninio's program is convenient for RNA molecules not longer than about 150 nucleotides when used under the most general conditions: odd pairs allowed and a minimum helix length of two base pairs. This is due partly to its combinatorial nature, and partly to the complexity of the underlying energy model. However, the program can be used in a less rigorous manner on longer molecules. A first run with odd pairing prohibited and a larger minimal helix length furnishes the major folding domains of the molecule. A second run in which some of the main first step helices are enforced (see Section 5.2) furnishes a refined structure prediction.

3.2 Recursive algorithms

A widely different class of secondary structure prediction algorithms is that of recursive methods (also called dynamic programming methods). Several authors have, independently in some cases, published this type of algorithm $(15-18)$ and the methods have been widely used. Although the early programs of Nussinov suffered from slight inconsistencies that sometimes resulted in inaccurate treatment of ss regions, particularly multiple loops (19), these authors eventually converged on nearly identical methods hereafter collectively named Zuker $-$ Nussinov $-$ Sankoff (ZNS) algorithm $(4,6,19-23)$.

3.2.1 *The Zuker $-$ Nussinov $-$ Sankoff algorithm*

The algorithm has been published in detail by Zuker and Stiegler (18) for the particular case where multiple loops are given no free energy contributions, and in a formal and general way by Sankoff *et al.* (4). It is presented here at a conceptual level.

Basically, the ZNS algorithm derives the best structure of a given fragment of the molecule from the best structures of smaller fragments included in it. The shortest fragments whose structures are evaluated are pentanucleotides because five nucleotides is the range of the folding with minimum length: a 3-base hairpin loop closed by 1 bp. More precisely, for all couples of nucleotides i,j $(i + 3 < j)$, the program computes two quantities:

(i) $W(i,j)$, the minimum free energy of all possible, closed or open, structures on sequence fragment (i,j).

(ii) $V(i,j)$, the minimum free energy of all possible structures on fragment (i,j) in which nucleotides i and j are paired with each other.

Here and below, if i and j cannot pair, $V(i,j)$ = infinity. Now the algorithm reduces to the computation of quantitites $V(i,j)$ and $W(i,j)$ for a given pair (i,j) from quantities $V(k,l)$ and $W(k,l)$ with $i \leq k < l \leq j$. As illustrated in *Figure 4* it is apparent that, whenever i and j can pair with each other, $V(i,j)$ is the minimum of three quantities $VH(i,j)$, $VI(i,j)$ and $VM(i,j)$ where:

$$VH(i,j) = \text{free energy of hairpin loop closed by base pair } i,j;$$

$$VI(i,j) = \text{minimum } [e(i,j,i',j') + V(i',j')]$$
$$i<i'<j'<j$$
$$i'.j' \text{ paired}$$

where: $e(i,j,i',j')$ = free energy contribution of the substructure extending from base pairs $i.j$ to $i'.j'$ (can be a bulge or internal loop or a 2-bp stack).

(1)   (2)   (3)

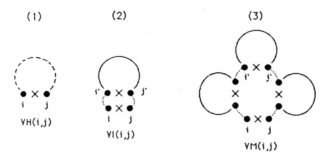

Figure 4. Computation of V(*i,j*) from previously computed V values. Dotted lines indicate ss regions, possibly of null length; solid lines may be of any folding; a cross between large dots indicates paired residues that appear in the V(*i,j*) computation. Base pair *i.j* can either (**1**) close a hairpin loop; (**2**) close an internal or bulge loop or extend a helix; (**3**) close a multiple loop (depicted here arbitrarily with four closing pairs).

$$VM(i,j) = \text{minimum } [e(M) + \Sigma \; V(i',j')]$$
$$i.j \text{ closes a} \qquad i'.j' \text{ closing pairs}$$
$$\text{multiple loop M} \qquad \text{of M except } i.j$$

where: e(M) = free energy contribution of multiple loop M.
  The W(*i,j*) recurrence equation is:

$$W(i,j) = \text{minimum} \begin{bmatrix} 0 \\ V(i,j) \\ \text{minimum } [W(i,h) + W(h+1,j)] \\ i<h<j-1 \end{bmatrix}$$

At each step, the molecular substructure which has the minimum free energy is recorded. Then backtracking through W and V matrices yields the final optimal molecular folding.

The ZNS algorithm, in the form just described, is able to use any energy model following the additivity principle (Section 2.2). Existing programs, though, do not have this generality. They have been written for Salser's energy model making use of several of its characteristics, particularly for multiple loops (18,20).

3.2.2 *Computation time dependence on sequence length*

The dependence of computation time on sequence length has been carefully studied by Sankoff *et al.* (4). Their results can be summarized in three points. First, in its most general form, this algorithm has an exponential time dependence on sequence length. This limits practical application to sequences shorter than about 200 nucleotides. Treatment of multiple loops is responsbile for the most time-consuming part of the algorithm. Second, these authors evaluated the effect of a simple form of multiple loop treatment in which the energy contribution of a loop depends only on the closing base pairs and on the number of ss residues but not on the loop topology. For instance, a multiple loop with three closing base pairs and six unpaired residues has the same contribution under the simple treatment wherever the six unpaired residues are located relative to the three base pairs. This yields a time dependence on the fourth power of molecule length. Third, similar simple treatment for all loops (multiple but also hairpin, bulge and internal) produces a time dependence on the third power. Sankoff *et al.* (4) consider that the simple form is acceptable for multiple loops only; this therefore limits practical usage of recursive algorithms to sequences of a few hundred bases. The Zuker and Stiegler program and Salser's energy model (*Table 2*) treat multiple loops in the simple form, that is according to total loop length but not to loop topology.

Zuker and Stiegler (18) and Nussinov and Jacobson (17) claim that the time dependence of their algorithms is on the third power. This probably results from a limit imposed on loop length in both programs (with a user-modifiable value of 30). These programs are therefore heuristic, because loops longer than 30 nucleotides are eliminated *a priori*. In practice such a threshold is quite acceptable.

Attempts to treat longer sequences have sometimes been made. Nussinov *et al.* (19) tried to predict the secondary structure of SV40 late precursor mRNA of length 2600 bases. After 3 months, they interrupted the computation which was then at the stage of studying subfragments of nucleotides in length 1790. Technical improvements at the programming level were proposed that speed computation but tend to make the program dependent on a particular computer architecture (6,22).

3.3 Differences between combinatorial and recursive methods

The main differences between recursive (ZNS) and combinatorial (Ninio's) algorithms are, from the theoretical viewpoint, 2-fold. Firstly, there is a more extensive usage in ZNS algorithms of the principle of free energy additivity (Section 2.2). Specifically, recursive algorithms require that helix contributions equal the sum of independent base pair contributions, while combinatorial algorithms use whole helix stabilities before adding them to ss region contributions. Secondly, recursive algorithms have been implemented to find one optimal structure rather than a few best foldings. On the other hand, they furnish, for the same work, the optimal folding of any fragment of the molecule being studied.

Obtaining several foldings is important because it allows one to assess whether widely different structures with similar free energies exist, or if the best structures are close variants of each other, or if the very best one is clearly energetically favoured with respect to its competitors. In the first case it would not be justifiable to choose the structure with lowest free energy because the underlying energy model is only approximate. Structural RNAs, especially those in the ribosome have a functional role which probably involves various conformational transitions (24,25). Such transitions are expected to occur between various structures of similar stabilities. Therefore, consideration of several foldings may also be a source of important indications about molecular function. Combinatorial methods are significantly slower than recursive methods, and limited to treatment of shorter molecules (~ 150 bases versus a few hundred).

Treatment of knotted structures (see *Figure 2*) also distinguishes the recursive and the combinatorial methods. Indeed, combinatorial algorithms may, at least theoretically, generate helix combinations giving a knotted structure to the molecule, while recursive algorithms are unable to take into account such conformations. Allowing knotted structures significantly increases the number of potential foldings to be considered. This would probably make the algorithm too slow for practical use.

Table 6. Percent success of energy models in predicting the correct folding of tRNA and 5S rRNA.

Sample	tRNA	5S RNA
Salser's model	26%	18%
Ninio's model	72%	68%
No. of sequences	200	100

Some examples of the application of the ZNS algorithm are to the 5′ end of SV40 late mRNA (19−21), to the *Escherichia coli* 16S rRNA (25,26) and to the *Tetrahymena* rRNA self-splicing intron (7). The work on 16S rRNA was done by dividing the molecule into four independent folding domains. Extensive use was also made of data on sensitivity to nuclease cleavage and to conservation of base pairings by comparison with 16S rRNA molecules from other species. Additional data were also used in the work on the self-splicing intron. For such long molecules, structure prediction based only on the nucleotide sequence is beyond the power of the ZNS methods (see Section 9).

Ninio's energy model and program were first applied to prediction of the 'cloverleaf' tRNA folding (1). In a refined form (that of Section 2.2.2), it has been applied to the analysis of folding of 100 5S rRNA molecules (8). The model has also been used to compute the overall stability of several molecular foldings: 5.8S rRNAs (27) and 5S rRNAs (24). *Table 6* compares predictions of tRNA and rRNA consensus secondary structures by Salser's and Ninio's models (8).

3.4 Overview of two heuristic methods

In an attempt to quickly fold large sequences, some heuristic methods have been proposed. Martinez's algorithm is based on the idea that molecular folding is a process of growth in which new helices form successively without modifying those previously formed (28). Martinez roughly estimates the probability that any candidate helix folds: stable helices have high folding probability, but among several helices of comparable stability, those whose formation give the molecule the least decrease in entropy are favoured. This algorithm has a time duration proportional to the square of sequence length, is written in C language for DEC VAX computers, and is available upon request. A similar folding process has also been considered by Stüber in his FALTUNG prediction program, also available upon request (29).

Yamamoto *et al.* (30) introduced statistical considerations in secondary structure prediction by estimating the probability of helix formation. In essence, high probabilites are given to helices that, when formed, leave most room for other helices to fold. The program, written in FORTRAN for VAX computers, has been applied to two precursor tRNA molecules and four U1 small nuclear RNAs.

4. USE OF THE ZUKER−STIEGLER SECONDARY STRUCTURE PREDICTION PROGRAM

The Zuker−Stiegler program has been chosen here as representative of ZNS algorithms because it is the most widely available secondary structure prediction program and is easily implemented on a variety of computers (18).

4.1 **Program distribution**

The Zuker–Stiegler program, called RNAFOLD and written in FORTRAN 77, is distributed upon request on magnetic tape by M.Zuker. Salser's data for secondary structure free energy computations and a user guide are also supplied. The hardware requirements are mainly in terms of main memory:

Molecule length (nucleotides)	100	200	500	L
Memory requirements (kbytes)	53	143	775	$23 + 3.L^2$

No special printing or plotting device is required because the predicted secondary structure models can be drawn by a line printer.

4.2 **Options**

The program provides a number of run-time options for greater flexibility.

4.2.1 *Selection of the sequence fragment to be folded*

The user may delimit a region of the molecule for folding, or excise a fragment from the molecule, possibly by a closed excision that replaces the fragment by three non-pairing bases forming a hairpin loop.

4.2.2 *Constraints on specific nucleotides to be paired or single-stranded*

Experimentally, secondary structure is usually investigated by measurements of nucleotide sensitivity to chemical attack or enzymatic cleavage. This results in data indicating specific nucleotides to be in either single-stranded or helical regions. Therefore the capability of a prediction program to take into account such data is very important. Two options serve this purpose. One forces one or several contiguous bases to pair. The other forces one or several nucleotides to be single-stranded. Note however that, due to the algorithm, the program is only slightly faster when using these options.

4.2.3 *Modification of the standard energy model*

Options allow one to modify by a constant factor the free energies attributed to base pair stacking, and bulge, internal, hairpin or multiple loops. Other options allow one to alter the maximum lengths of bulge or internal loops, or the minimum length of hairpin loops.

4.2.4 *Short range folding*

The distance between two paired nucleotides may have an upper limit, this restricts predicted structures to short range folding only. Moreover, closed multiple loops may be prohibited. The output is in this case a series of hairpins, bulges and internal loops (thus making RNAFOLD very similar to the SEQL program, see Section 6).

4.2.5 *Time-consuming runs*

When sequences above ~200 nucleotides are to be analysed the computation time may

```
     A sequence file may contain many sequences. One sequence file
added to the end of another yields a valid sequence file. Any number
of comment lines may precede each sequence, as long as the character
"(" never occurs in column one. As soon as a "(" occurs in the first
column of a line, that line is recognized as the format for the
sequence. The next line contains the number of nucleotides and the
sequence label. The format for this line is
     (I4,5X,50A1) .
Subsequent lines must contain the sequence according to the specified
format.
(60A1)
 120     Drosophila melanogaster 5S rRNA
GCCAACGACCAUACCACGCUGAAUACAUCGGUUCUCGUCCGAUCACCGAAAUUAAGCAGC
GUCGGGCGCGGUUAGUACUUAGAUGGGGGACCGCUUGGGAACACCGCGUGUUGUUGGCCU
```

Figure 5. Example of input to program RNAFOLD. *D. melanogaster* 5S rRNA sequence from (2).

become too great for interactive working. Users may employ an option that will restrict the program to doing the time-consuming computation of matrices V and W (Section 3.2.1) and writing the results to disc. This task may be run in 'batch' mode. Later, users may instruct the program to read the minimum energy matrices and do the final rapid interactive part of the task, that of backtracking through the matrices to obtain the optimal folding. The first part is called a 'save' run, the second a 'continuation' run.

4.2.6 *Output options*

Apart from giving a line printer drawing of the predicted best structure, the program is capable of writing results in a form suitable for treatment by two secondary structure drawing programs by Shapiro (31) and Feldmann (32) (see Section 8). Any combination of these three outputs may be required.

4.3 **Example**

An example of the sequence input format needed by the program RNAFOLD is given in *Figure 5* for the *Drosophila melanogaster* 5S rRNA sequence. An input file may contain several sequences. The program dialog generated by the computation of *D. melanogaster* 5S rRNA secondary structure is given in *Figure 6*. Bases are numbered in multiples of ten. Helices appear on adjacent lines while ss regions are written on lines shifted above or below the lines used for helices. Bulges and asymmetrical internal loops are filled with dashes on the shortest branch. Each multiple loop determines a two-space insert in the drawing (in *Figure 6* between A13 and C14) and each branch is plotted at successively lower positions vertically aligned with the insert. Dotted lines have been added to *Figure 6* for clarification of the multiple loop topology.

5. USE OF THE NINIO SECONDARY STRUCTURE PREDICTION PROGRAM

5.1 **Program distribution**

Ninio's program, CRUSOE, is distributed upon request on magnetic tape by the present author. A detailed installation and user guide is also furnished. CRUSOE, written in

```
$RUN RNAFOLD
ENTER ENERGY FILE NAME
SALSER
ENTER C FOR A CONTINUATION RUN, S FOR SAVE (FUTURE CONTINUATION)
AND <CR> FOR A REGULAR RUN.
<carriage return>
ENTER SEQUENCE FILE NAME
DROSO5S
Drosophila melanogaster 5S rRNA                          120 NUCLEOTIDES
OUTPUT TO TERMINAL? Y OR N
Y
CT FILE GENERATION? Y OR N
N
ENTER ENDPOINTS OF FRAGMENT TO BE FOLDED. (DEFAULT  =  1,N)
<carriage return>
ENTER: F BEGIN FOLDING, T TERMINATE, A AUXILIARY INFORMATION,
P PARAMETER DEFINITION
F
FOLDING BASES    1 TO  120 OF Drosophila melanogaster 5S rRNA
ENERGY  =     -51.7


        10            20          30
   --      CAUA    -      ---   ACA    -- UC
    GCCAACGAC    CC ACGCUG   AAU   UCGGU  UC  G
    CGGUUGUUG   ,GG UGCGAC   UUA   AGCCA  AG  U
 UC        ---U : C      GAA   --A     CU  CC
 .        110    :  :    60         50        40
               :  :
               :  :  70       80        90
               : 'G       UA  ACUUAGAUGGGGGA   GC
               : CGCGGU   GU                   CC
               : GCGCCA   CA                   GG  U
               '.__:      --   -------------A  GU
                          100
ENTER: T TERMINATE, NS NEW SEQUENCE, NF NEW FRAGMENT, O OUTPUT PARAMETER
DEFINITION, OR THE ENDPOINTS OF A SUBFRAGMENT BETWEEN    1 AND    120.
END WITH 1 TO FORCE ENDS TO BASEPAIR.
T
FORTRAN STOP
Charged CPU time:    0 00:01:08.56
```

Figure 6. Computation with program RNAFOLD of *D. melanogaster* 5S rRNA secondary structure. All user input is underlined. The file containing the free energy values is called SALSER. Continuation/Save operations are described in Section 4.2.5. The output can be sent to the terminal as here, or to a disc file for subsequent printing. The CT file interfaces RNAFOLD with Feldmann's drawing program (Section 4.2.6). Options are entered after the query "Enter: F begin folding…" and begin with character "A". The user's manual fully describes the option syntax. Finally, the total free energy (kcal/mol) precedes the computer-drawn predicted structure. The folding generally follows de Wachter's consensus model (2) except for region 73−102: bases 78−86 should pair with 91−98, A83 being a bulge, but this involves the odd pair U80.U96 which cannot be found by the program. CPU time is for a DEC VAX 11/730 computer.

FORTRAN 77, has proved easy to implement on several types of computer. No special graphic device is needed because secondary structure outputs are for a line printer. The limitations derive mainly from core memory requirements. The nature of the algorithm is such that memory usage is governed by the number of potential helices combined rather than by sequence length (typically, a 5S rRNA sequence will produce ~ 800 potential helices of length two or more).

Number of helices	400	800	H
Memory requirements (kbytes)	32	104	$H^2/8 + 28.H + 27$

5.2 Options

Several options are under user control.

5.2.1 *Minimum length of helices*

The minimum length of helices the program will consider can be set to two or more. A larger value (four) can be used in a first run to obtain an approximation to be refined later using lower values. Consideration of helices of length two proves necessary for optimal prediction.

5.2.2 *Maximum number of helices per structure*

For internal reasons, the program requires that the number of helices per secondary structure be limited. However, because the computing time is insensitive to this limit, a rather large value (15) can be used. It is good practice to always have the limit at least one unit higher than the maximum number of helices actually found, in order to be sure not to miss any potential structure.

5.2.3 *Number of best structures computed*

The algorithm furnishes several best structures rather than one only. The desired number can be fixed between 1 and 20.

5.2.4 *Output of helix list*

The full list of potential helices can be printed. The program can be restricted to the computation of the helix list without making the structure prediction. This gives an estimate of the duration of a full computation.

5.2.5 *Odd pairs allowance*

Potential helices can be allowed to contain odd pairs, or be limited to G.C, A.U and G.U pairs, or even to G.Cs and A.Us only.

5.2.6 *Energy set*

The standard data set of pairing and stacking energies (*Table 3*) can be modified.

5.2.7 *Confidence free energy interval*

The confidence free energy interval function has been described in Section 3.1.2. Its default value (4 kcal/mol) is usually sufficient but may need to be increased when the required number of best structures is not found in a program run.

```
/L  2/B 10/A  2
   Bacillus pasteurii 5S rRNA
UNUGGUGGCGAUAGCGAAGAGGUCACACCCGUUCCCAUACCGAACACGGAAGUUAAGCUC
UUCAGCGCCGAUGGUAGUUGGGGUGUUAGCCCCUGCAAGAGUAGGACGUUGCCANGC
```

Figure 7. Example of input to program CRUSOE. *B. pasteurii* 5S rRNA sequence from (2). The user's manual fully describes all option syntax.

5.2.8 *Forced helices and ss regions*

It is possible to force one or several helices to occur in the predicted structures. Note however that such a helix will not be extended by the program even when this is possible. Specific residues may be forced to remain single-stranded by coding them with the character 'X' in the input sequence data. Unlike RNAFOLD, computation time is greatly reduced by the use of forced helices or ss regions.

5.2.9 *Stability computation of a supplied secondary structure*

A simpler program called STABIL contains the code for stability computation but not that for structure prediction. It may be used to obtain the overall free energy of any supplied secondary structure.

5.3 **Example**

An example of the input format for program CRUSOE is given in *Figure 7* for the *Bacillus pasteurii* 5S rRNA molecule. Input data is structured as follows. The first line contains the values of the options. Each option has a default value and only options given a different value are required on this line. Options start with the character ''/'' followed by a letter that identifies them and possibly by a number or a sign. The second line contains the sequence title. Optionally, one line for each forced helix is used after the title line. Finally, the sequence is written on an arbitrary number of lines of any length. Sequences may contain characters A,C,G,U (or T), or X to indicate a non-pairing residue. The results produced are shown in *Figure 8*. In each secondary structure model the sequence is numbered in multiples of ten bases. The structure is symbolized in the line above each sequence: a nucleotide series crowned by a given character pairs with the other series crowned by the same character. The bottom line shows the facing nucleotide under each paired residue thus allowing base pairs to be visualized. Three odd pairs occur in this secondary structure: G22.G57, G70.A103 and U79.U94.

6. DETECTION OF LOCALLY STABLE SECONDARY STRUCTURES

A less ambitious goal than secondary structure prediction is to identify all locally stable structures in a sequence. A *locally stable secondary structure* is a sufficiently stable connection of double helices in a localized region of the sequence, bounded by one hairpin loop, and possibly containing bulges and internal loops, but without multiple loops. This approach is proposed by Kanehisa and Goad (with their program SEQL) as an alternative both to recursive methods, which are limited to the prediction of one structure only, and to combinatorial methods, which are limited by their comparatively slow algorithm (33).

```
***      Bacillus pasteurii  5S rRNA

NUCLEOT : 117 , SEGMENTS : 805 ( 851 ) , ILOTS FIXES:   15
RANGE   : 4.0 , FAMILIES : 274 ( 559 ) , DEPTH ':  10
L.MIN.  2 , DEGEN. YES , B.P. ODD , MODIF. YES , SET Def

     FINAL STAB. :  -40.2 KCAL (  -57.4 ) ,STEMS USED :  8
     +   1   +   2   +   3   +   4   +   5   +   6   +   7   +   8   +   9   +  10   +  11   +
     FFFFFFFF  HHAAAAAAAA  CC  BBBBDD      DD  BBBB  CC  AAAAAAAA  HH  GGGGG     EEEEEEE     EEEEEEE    GGGGG  FFFFFFFF
     UNUGGUGGCGAUAGCGAAGAGGUCACACCCGUUCCCAUACCGAACACGGAAGUUAAGCUCUUCAGCGCCGAUGGUAGUUGGGGUGUUAGCCCCUGCAAGAGUAGGACGUUGCCANGC
     ACCGUUGC   CGCUUCUCGA  UG  GGCAAG      CU  UGCC  CA  UGGAGAAG  CG  GGAUG     CGUCCCC     GGGGUUG    UAGCC  GCGGUGGU

     FINAL STAB. :  -39.9 KCAL (  -57.4 ) ,STEMS USED :  8
     +   1   +   2   +   3   +   4   +   5   +   6   +   7   +   8   +   9   +  10   +  11   +
     FFFFFFFF  HHAAAAAAAA  CCBBBBDD        DD  BBBB  CC  AAAAAAAA  HH  GGGGG     EEEEEEE     EEEEEEE    GGGGG  FFFFFFFF
     UNUGGUGGCGAUAGCGAAGAGGUCACACCCGUUCCCAUACCGAACACGGAAGUUAAGCUCUUCAGCGCCGAUGGUAGUUGGGGUGUUAGCCCUGCAAGAGUAGGACGUUGCCANGC
     ACCGUUGC   CGCUUCUCGA  UGGGCAAG       CU  UGCC  CA  UGGAGAAG  CG  GGAUG     CGUCCCC     GGGGUUG    UAGCC  GCGGUGGU

SEARCH :  296663 , CALLED :  78992 , ACCEPT :2145 , INSERT :1905
SETUP TIME :   2m 38s , SEARCH TIME :  44m  5s
FIN NORMALE
```

Figure 8. Output of program CRUSOE applied to *B. pasteurii* 5S RNA. The first line shows the sequence title. Then, the sequence length (117), the numbers of potential helices (805) and of islets in which helices have been placed (15) are printed. Next, option settings are repeated. Range (4.0 kcal) is the confidence free energy interval (Section 5.2). Families (274) is the number of maximum length potential helices. Depth (10) is the maximum allowed number of helices per structure. L.MIN. (2) is the minimum helix length. B.P. states whether odd pairs are allowed (odd) or not (wob). Set equals Def when the default energy model (Ninio's) has been used. Then, as many best structures as required (here two) are displayed. For each, the overall free energy (-40.2 kcal for the first one) is given with, in parenthesis, the approximate stability (Section 3.1.2). The best predicted structure perfectly follows de Wachter's model (2). Finally, the durations of the two phases of the algorithm appear: setup, potential helices identification and compatibility sets computation; search, combinatorial search for optimal structures. CPU time is for DEC VAX 11/730 computer.

In essence, the SEQL algorithm can be viewed as a particular case of the ZNS algorithm (Section 3.2.1): it computes the best folding free of multiple loops for successive parts of the molecule. More precisely, given a user-chosen range LEN, the program searches for structures involving at most LEN bases on successive portions 2 × LEN bases long shifted by LEN bases with respect to the previous portion. Stability computations are made using Salser's energy model (Section 2.2.1) except that the contribution of ss regions is linearly approximated for reasons of computational efficiency. This overestimates the destabilizing effect of loops longer than about seven nucleotides. The minimal helix length is 3 bp. A sample program output is given in *Figure 9*.

Several parameters are under user control. The maximum free energy (or minimum stability) for a local structure to be displayed may be changed (default value is -10 kcal/mol). The maximum lengths of hairpin, internal and bulge loops (default values 20, 10 and 5, respectively) and LEN, the above-mentioned maximum width of local structures (default value 100), may be modified.

SEQL, written in FORTRAN, is available upon request from its authors. It requires a memory storage of about $LEN^2/2$; its computation time dependency is proportional to LEN^2. This is a fairly rapid method because the most time-consuming part of the ZNS algorithm, devoted to multiple loop treatment, is bypassed and computations are restricted at all times to small parts of the molecule.

7. COMPUTER-AIDED SECONDARY STRUCTURE MODEL BUILDING

The approach chosen by Quigley *et al.* (34) is to use the computer as a help for handling many alternative secondary structures. Here the user himself chooses one particular structure employing his biological knowledge about the molecule; the computer mere-

LOCALLY STABLE SECONDARY STRUCTURES IN ECOTRMF

```
#1(12) -21.3                                    #2(8) -10.2
         10        20                            10         20
CG      CGGGGUG GAGCAGCCUGGUAG                  GGAGC    AGCCUGGUAG
::      ::::    ::::    ::                      ::       ::::.:    C
GCUGGAAGCCCAAUACUCG   GGCUGCUC                  CCAAUACUCGGGCUGCU
          40        30                          40         30


#3(20) -40.1
         10        20                    30
GCGGGGUGGAGCAGCCUGGU    AGCUCGU          CGGGCUC
:::::  ::       :: ::       :.: ::       ::::   A
CGCCC  CC    GG CCUAAACUUG GCUGCUGGAAGCCCAAU
  70              60        50        40


#4(18) -23.8                                    #5(9) -13.7
  20        30          40                                50
UGGUAGCUCGUCGGGCUC AU    AACCCGAAGG             CGAAGGUCGUCGGUUC
::::: ::    ::::    ::    :: ::::               ::  ::  :.::: A
ACCAACGCC    CCCGGCCUAAACUU GGCUGCU             GCC CC  CGGCCUAA
      70          60        50                  70        60
```

SEQUENCE ECOTRMF

```
          10        20        30        40        50        60
CGCGGGGUGGAGCAGCCUGGUAGCUCGUCGGGCUCAUAACCCGAAGGUCGUCGGUUCAAA
                  ^^^^^^^^^^^^^^^ ^^^^^^^^^   ^^^^^^^^^  ^^^^^^^^
                 #1               #3          #4         #5
                      ^^^^^^^^^^^^
                 #2


         70        80
UCCGGCCCCCGCAACCA
^^
#5
```

Figure 9. Output of the SEQL program applied to *Escherichia coli* initiator Met-tRNA (9). Hairpin loops are at the right end of each structure. Parenthesized values (number of base pairs per local structures) are followed by the net free energy in kcal/mol.

ly helps him to find his way amid a forest of potential foldings.

The central tool is here the *dot matrix* (35). Given an RNA sequence, its associated dot matrix is obtained by filling a squared matrix with a dot at the intersection of row i and column j each time nucleotides i and j of the sequence may pair (only G.C, A.U and optionally G.U pairs are considered). A sample dot matrix is given in *Figure 10*. Since helices are series of paired nucleotides, any potential helix is seen in the dot matrix as a couple of short diagonals at symmetric positions relative to the main matrix diagonal (labelled a−e in *Figure 10*). Therefore the dot matrix may be used as a graphical display of all potential secondary structures in the molecule.

Dot matrix interpretation is governed by several rules that require a little practice to be assimilated. These rules are defined relative to two squares associated to each helix, an inner and an outer square, both drawn for helix b in *Figure 10* as an example.

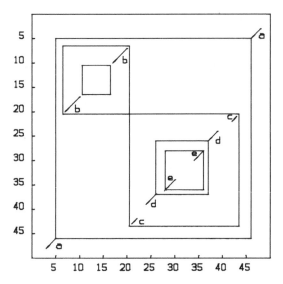

Figure 10. Dot matrix plot for the sequence of *Table 5* folded as in *Figure 1*. Helix labelling as in *Figure 1*. Only dots corresponding to base pairs that occur in molecule folding are represented.

A hairpin loop appears as a pair of diagonals (labelled b in *Figure 10*) with no other dot in the associated inner square. Series of helices separated by bulges or internal loops appear as series of interrupted, nearly co-linear diagonals (labelled c, d, e in *Figure 10*). Multiple loops occur when the inner square associated with a helix (labelled a) encompasses several non-intersecting outer squares for other helices (b and c). In this situation, the loop-closing helix is the outermost one. No nucleotide can belong simultaneously to more than one helix, so that any two helices represented by diagonals simultaneously crossed by a horizontal or vertical line are mutually exclusive. Indeed no horizontal or vertical line crosses any two diagonals in *Figure 10*. There are two cases when two helices are compatible, that is can exist simultaneously in a secondary structure:

(i)   outer squares associated with each helix do not intersect (helices b and c);
(ii)   the inner square for one helix totally encompasses the outer square for the other (helices d and e).

   Quigley's method provides several filters for dot matrices. A filter is the selection of a reduced number of potential base pairs to be drawn in the dot matrix according to a specific criterion. Filtering is necessary because raw matrices are generally too crowded with dots to be easily interpreted. Filtering criteria are:

(i)   helix minimal length;
(ii)   helix minimal stability (computed from very rough, stacking independent values);
(iii)   experimental evidence of the ss or ds nature of various nucleotides.

   The capability is provided to remove quickly from a dot matrix any helix incompatible with a set of other helices.

   With these two basic tools, a general procedure may be followed.

(i)   Enter input data: sequence and results of enzyme digestions, if any.

280

(ii)   Start with a stringent filtering: have long, stable helices appear in the dot matrix.

(iii) Examine the filtered dot matrix for potential helices; select with first priority short-range helices.

(iv)   Have all helices incompatible with the selected ones removed from the matrix.

(v)   Repeat steps (ii)−(iv) with progressively less stringent filterings, taking particular note of pairing regions brought close by previous pairings and of domain structures (i.e. helices closing off a number of paired regions).

The authors of this method claim the frequent occurrence in natural molecules of alternate secondary structures representing either static heterogeneity or dynamic molecular 'breathing'. The structure modelling approach, unlike recursive prediction methods, can take these possibilities into consideration because it has the ability to display different, mutually exclusive foldings. The program is written in FORTRAN 77 for DEC VAX computers and is partly device specific. It is available from the authors but is unsupported.

8. SECONDARY STRUCTURE PLOTS AND DRAWINGS

Particularly for long molecules, drawing a representation of RNA secondary structure is not easy. Hand drawing is an error-prone process that requires a lot of time. Moreover, making an overlap-free plot can be a difficult task for complex structures. Another source of difficulty arises because topologically identical foldings can be drawn in a large number of different ways, so that, in general, structural similarities in a series of folding plots are not visually identifiable. Several authors have written computer tools that help to visualize molecular folding models and structural similarities.

8.1 Stüber's secondary structure drawing program

Stüber's CLOVER program produces a two-dimensional layout of RNA secondary structure suitable for a normal line printer (29). Branch overlaps are removed from the plot by suitable elongation of ss regions with hyphen series insertions. Drawings otherwise have the conventional aspect, that of *Figure 1*. Written in Pascal, and, according to its author, adaptable to many computer systems, CLOVER is available upon request from K.Stüber.

8.2 Shapiro's displays of secondary structures

A sophisticated program generating non-overlapping displays of secondary structures has been presented by Shapiro *et al.* (31). This program exists in two versions, one written in SAIL for DEC System-20, the other in Pascal. The Zuker−Stiegler prediction program has an option that produces output suitable as input for Shapiro's program (Section 4.2). Foldings are displayed on Tektronix 4000 series, DEC VT100 video terminals or on Zeta plotters for hard copy. The program automatically produces a conventional display that is nearly completely non-overlapping (for sequences under 500 nucleotides the result is usually free of overlaps). In a final step, the user completely untangles the molecule with the help of a rotation capability provided by the program. Moreover, due to the drawing algorithm, molecules that have similar secondary structures tend to have similar shapes on the displays. This allows image com-

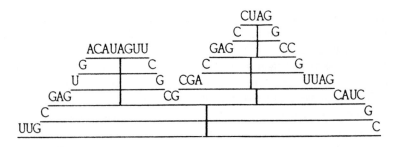

Figure 11. Hogeweg and Hesper's (23) secondary structure representation applied to sequence of *Table 5* as in *Figure 1*.

parisons to be made for seeking structural similarities. Another program capability is the zoom facility which allows users to select part of the drawing and have it enlarged. This, combined with the rotation facility, makes similarly shaped structural domains of different molecules visually identifiable.

8.3 An unconventional representation of secondary structures

The conventional representation of secondary structures (that in *Figure 1*) suffers from two main disadvantages. It suggests higher order proximity relationships which are meaningless and does not facilitate the identification of common substructures amongst several molecules.

Hogeweg and Hesper (23) have proposed an alternative representation that, according to them, overcomes the above problems. This very easily programmed representation, shown in *Figure 11*, is constructed as follows:

(i)   all nucleotides are horizontally positioned at regular intervals;
(ii)   helices are vertically expanded while retaining the normal horizontal position of any nucleotide;
(iii)   ss regions are not vertically expanded;
(iv)   base pairing bonds are represented by connecting lines;
(v)   vertical lines link the centres of the horizontal connecting lines.

Vertical lines bring out the tree-like structure of conventional displays. Internal and bulge loops are seen as interrupted verticals, and multiple loops seen as tree branchings. The representation is unique for a given folding, so that identical substructures differ only by level changes, regardless of how they are embedded in the global structure. Because the plot is essentially linear, different sequences can be easily aligned in order to identify corresponding parts.

The program, written in PL/I, is integrated in BIOPAT, a package for bioinformatic pattern analysis.

9. CONCLUDING REMARKS

Obviously RNA secondary structure prediction is not yet at a stage where easy-to-use methods can be employed, except for short molecules. This fact has been recognized by all authors of prediction methods as is stressed in the following quotation of the article by Zuker and Stiegler presenting their algorithm: 'A program based solely on conformational rules and thermodynamics will not yield a biologically meaningful folding of a molecule on its own... More and different kinds of additional information must be incorporated into the algorithm as well' (18). Users will have to identify among several methods, each with their own advantages and drawbacks I have tried to detail here, those that best fit their particular problem.

The limitations of prediction methods mainly result from five causes.

(i)   Prediction of one 'best' folding is insufficient because there may exist several different structures with similar energies.

(ii)   The underlying energy model is inadequate: much more experimental data are needed on the thermodynamics of RNA chain folding.

(iii)   Tertiary interactions (i.e. three-dimensional folding) should be taken into account.

(iv)   Inter-molecular interactions, especially for rRNA molecules that interact with ribosomal proteins, should also be taken into account.

(v)   Long RNA molecules do not fold into the structure of absolute minimum free energy; it is generally thought that they progressively fold during their $5' \rightarrow 3'$ synthesis into a locally optimal conformation.

Ninio's energy model and prediction algorithm is a significant step towards combating the first two limitations listed above. This is achieved at the cost of an increased computational load relative to the recursive methods. Several attempts have been made to incorporate the idea suggested in the fifth point, above, but they are not yet at the stage of being practical, extensively tested prediction methods (4,6,14,20).

One of the most powerful kinds of supplementary information that may be used in secondary structure analysis is given by the comparative study of homologous molecules from different organisms. It is highly probable that potential foldings preserved during evolution are structurally and functionally meaningful. There have been some attempts to have prediction tools take into account several sequences and their homology (36,37). However these methods cannot yet be considered as general purpose tools. They are not easily adaptable from one family of molecules to another, nor are they, to the author's knowledge, widely available. Therefore no program of this kind has been described in this chapter.

10. ACKNOWLEDGEMENTS

I would like to thank Dr. Jacques Ninio for his diligent help throughout the writing of this chapter and Dr Richard Grantham for critical reading of the manuscript. Dr Patrick Stiegler kindly sent me information on the Zuker—Stiegler program.

11. REFERENCES

1. Ninio,J. (1979) *Biochimie,* **61**, 1133.
2. Erdmann,V.A., Wolters,J., Huysmans,E. and de Wachter,R. (1985) *Nucleic Acids Res.,* **13**, r105.
3. Rietveld,K., Linschooten,K., Pleij,C.W.A. and Bosch,L. (1984) *EMBO J.,* **3**, 2613.
4. Sankoff,D., Kruskal,J.B., Mainville,S. and Cedergren,R.J. (1983) In *Time Warps, String Edits, and Macromolecules: The Theory and Practice of Sequence Comparison.* Sankoff,D. and Kruskal,J.B. (eds), Addison-Wesley, p. 93.
5. Salser,W. (1977) *Cold Spring Harbor Symp. Quant. Biol.,* **42**, 985.
6. Jacobson,A.B., Good,L., Simonetti,J. and Zuker,M. (1984) *Nucleic Acids Res.,* **12**, 45.
7. Cech,T.R., Tanner,N.K., Tinoco,I., Weir,B.R., Zuker,M. and Perlman,P.S. (1983) *Proc. Natl. Acad. Sci. USA,* **80**, 3903.

8. Papanicolaou,C., Gouy,M. and Ninio,J. (1984) *Nucleic Acids Res.*, **12**, 31.
9. Sprinzl,M., Moll,J., Meissner,F. and Hartmann,T. (1985) *Nucleic Acids Res.*, **13**, r1.
10. Sprinzl,M., Vorderwülbecke,T. and Hartmann,T. (1985) *Nucleic Acids Res.*, **13**, r51.
11. Pipas,J.M. and McMahon,J.E. (1975) *Proc. Natl. Acad. Sci. USA*, **72**, 2017.
12. Studnicka,G.M., Rahn,G.M., Cummings,I.W. and Salser,W.A. (1978) *Nucleic Acids Res.*, **5**, 3365.
13. Dumas,J.-P. and Ninio,J. (1982) *Nucleic Acids Res.*, **10**, 197.
14. Gouy,M., Marlière,P., Papanicolaou,C. and Ninio,J. (1985) *Biochimie*, **67**, 523 (in French).
15. Waterman,M.S. and Smith,T.F. (1978) *Math. Biosci.*, **42**, 257.
16. Nussinov,R., Pieczenick,G., Griggs,J.R. and Kleitman,D.J. (1978) *SIAM J. Appl. Math.*, **35**, 68.
17. Nussinov,R. and Jacobson,A.B. (1980) *Proc. Natl. Acad. Sci. USA*, **77**, 6309.
18. Zuker,M. and Stiegler,P. (1981) *Nucleic Acids Res.*, **9**, 133.
19. Nussinov,R., Tinoco,I. and Jacobson,A.B. (1982) *Nucleic Acids Res.*, **10**, 351.
20. Nussinov,R. and Tinoco,I. (1981) *J. Mol. Biol.*, **151**, 519.
21. Nussinov,R. and Tinoco,I. (1982) *Nucleic Acids Res.*, **10**, 341.
22. Comay,E., Nussinov,R. and Comay,O. (1984) *Nucleic Acids Res.*, **12**, 53.
23. Hogeweg,P. and Hesper,B. (1984) *Nucleic Acids Res.*, **12**, 67.
24. de Wachter,R., Chen,M.W. and Vandenberghe,A. (1982) *Biochimie*, **64**, 311.
25. Stiegler,P., Carbon,P., Zuker,M., Ebel,J.-P. and Ehresmann,C. (1980) *C.R. Acad. Sci. Paris (Sér. D)*, **291**, 937 (in French).
26. Stiegler,P., Carbon,P., Zuker,M., Ebel,J.-P. and Ehresmann,C. (1981) *Nucleic Acids Res.*, **9**, 2153.
27. Ursi,D., Vandenberghe,A. and de Wachter,R. (1982) *Nucleic Acids Res.*, **10**, 3517.
28. Martinez,H.M. (1984) *Nucleic Acids Res.*, **12**, 323.
29. Stüber,K. (1985) *CABIOS*, **1**, 35.
30. Yamamoto,K., Kitamura,Y. and Yoshikura,H. (1984) *Nucleic Acids Res.*, **12**, 335.
31. Shapiro,B.A., Maizel,J., Lipkin,L.E., Currey,K. and Whitney,C. (1984) *Nucleic Acids Res.*, **12**, 75.
32. Feldmann,R.J. (1981) *Manual for Programs NUCSHO and NUCGEN of Nucleic Acid Structure Synthesis and Display*.
33. Kanehisa,M.I. and Goad,W.B. (1982) *Nucleic Acids Res.*, **10**, 265.
34. Quigley,G.J., Gehrke,L., Roth,D.A. and Auron,P.E. (1984) *Nucleic Acids Res.*, **12**, 347.
35. Maizel,J.V. and Lenk,R.P. (1981) *Proc. Natl. Acad. Sci. USA*, **78**, 7665.
36. Studnicka,G.M., Eiserling,F.A. and Lake,J.A. (1981) *Nucleic Acids Res.*, **9**, 1885.
37. Michel,F., Jacquier,A. and Dujon,B. (1982) *Biochimie*, **64**, 867.

NOTE ADDED IN PROOF

The ZNS secondary structure prediction method has recently been improved for finding the best few alternative secondary structures of an RNA molecule (38).

38. Williams,A.L. and Tinoco,I. (1986) *Nucleic Acids Res.*, **14**, 299.

CHAPTER 12

Protein structure prediction

WILLIAM R.TAYLOR

1. THE PROTEIN FOLDING PROBLEM

The *in vivo* synthesis of a protein from the information stored in DNA is a very complicated process involving large assemblies of proteins that process the message and synthesize the peptide bonds. Despite the complexity of the cellular apparatus through which the message is processed, the resulting protein chain maintains essentially a linear relationship to the original DNA. Although recent research has shown this not to be strictly true, and that a certain amount of editing occurs in between, the simple one-to-one relationship is a sharp contrast to the further transformation of the message into a functional entity.

Whether the protein performs a structural or enzymatic role in the cells, it does so, normally not as an extended chain, but as a compact and folded chain. The fold of the chain is the same for all copies of the protein and serves to maintain a particular shape, perhaps of structural importance, or to bring together certain amino acids such that their chemical groups are positioned to effect catalysis. The transition between the initial conformation, as synthesized on the ribosome, and the functional conformation appears to be a spontaneous folding, unaided by any cellular apparatus; again, a distinct contrast to the earlier history of the message.

Not only is the folding process unaided but there is no obvious relationship between the chemical nature of the protein sequence and the folded structure it adopts. It is certainly not a simple one-to-one, but a many-to-one relationship in which sequentially remote amino acids contribute to any given part of the structure. The relationship may indeed be an all-to-one relationship in which every amino acid in the sequence contributes, to a greater or lesser extent, to the maintenance of each sub-structure.

1.1 Theoretical importance

How the sequence determines a specific structure has been a constant source of fascination and speculation since the problem was identified. But, as yet, no one had described the transformation 'formula' to convert sequence to structure. This problem, generally referred to as the 'folding problem', was quickly identified as one of the major outstanding problems in fundamental molecular biology being referred to by Jacques Monod as an unidentified protein 'assembly law' (1).

1.2 Practical importance

1.2.1 *Nucleic acid sequence data*

We may not be much closer to defining Monod's 'assembly law', but nucleic acid research has progressed to such an extent that describing the sequences has become almost a routine occupation. As many of the contributions to this volume will bear witness, there is now such a large body of DNA sequence data that even storing it efficiently can create problems. Much of this sequence data is either known, or can be inferred to code for protein sequences and many, besides their own intrinsic interest, are associated with various disease states, such as cancer or viral disorders. Indeed, these connections often provided the original motivation to determine the sequence and if a three-dimensional structure could be inferred from the sequences then further insight might be gained about the mechanism or possible cure for the disease.

1.2.2 *Protein engineering*

The techniques used in nucleic acid research not only allow the DNA sequences to be read but also enable specific changes to be induced in the DNA of living cells and modified (or mutant) proteins produced. This process of site-specific mutagenesis has many wide ranging applications, only a few of which can be alluded to here. Effectively the technique allows existing proteins to be engineered to tackle new or modified functions or to perform existing functions more efficiently. The limiting factor in this rapidly emerging field is not the nucleic acid techniques but the difficulty in anticipating the effect that the modifications will have on the protein structure and activity. To overcome this difficulty requires a good understanding of how proteins maintain and use their natural (or native) structure. This is effectively the protein folding problem rephrased; for if it were possible to predict a structure from sequence, the structure of the modified sequence could also be anticipated.

These recent developments in nucleic acid research have transformed protein folding from a rather obscure pursuit to one of central importance in many lines of research. In what follows I hope to show that the elusive folding law or formula envisaged by Monod and many others cannot be a practical reality but that, despite this, there is a real possibility of predicting protein structure from sequence in a way that will be much more accessible to the new generation of biochemists and protein engineers. To appreciate this, however, requires a background of the principles of protein structure which the following section attempts to provide.

After this, the practical issues of using established protein structure prediction methods are discussed together with some more recent techniques which have been developed in order to improve predictive accuracy.

2. PROTEIN STRUCTURE

Almost all that is known of the details of protein structure is derived from an analysis of X-ray diffraction patterns of protein crystals. Under favourable conditions, crystallographic analysis allows atomic resolution of protein structure. Unfortunately, such an analysis often takes years and despite significant improvements such as the use of synchrotron radiation sources for data collection and supercomputers for data processing, it is unlikely that in the near future the determination of protein structures will become routine. Further limitations to protein structure determination are the ability to isolate and crystallize the protein. Although the former aspect will soon be greatly aided by the isolation and cloning of the gene that encodes the protein, the ability of the protein to crystallize will remain a fundamental limitation. Indeed, even now this is rapidly becoming the rate limiting step in protein structure determination. Whatever the rate at which structures are determined it is certain that this will lag several orders of magnitude behind the rate at which protein sequences are determined, leaving a large backlog to be tackled by a faster, if less certain, approach.

2.1 Structure representations

In what follows I will concentrate not on details of the now considerable number of structures that have been determined by protein crystallography (about 200), but rather on ways in which these structures can be represented and analysed to emphasize underlying features of importance to structure prediction.

For the more basic principles of protein structure the reader who is unfamiliar with these is advised to consult one of the following references (2−4).

A typical protein structure contains roughly 2000 atoms whose positions are specified in space (by crystallography) and whose chemical connectivity is generally inferred from knowledge of the amino acid sequence. Such a structure contains too much information to be readily comprehensible in any representation (even with fast, colour, interactive, three-dimensional computer graphics). Much of the comparative analysis and appreciation of protein structure is thus dependent on finding suitable simple representations. Fortunately, analysis of the known structures has revealed that protein structure is organized in a hierarchical way, with large structures being composed of identifiable sub-structures which themselves are composed of sub-structures. Such a hierarchy lends itself very well to representation at different levels depending on which aspects are to be emphasized. Unfortunately, there is no standard way of representing them, and a variety of conventions will be found throughout the literature. The brief explanation below should, however, enable the more common representations to be identified in their different guises.

An initial step in the process of simplification is to neglect the detailed structure of the amino acid side chains and concentrate only on the chain traced by the protein backbone or even more simply just a chain of 'virtual bonds' created by connected α-carbon atoms (i.e. only one point per residue).

2.2 Structural hierarchy

2.2.1 *Secondary structure*

The first successful prediction of protein structure was made by Pauling in 1951 (5) who, from consideration of the steric restraints implied by the peptide bond, deduced that protein chains would naturally adopt two regular compact structures. One of these was a helix and the other consisted of strands of protein chain aligned to form a sheet (*Figure 1*). Both are maintained by hydrogen bonds. These two states were identified with two known distinct physical states observed in fibrous proteins, one being called the α-helix and the other the β-sheet. They were generally referred to as secondary structures in a hierarchical classification that defined the sequence as primary and the whole three-dimensional structure as tertiary. Their occurrence has since been confirmed in almost every globular protein structure.

The dominance of protein structure by only two major sub-structures is a great ad-

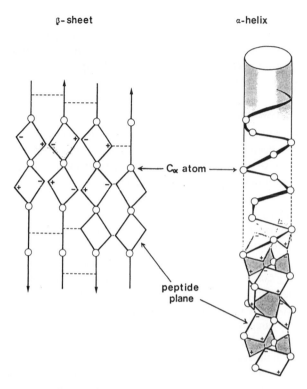

Figure 1. **(a)** β-sheet formed from extended parallel (or antiparallel) aligned strands of protein chain. The strands need not be sequentially consecutive. **(b)** α-helix shown in progressively simpler representations towards the top. The small circle represents the α-carbons (C_α) to which the amino acid side chain (if any) is attached. The atoms between C_αs have been simplified as a diamond shaped plane emphasizing that there are no rotatable bonds in this region. With this constraint, the α- and β-structures are two of a very limited set of structures that can internally satisfy all the main chain hydrogen bonds (indicated as $\oplus$ . . . $\ominus$).

vantage for simplification, as a structure need only be defined by the relative location and connection of its secondary structures. Commonly, α-helices are represented as idealized helical ribbons or solid cylinders. β-strands tend to retain their extended character and are usually drawn as arrows or, in slightly more detail, as twisting ribbons with an arrow head indicating the chain direction. So useful is such a representation that Richardson (4) has represented most of the known protein structures in this idealized 'cartoon' way and her review is to be recommended both for a summary of basic principles and an overview of the variety of structures adopted by proteins (see *Figure 2*).

2.2.2 Super-secondary structures

As the analysis of protein structure became more sophisticated, structural organization was identified in the region between the secondary and tertiary levels. These structures, which were termed super-secondary structure (6), consist of a few elements of secondary structure that are generally sequentially adjacent and pack together in a regular way. These recurrent motifs, which dominate tertiary folds of most globular proteins, are probably energetically favourable and possibly provide stable nucleation centres in

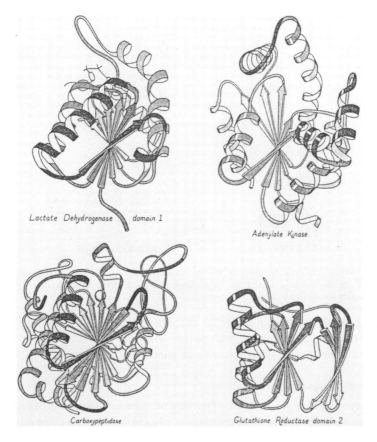

Figure 2. Idealized 'cartoon' representations of protein structures (4). α-helices are represented as helical ribbons and β-sheets as aligned (slightly twisted) arrows.

folding. The super-secondary structures identified so far include the $\beta\alpha\beta$ unit (6,7), the β-hairpin and the β-meander (3) and the four-helical bundle (8,9).

2.3 **Protein topology**

2.3.1 *'Bundle of rods'*

Because of their larger scale, the presence of super-secondary structures can be more easily appreciated if a further simplification is made to the representation of protein structure. A natural progression in this direction is to neglect the structure of the loop regions that connect secondary structures. In addition, both α-helical and β-strands are regular linear structures which allows them to be approximated by a single line or rod. If the structure of proteins is considered as a packing together of rods, then it is found that the rods can often be considered as lying parallel as would naturally occur in a bundle of sticks. Such packing is constrained between the β-strands by interstrand hydrogen-bonding to form a sheet. Between helices and sheets a parallel alignment is often imposed by the necessity to maintain a tightly packed structure. Similarly, if two

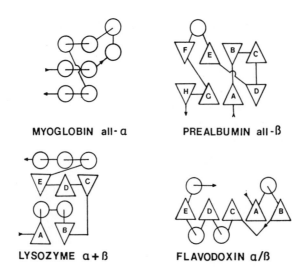

MYOGLOBIN all-α   PREALBUMIN all-ß

LYSOZYME α+ß   FLAVODOXIN α/ß

Figure 3. Structural classes of the globular proteins in two-dimensional circle and triangle representation (7). The circles represent α-helices; the up-pointing triangles represent approaching β-strands while the down-pointing triangles are receding β-strands.

sheets pack together, the requirement for efficient packing of their faces again imposes an effectively parallel alignment. Such packing constraints are not so severe between helices, but these too are often found to lie parallel (10).

2.3.2 *Two-dimensional representation*

With the assumption of general parallel packing of secondary structures, a gross simplification can be made in the representation of protein structures as one dimension is now redundant and the structure can be represented in two dimensions by specifying only the relative location and connection of the secondary structures. Although different symbols may be used to represent helices and strands in such diagrams, I will adhere to those of Sternberg and Thornton (7) who represent a helix as a circle and a strand as a triangle (see *Figure 3*).

At this extreme level of abstraction, the several thousand numbers required to specify the protein structure have been reduced to about 20. It must therefore be remembered that such representations can contain large distortions, especially if the protein structure itself is large. Nevertheless, the simplification achieved facilitates the description, and perhaps even the discovery of topological features of the chain fold. Most importantly, such abstract representations allow the structural similarity of apparently otherwise unrelated structures to be appreciated. An example of this was the discovery of a pattern of β-strand arrangement in some structures that was equivalent to the decorative motif, or key, traced around many classical Greek vases (11). This Greek key pattern was observed in the folds of two otherwise unrelated structures (12).

2.4 **Domains**

The term domain is used variously to refer to structural entities larger than secondary structures. Common usage now tends to restrict its application to structures generally larger than super-secondary structures that are either functionally or structurally intact. In other words a domain would be expected to maintain its structural identity in isolation. No automatic method of defining domains has yet produced a set of definitions that is widely agreed on.

Domains have a strong tendency to fall into well defined structural classes, characterized by their secondary structures. The structural classes, initially identified by Levitt and Chothia (13) and elaborated by Richardson (4), include domains dominated by only one type of secondary structure (all-α and all-β) and those in which α-helices and β-strands alternate along the sequence (β/α). A pattern of mixed secondary structures, segregated along the sequence ($\alpha+\beta$) also occurs but is more susceptible to interpretation as two domains. There are many 'unclassifiable' structures but these tend to be small structures dominated by the binding of a large ligand such as haem or iron − sulphur cage.

3. A MECHANISTIC APPROACH TO FOLDING

Many attempts have been made to predict the tertiary structure of a protein from its sequence. These fall into two distinct approaches.

(i)   One is to set up a realistic mechanical model of the protein chain (preferably in a computer) and to simulate the folding process. This is, perhaps, closest to defining the folding laws that Monod aspired to and is essentially the approach of the physicist who expects to derive complex results by extrapolation from fundamental laws.

(ii) Other approaches are empirical as they proceed by inference from known tertiary structures to unknown.

Both these approaches have inherent limitations that will be discussed fully in the following sections.

3.1 Energy minimization

An underlying assumption to much of the work that attempts to simulate the folding process is that it is possible to define a function which is at a global minimum when the structure is in its native form. This function has, naturally, been equated with the Gibbs free energy of the system. Thus most approaches have centered on attempting to describe the protein's chemical structure by a realistic energy function. Because such a function is far from linear, its minimization must proceed in discrete steps, with the energy and its first two derivatives being evaluated at each step. Each step is calculated from the previous such that each time a lower energy state is achieved. The minimization should thus proceed with energy decreasing in a monotonic way to a minimum energy state. [See (14) for a review of basic principles.]

3.1.1 Local minima problem

For simple energy functions that have a smooth continuous surface, any of a variety of techniques can efficiently find the global minimum. However, from even a superficial consideration of basic protein structure, it should be apparent that the surface of its energy function over which minimization must occur is far from smooth or continuous. A typical protein has perhaps 1000 rotatable bonds, the rotational freedom of which are interdependent. Minimization must therefore take place in a space of corresponding dimensionality (i.e. 1000-D). With modern computers such a problem could be solved if there existed a distinct minimum energy that corresponded to the correct structure and a series of progressively lower intermediate states leading to it. The real situation, however, could not be further from this ideal and an energy surface has been graphically described as a very large egg box. In other words, the minimization surface is pitted by deep holes, or local minima, between which there is effectively no energetic distinction. Most of these local minima undoubtedly correspond to a compact, protein-like globule. But in each globule the chain fold may be quite different. Techniques have been adopted to circumvent the local minima problem and are equivalent to 'heating up' the protein or giving it a 'kick'. These however only cause a jump into the next local minimum and do little to solve the essential problem.

Recently, Robson and Platt (15) have described a powerful approach based on a SIMPLEX minimization strategy that allows many minima to be sampled. Most significantly, however, they evaluate the structures at the minima not just by their energy but by a variety of global characteristics including hydrophobic packing and chain fold patterns.

3.1.2 *Potential functions*

Further difficulties are encountered with energy minimization, some of which are practical but others have a more fundamental nature. One that occupies much of the attention of those involved is the definition of a suitable potential function with which to evaluate the energy. Ideally, the potentials should be described by Gaussian approximations to the quantum mechanical wave equation. However, even in small molecules this description approaches the limits of what is practically computable. A representation commonly adopted in molecular dynamical simulations of protein structure (which hope to elucidate aspects of protein motion) is one containing some mixture of classical electrostatic and Van der Waal's potentials. Such simulations often run for several hours on the world's fastest computers to simulate protein motion in time spans usually measured in picoseconds (10^{-12} sec). However, *in vivo*, proteins often take seconds to attain their native structure and therefore to run a realistic simulation of this length is clearly impossible. Attempts to speed up the simulation so that useful time periods can be observed have entailed simplifications in the representation of the protein structure, such as that used by Levitt (16) who reduced each amino acid to a sphere on a stick (the lollipop model) and connected them by single rotatable bonds. Simplifications have also been made in the energy functions. For example, Robson and Osguthorpe (17) restricted rotational freedom of the two backbone bonds to an ellipse in their joint space. However, all such compromises gain speed at the expense of realism and this has the dangerous effect that it blurs any distinction that might have existed between the global minimum and the multitude of false minima.

Practical considerations of computability also make it necessary to neglect the effect that the surrounding water has on protein folding. This is perhaps the most fundamental objection to the whole approach as it is generally recognized that water and its associated hydrophobic effect on some amino acids is probably the principal force in the determination of protein structure. To run a simulation long enough, with sufficient simulated water molecules so that these mainly entropic effects are modelled effectively is, for the foreseeable future, wholly impractical.

3.2 Future use of energy minimization

Considering the above critique, it is perhaps not surprising that none of the methods using energy minimization have produced a structure that is significantly close to a known structure as measured by the rigorous criterion of Cohen and Sternberg (18). These past failures have generally discouraged further *ab-initio* folding simulations. Nevertheless, I have considered this topic in more detail than is justified by its success because the techniques are very effective at solving more limited problems. If other methods can determine the overall fold of the protein chain and, in effect, solve the global search problem, then energy minimization methods will prove very useful in refining these rough folds towards well packed protein structures. The final energy of these structures may, perhaps, provide a measure to distinguish amongst competing folds. In this situation the minimization problem is restricted to finding a solution that is within its relatively small radius of convergence.

The combination of empirical methods with those of energy minimization, either as separate cycles or in more integrated methods, seems to hold most promise for the future.

4. EMPIRICAL METHODS OF STRUCTURE PREDICTION

The failure of energy minimization to fold a protein sequence even remotely near its native structure has led workers increasingly to seek a solution to the folding problem by studying known structures and predicting by inference. These approaches range from very accurate methods that require a close sequence homology with a known structure, to more general methods that indicate preferences for regions of sequence to adopt secondary structure. In the following sections I will begin with the most general methods and progress towards the specific as this tends to coincide with their historical development.

The approach to structure prediction discussed in the preceding section has the innate (possibly fatal) attraction that, in theory, it requires no foreknowledge of protein tertiary structure (or even of proteins). If it were to be successful it could be applied uniformly to all sequences. By contrast, all methods based on inference from known structures are inherently limited in their applicability. Their validity is limited to the prediction of structures that are similar to those which were used in the inference process. This involves circularity in that it often cannot be determined if the inference rules are valid until a structure has been predicted! Fortunately, there are often biophysical or biochemical clues that aid in this decision and they will be discussed in a later section. These limitations are often forgotten by those who apply the methods and it is not uncommon to find in the literature predictions of structure for small peptides based on a method that was derived from a database of relatively large globular proteins.

4.1 Secondary structure prediction

4.1.1 *Statistical approaches*

Statistically significant correlations are readily apparent between the amino acid composition of the local sequence and the secondary structures they adopt. The analysis of such correlations has provided the basis for many methods that attempt to predict the location of secondary structure along the sequence. Some of these have gained wide popularity, sometimes as a result of their ease of application rather than any claim to outstanding predictive accuracy. In this section I will briefly survey the principles underlying a few of the more popular approaches and others that have provided a base for recent developments. For details of all these methods and references to the less well known, the reader is recommended to consult the relevant chapter of Schultz and Schirmer (3). Although this work is becoming a little dated, most of the methods have not changed in the intervening time. This stagnation reflects the increasing obsolescence of these methods as most have not been revised in the light of the growing database of protein structures.

(i) *The method of Chou and Fasman.* Few of the methods referred to above can be applied without the aid of a computer. An exception is the method of Chou and Fasman (19) which can be applied, if somewhat tediously, by hand. In its simplest formulation the method calculates a moving average of values that indicate the probability or propensity of a residue type to adopt one of three structural states, α-helix, β-sheet and turn conformation. The propensities are simply the frequencies of a given residue type to be observed in a particular secondary structure, normalized by the frequency expected by chance. From this, initially simple, approach the method then requires the application of several *heuristic* rules that attempt to determine the exact ends of secondary structure elements. These rules suffer from the defect of appearing somewhat arbitrary and even ill-defined (20). They are easily applied by hand but their implementation in a computer is more difficult and variations in the extent of their implementation is found in different programs.

(ii) *The method of Garnier, Osguthorpe and Robson.* The method of Garnier *et al.* (21) (sometimes referred to as the Robson method) has a much more sophisticated theoretical basis than the Chou and Fasman method. Its development began with a series of papers by Robson and Pain in which the concepts of information theory were applied to the folding problem in the hope of extracting all available information from the sequence. Despite this difficult background, the resulting method is remarkably simple and almost trivial to implement as a computer program since it requires only addition and contains no complicated heuristic rules.

The method considers the effect that residues within the region eight residues N-terminal to eight residues C-terminal of a given position have on the structure of that position. Thus for each residue type there exists a profile (17 residues in extent) that quantifies the contribution the residue type makes towards the probability of one of four states, α-helix, β-sheet (referred to as extended), turn and coil. Thus at each position the value of the profile for each residue type ± 8 of the current position is selected at a corresponding displacement and added to the mid-profile value of the current residue

a

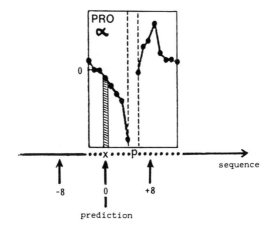

b

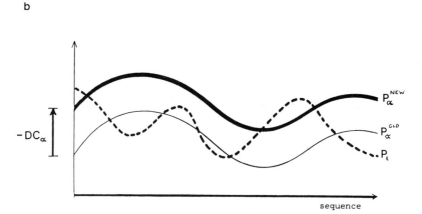

Figure 4. (a) The Garnier *et al.* (21) prediction method showing how the prediction of structure for position X is affected by a proline residue five positions C-terminal to X. A profile indicating the local effect of proline on the prediction of helical structure is centred on the nearby proline and the value of this profile at position X is added to a total of all other representation profiles within eight positions. **(b)** The effect of biasing the prediction of structure by a decision constant (DC) is shown. This simply raises or lowers the entire profile to which it is applied resulting in a new definition of structure defined by the highest profile at each position along the sequence.

type (*Figure 4a*). This can be confusing to humans, and care must be taken in not applying the displacement backwards, but for computers the process is simple and fast. The result is four 'probability' profiles and the only rule in the procedure is that the highest at any position constitutes the prediction of structure. If only the mid-point value of the information profiles is taken, the values can be used as in the Chou and Fasman method to calculate a moving average. Garnier *et al.* (21) go to considerable lengths to demonstrate that such an approach is not as good as using their 'directional' information. However, as many of the information profiles die away rapidly and symmetrically from their mid-points, the difference between the methods amounts to only a few percent.

Since no post-processing of the probability profiles is required as part of the prediction, this lends to the Garnier, Osguthorpe and Robson method a simplicity that allows overall biases to be imposed on the profiles. This was exploited by Garnier *et al.* (21) to tackle the identification and prediction of different structural classes of protein. Because the structural classes (discussed in Section 2.4) included types of structures that are composed almost entirely of one secondary structure type, then, if the predicted structure tends strongly towards one secondary structure, it is advantageous to bias it further towards that one structure. This is effectively an attempt to *boot-strap* out of the circularity mentioned previously as it is performing a prediction to determine the validity of a further, more specific prediction.

The Garnier, Osguthorpe and Robson approach is to simply raise or lower one or more of the probability profiles uniformly over the entire sequence by the addition of a constant value refered to as a *decision constant* (DC). For reasons of consistency with information theory nomenclature, a positive DC is subtracted from the profile, not added as might be intuitively expected. The new prediction is then redefined by the new highest peaks (*Figure 4b*).

The approach works well for the all-α and all-β type of protein structure but is obviously of limited application to the β/α type of structure where the two types alternate. Despite this, the paper of Garnier *et al.* (21) specifies recommended DCs for use with the β/α type of structure. The reason for this and its validity was considered more closely by Taylor and Thornton (22).

In an analysis of the β/α class of structure and its prediction we found that there was effectively no structural distinction between a turn and a coil. This results from the relatively long connections between β-strands and the following α-helix. Similarly, the Garnier, Osguthorpe and Robson. method was found to predict either turn or coil conformations for these loop regions with almost equal frequency. These observations were confirmed by correlating the Garnier, Osguthorpe and Robson probabilities with the observed conformation. The resulting 4 $\times$ 4 correlation matrix (*Table 1*) should be positive on the diagonal and negative elsewhere. We found this to be so for the β and α correlations but both the turn:coil and coil:turn elements of the matrix were positive, indicating almost no distinction between these states. It then made sense to treat β/α structure prediction as a three state prediction with the states being α, β and unstructured.

The change from four to three states, however, has a subtle effect on the choice of optimal DC. If four states are predicted, two of which are redundant, it is advantageous, in terms of correctly predicted residue states, to over-predict the less redundant β and α structures at the expense of the poorly correlated turn and coil predictions (many of which will be wrong anyway). This results in the DC_α and DC_β values recommended by Garnier *et al.* (21) biasing the predicted structure towards too much secondary structure. We consequently re-optimized the DC values to maximize the prediction of three states. This process converged towards an almost unbiased prediction ($DC_\alpha = 20$, $DC_\beta = 0$) which, bearing in mind the different structure data sets used, was insig-

Table 1. Correlation (43) matrix of observed secondary structure with structure predicted by the (unbiased) Garnier *et al.* (21) method for *βαβ* regions. The matrix shows reasonable correlation for *α:α* and *β:β*, but poor correlation for the turn and coil regions indicating that there is little predictive discrimination between these states. For the *β/α* class of protein a three state prediction is thus recommended and this may also be true for other structural classes

	CORRELATION MATRIX			
Predicted structure	*Observed structure*			
	α	*β*	*t*	*c*
α	0.320	−0.246	−0.076	−0.028
β	−0.257	0.476	−0.149	−0.124
t	−0.061	−0.089	0.141	0.043
c	−0.057	−0.149	0.144	0.130

nificantly different to the unbiased DCs of the Garnier, Osguthorpe and Robson method.

We did not investigate the significance of this effect on the other structural classes, but care should be taken, especially in the all-α type of structure where the loop lengths between secondary structures are equally as long as the β/α class.

(iii) *Multiplet frequency analysis.* Several methods were developed on the assumption [like that of Garnier *et al.* (21)] that neighbouring residues have an effect on each others conformation. This, not unreasonable, supposition prompted analyses of the conformation adopted by all possible combinations of sequential amino acid pairs and even triplets (e.g. AAA, AAC, ACA . . . . YWY, YYW, YYY). The predictive power of this approach proved to be less than expected and, indeed, methods based on it were no more successful than the simpler approach of Chou and Fasman that is based on singlet statistics. An inherent problem of the approach, whether using doublets or triplets, was the large number of combinations that statistics had to be gathered for. This led to a thin spread of the available data, resulting in poor statistics for many of the combinations (3).

This approach is mentioned here as it has had a recent revival in a new prediction method of Kabsch and Sander (23). They considered the occurrence of pentapeptides, and in an initial survey (24) were surprised to find two identical pentapeptides, one with an α-helical conformation and the other with a β-strand. Despite this they derived pentapeptide statistics, but to maintain the significance of the values they considered the 20 amino acid types to consist essentially of just five distinct types, thus greatly reducing the possible combinations. [This choice was made on the basis of a previous analysis of redundancy in amino acid usage (25)]. The results of this method are inseparable from a complex post-processing operation that will be described in a later section along with other similar approaches.

(iv) *Statistical-mechanical approach.* A method for calculating helix/coil transitions in polymers, due to Zimm and Bragg [see (3), appendix 4, for a simple exposition], has similarly been revived and applied more generally to secondary structure prediction with reasonable success (26).

4.2 **Hydrophobic plots**

The border-line that divides methods that rely on statistically derived parameters and those that rely on empirical occurring measures is not one that greatly affects the techniques used to subsequently predict conformational state. Many aspects of amino acid properties, such as charge or size, could be investigated for their predictive value, but the amino acid property that has attracted most attention is the hydrophobicity of the side chain, which is one of the dominant factors determining protein structure. Although such approaches often claim to be non-empirical, they are all based on the fundamental observation that protein structures have a hydrophobic interior shielded from solvent by a hydrophilic shell.

4.2.1 *Hydrophilic turn predictor*

Considering the bundle of rods model of protein structure (discussed in Section 2.3.1), it is clear that generally the protein chain will tend to pass through the hydrophobic interior and double back again via a turn that will be exposed to solvent on the protein surface. This model underlies the method of Rose (27) for the prediction of turns in proteins. It entails, simply, a moving average of a physico-chemically based measure of amino acid hydrophobicity. The correspondence of the peaks in the resulting hydrophilicity profile (hydrophobic minima) with observed turns was very good, rivalling equivalent predictions made by conventional secondary structure prediction methods.

4.2.2 *Hydrophobic patterns*

Other workers have attempted to utilize the predictive power of variations in hydrophobicity along the sequence and combine it with standard structure prediction techniques (e.g. 22,28). However, because of the redundancy of information between hydrophobicity and some statistically derived measures, the gain in predictive power has been slight.

(i) *Pattern convolution.* In a study of the β/α class of protein, Taylor and Thornton (22) found that the *convolution* of a small peak (equivalent to a positionally weighted moving average) with a simple measure of hydrophobicity was a surprisingly good predictor of buried β-strands. The hydrophobic pattern associated with an amphipathic (half-buried) α-helix was also matched against the sequence. In this case the pattern (or probe sequence) that was convoluted with the hydrophobic profile of the amino acid sequence was chosen to correspond to that observed by Chothia *et al.* (29) and Cohen *et al.* (30). The value of this convolution oscillates as the probe sequence goes in and out of phase with the patches of hydrophobicity found in helical regions. Thus to obtain a useful measure, the root mean square amplitude of this function was taken to indicate the presence of a helix, rather than the value of the convolution itself. Using a correlation measure to evaluate the predictive value of the result [see Section 4.1.1(ii)] this α-helical hydrophobicity measure was found to be better than the P_α of Garnier *et al.* (21). However, on combining these hydrophobic measures with the Garnier,

Osguthorpe and Robson probability profiles the prediction obtained was only better by 2%.

(ii) *Hydrophobic moments.* The predictive significance of hydrophobic periodicities was first recognized by Schiffer and Edmundson (31) and by Palau and Puigdomenech (32) who represented the residues associated with an α-helix on the circle that would be seen looking down the axis of the helix (a spiral representation has also been suggested). In such a representation it can easily be seen if the helix has a hydrophobic side. In many cases it is often more informative simply to represent the helix as a net (33) and mark hydrophobic patches.

A more sophisticated formulation is due to Eisenberg *et al.* who described amphipathicity by a hydrophobic *moment* (34). Rather than performing convolutions, as crystallographers, they naturally expressed their hydrophobic moment in more versatile representation as the *modulus* of the Fourier transform of the (one-dimensional) hydrophobicity function. This allowed them to identify peaks in the resulting *power spectrum* at any frequency. They found the α-helical periodicity of 3.6 to be clearly defined.

A moving (local) Fourier transform thus provides a predictor of regular structures that have a hydrophobic and a hydrophilic side and has been used predictively (35).

Eisenberg also devised a simple measure based on the hydrophobic moment that distinguishes transmembrane helices from amphiphilic helices (36), but care must be taken in applying this as transmembrane helices in large assemblies may be amphipathic.

All the above approaches have much in common with the secondary structure prediction method of Lim (37).

4.3 Pattern recognition methods

4.3.1 *Lim's method*

An increasing number of secondary structure prediction methods are appearing that search for predefined linear patterns of residues and residue properties (e.g. hydrophobicity) along the protein sequence. The methods, to a large extent, follow the approach of Lim (37).

Lim's method concentrates principally on the identification of sequentially local patterns of hydrophobicity. These typically correspond to the hydrophobic positions expected with amphipathic elements of secondary structure. For example, a half-buried α-helix, in which the residue at position i points towards the core, would be expected to have hydrophobic residues at positions i, $i+3$ and $i+4$ or its reverse (i.e. i, $i-1$ and $i-4$). A large body of such rules was defined by Lim specifying not only patterns of hydrophobicity but also the size of side chains allowed and other properties.

Lim's rules were derived from all the then known structures, but despite this moderate sized database, many of the rules now seem arbitrary in the light of current knowledge of structures and structural analysis. Despite this, the method remains surprisingly effective and provides a useful contrast to statistically based methods. An early criticism of Lim's method was the larger number of parameters implied by its numerous rules. As these parameters were derived (or optimized) on all the then known structures it was difficult to assess how specific the parameters were to this data set. To answer this, some modern comparative assessments of Lim's method are considered in Section 4.4.

A further problem with Lim's method was the difficulty of encoding it as a computer program. The complex interactions of the many rules could not be handled effectively in the FORTRAN language. Currently an attempt is being made to encode the Lim rules in the artificial intelligence oriented language PROLOG (38). It is to be hoped that such a formulation will render the method more accessible and perhaps enable revisions.

4.3.2 *Method of Cohen, Abarbanel, Kuntz and Fletterich*

The suitability of more modern computer languages for pattern matching has encouraged Cohen *et al.* (39) to implement a larger number of Lim-like rules in the programming language C. The method is specific to the β/α class of structure and some of the rules specified derive from a previous analysis of that type of structure (30). Some of the other rules are those described by Richmond and Richards (40) to identify the packing faces of α-helices [see Section 5.1.1(i)] while others are embodiments of Lim's principles or are intuitively derived.

An important part of the method is its further treatment of the pattern locations once they are identified. The success of this post-processing is largely attributed to the accurate prediction of turn regions. The initial prediction of turns is based on an assessment of hydrophobilic peaks in the manner of Rose (27) (see Section 4.2.1) but is made contingent on the prediction of other structures by a series of conditions and cutoffs. Although it is difficult to assess the implicit degrees of freedom of these rules it should be considered whether they are not so many that they could allow any database, especially a restricted one, to be fitted well. Following the assignment of turn regions the structures of the intervening segments are permutated and a best selection made by a further series of rules that bias the prediction of structures towards the expected alternating β/α pattern.

The results of the method are again difficult to assess as they are presented as the best combination of possible segment assignments that match the observed. For a medium-sized protein there are only about 10 segments of structure that can be only one of two types (α or β). The number of permutations that are needed, on average, to correct mispredicted structures is difficult to estimate but cannot be many (no controls are reported). Without the percentage of residues correctly predicted by the top scoring combination it is difficult to compare the results of this method with others.

4.3.3 *Method of Taylor and Thornton*

When assessed objectively, the wide variety of secondary structure methods discussed in the preceding sections attain a level of accuracy that is not easily improved by either refinement of the method or extending the data set of structures from which parameters are derived. This difficulty almost certainly arises from the influence of sequentially long-range interactions and the restrictions required to form favourable chain fold topologies.

(i) *Super-secondary structure prediction.* Ideally a prediction of structure from sequence should incorporate the wealth of knowledge that has been gathered about long range interactions in the analysis of higher levels of protein structure organization. The most obvious interactions to consider are those in super-secondary structures, since interactions between sequentially adjacent secondary structures predominate.

Taylor and Thornton (41,42) analysed these interactions in the $\beta\alpha\beta$ unit. This widespread and well defined super-secondary structure is especially difficult to predict as the β and α structure predictions frequently coincide.

(ii) *Template matching.* The essence of this approach was the use of a template, derived from the analysis of all known $\beta\alpha\beta$ structures, to locate probable $\beta\alpha\beta$ sites on a sequence. Templates of different size were used to model the range of observed variation and each template was fitted to the probability profiles produced by the method of Garnier *et al.* (21). The areas of the probability profiles in the corresponding region of the template was measured and the product of these areas taken as a goodness-of-fit. Thus, if the alignment of the template segments corresponded with peaks in the profiles the fit was good, but if any of the required peaks was missing then the fit was poor.

Having read and stored the template data, the fitting program aligns every template at every position along the sequence and calculates a fit. After this the highest scoring templates are retained for further processing. For these templates, a table is generated which defines whether any pair of fits can logically co-exist in the same prediction. This uses the simple rule that if the β- and α-segments overlap then the pair cannot be simultaneously selected. The table is processed by a recursive procedure that searches the *tree* of possible combinations for the highest scoring, structurally consistent set of templates.

For structure prediction, the selected template fits were not taken directly as a prediction but were used only to modify the original probability profiles of Garnier, Osguthorpe and Robson. This was implemented using the strength of fit as a weight to moderate the effect of templates. Thus good fits alter the profiles most, while weak fits have least effect and where there are no fits the Garnier, Osguthorpe and Robson prediction is unaffected (*Figure 5*).

(iii) *Advantage of super-secondary structure prediction.* The method was applied to 16 β/α class proteins of known structure. The average improvement over the method of Garnier *et al.* (21) (which was optimized on the same 16 proteins) was 7.5%. The various methods for measuring success all concentrate on accuracy at the level of the individual residues. However, when considering the super-secondary structure of a protein, a higher level measure is useful for this coarser, yet topologically important, definition of structure. For this, Taylor's structure abstract (42) was used [see Section 4.4.1(iii)] and by this measure the location of structural units was improved by 15%. This is a significant gain over simple secondary structure prediction.

An important advantage gained from predicting super-secondary structure is that the units predicted have an implicit three-dimensional structure. For example, the $\beta\alpha\beta$ unit is always right-handed (7). Thus, their prediction is a step towards predicting tertiary structure of a protein. This aspect will be considered further in a following section.

4.3.4 *Method of Kabsch and Sander*

The Kabsch and Sander method of pentapeptide conformation prediction (23) segments the entire sequence into overlapping pentapeptides [see Section 4.1.1(iii)]. Many of these overlaps, like the templates of the previous method, will imply the overlap of incompatible structures. To avoid this the method considers all the possible combinations of segments and selects the one with the highest sum of scores that includes no structural inconsistencies.

4.4 Accuracy of secondary structure prediction

4.4.1 *Measures of accuracy*

(i) *Standard residue count.* A detailed consideration of the accuracy of secondary structure prediction has been delayed until this point in the text as it is not a simple matter to know what to measure. This problem is considered by Schultz and Schirmer (3) who have collected a bewildering variety of scoring methods that have been used to evaluate accuracy. Their varied use by different workers (often to the advantage of their own method) makes a comparison of the predictive accuracy of different methods very difficult. Schultz and Schirmer recommended that the simplest method of measuring accuracy be adopted uniformly and this has to some extent been adhered to in recent works. This measure is the percentage of residues correctly assigned to all the conformational states considered, including a coil or unstructured state whether this was specifically predicted or not.

(ii) *Correlation coefficients.* Use of the above measure should not, however, discourage the use of Matthews' (43) modification of the correlation coefficient. The standard correlation coefficient is a product of two probabilities, normalized by their standard deviations. If the probabilities are always one and zero (as when a structure either exists or does not) then the simplifications made to the evaluation of the correlation coefficient by Matthews are possible. If, however, only one probability distribution is discrete (one or zero) while the other is continuous, then only a partial simplification can be made. Such a measure was found to be useful by Taylor and Thornton (22) to evaluate the predictive potential of hydrophobic profiles. The measure has the advantage that it can be applied to the comparison of a single profile with the occurrence of a given

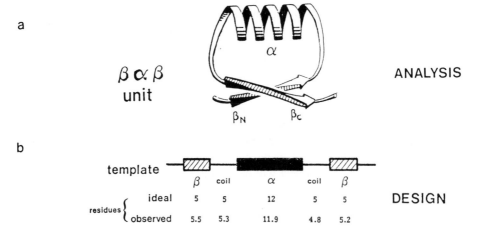

302

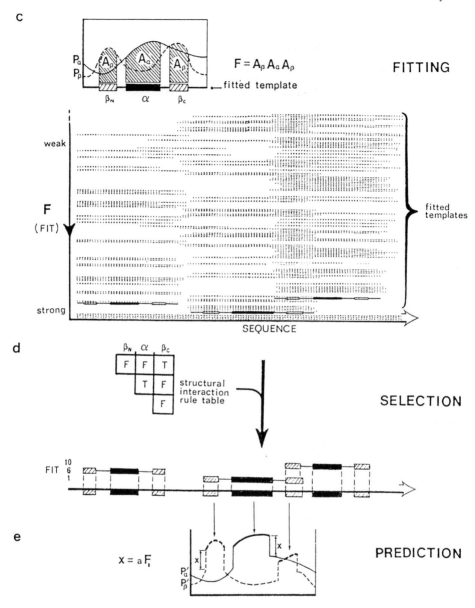

c

$$F = A_\beta A_\alpha A_\beta$$

fitted template

FITTING

d

SELECTION

e

$$x = aF_i$$

PREDICTION

Figure 5. Graphic outline of the template based structure prediction method of Taylor and Thornton (22,55,56). (a) ANALYSIS: Schematic representation of a $\beta\alpha\beta$ unit. Analysis of these structures provided the data from which a template was designed to locate the structures in the sequence. (b) DESIGN: Templates are designed to represent the observed $\beta\alpha\beta$ units. Only one (the average) is shown but 30 of different sizes were used to reflect the observed variation. (c) FITTING: Each probability is fitted at every position along the sequence and a goodness-of-fit calculated as the product of the peak areas of the probability profiles against which the template is matched. The strongest fits are shown for a small protein, rank ordered with the strongest fit closest to the sequence. (d) SELECTION: The fitted templates, along with a logical table of whether pairs can co-exist in the same prediction, are processed by a recursive procedure that searches the tree of possible template combinations and selects the highest scoring. (e) PREDICTION: The selected templates modify the original probability profiles in proportion to their strength of fitting. After this the highest modified probability profile constitutes a prediction of structure.

303

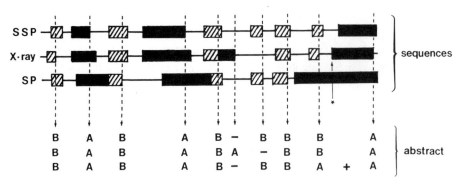

Figure 6. Three sequences of a small protein are shown with secondary structure indicated as α-helix solid, β-strands hatched and turn or coil as a line. SP is a secondary structure prediction (21) and SSP a super-secondary structure prediction (22), while X-ray is the secondary structure determined by crystallographic analysis. Broken lines mark each C-terminal edge of a structure where an entry in the abstract is made. These are coded as B=β, A=α and − =unstructured. The starred arrow indicates a point where the SP terminal helix is reconsidered for abstract entry (caused by the start of the terminal X-ray helix). The resulting double entry for this structure in the abstract is linked by a + indicating a continuous structure.

structure, avoiding the necessity of converting the profile to a prediction (which would involve the use and optimization of a cutoff level).

(iii) *Structure abstract.* When attempting to predict a protein chain fold, the number of predicted structures is often more important than their exact length. For example, in the bundle of rods model of a protein structure, if one of the rods is not predicted, then the direction of all the following rods relative to all those preceding will be reversed. Furthermore, the exact lengths of secondary structures is probably determined by the co-operative effect of sequentially remote segments of chain so it should not be expected that secondary structure prediction methods (which only consider local information) could accurately delimit structure. In predictions of known structures, the strands and helices are often 'correctly' predicted but are slightly displaced or of the wrong length. To circumvent this problem, I devised a simple algorithm that can easily be applied by hand to any number of aligned structure predictions (42). This procedure reduces the secondary structural units to single letters in what is referred to as a structure abstract (*Figure 6*).

The process begins by 'tidying' the prediction by making 'odd' single residues conform to the surrounding structure type (single residue breaks in structure were, however, retained as they are structural possibilities). After this any unreasonably short structures were discounted [although not using such severe criteria as Nishikawa (20)] and then at the C-terminal edge of every structure a record in the abstract is made of any overlapping structures. With a few extra conditions to allow for the unique status of the X-ray structure, this produces a reasonable summary of the correspondence between structures.

The measure is especially appropriate for methods that explicitly segment the structure prediction (i.e. those in the previous section) as these tend to be precursors of tertiary structure prediction techniques.

4.4.2 *Secondary structure definition*

Even if a standard measure is adopted, a serious problem remains as to the standard with which secondary structure predictions should be compared. The problem arises because no two protein crystallographers work with exactly the same idea of what constitutes a secondary structure. The basic hydrogen bonding patterns of secondary structure are not in dispute, but the hydrogen bonds themselves often are. This arises because the bonds vary continuously in length and angle and ill-defined cutoffs on these define their existence. Thus, one crystallographer might have a lenient definition of hydrogen bonds and define, say, an α-helix that would be broken in two by another.

An early attempt to automate secondary structure definition was made by Levitt (44) but this considered only C_α positions and was not wholly satisfactory. A recent attempt by Kabsch and Sander (45) was more successful as it considered a full backbone representation and calculates a rough energy for each bond. The method, unfortunately, still produces some serious discrepancies with the crystallographic definitions and further refinement may be necessary.

4.4.3 *Objective assessments of accuracy*

Nishikawa (20) objectively assessed the predictive accuracy of the three main secondary structure prediction methods (those of Chou and Fasman; Garnier, Osguthorpe and Robson; and Lim). He found a surprisingly low accuracy for all the methods of less than 55% for three states (β, α and coil) and less than 45% for four states (β, α, turn and coil). Strangely, this was 20% less than the accuracy reported in the literature as reviewed by Schultz and Schirmer (3). He concluded that some unconscious positive bias enters the interpretation of the results of the less rigorous methods.

Considering the same three methods, Kabsch and Sander (46) also carried out an objective assessment of predictive accuracy against their automatic definition of structure. They too found poor predictive accuracy with two of the methods with an accuracy of around 55% and the popular Chou and Fasman method worst at only 50%.

4.5 **Prediction of structural classes**

Garnier *et al.* (21) and more recently other workers (22,28,39) have demonstrated that the accuracy with which secondary structure can be predicted is improved if the structural class to which the protein belongs can be identified.

4.5.1 *Predicted structure composition*

(i) *Composition plot.* Following a series of decision constant (DC) [see Section 4.1.1(ii)] optimization trials on their data set, Garnier *et al.* (21) recommended different combinations of DC for α- and β-structure based on 20% and 50% thresholds in the amount of predicted secondary structure [see Table 6 in (21)]. Using an extended database of structures, Taylor and Thornton (22) investigated this relationship between known and predicted structural class. The secondary structure composition of each protein was plotted on a graph of percentage α-helix against β-sheet. This point was then connected to the corresponding secondary structure composition that was predicted by the Garnier,

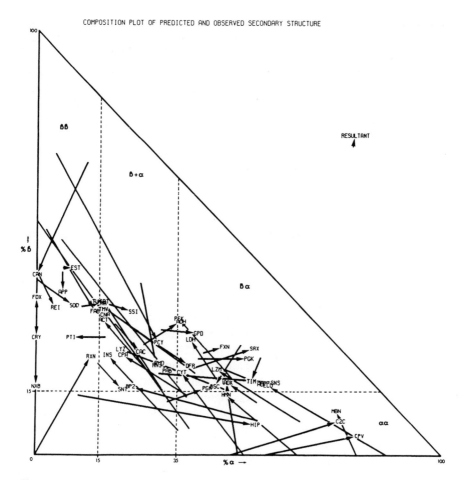

Figure 7. Composition graph of observed and predicted secondary structure. Lines join the observed composition of the protein to its predicted composition. The length of line thus indicates the error in the prediction. The Protein Structure Databank codes for the proteins are plotted at the end of the arrows. (Confused areas are resolved in *Table 2*.) The broken lines are the percentiles that best divide the proteins into their structural classes which are indicated in each sector.

Osguthorpe and Robson method with zero (unbiased) DC (*Figure 7*). The result was not intended to illustrate how poor secondary structure prediction is but it provides a very graphic representation of this. On the plot, each protein should be represented by a dot if the prediction method were 100% accurate. The length of the lines thus indicates the error in predicted composition and these have root mean square errors of about 16% each. This is not a result of systemic biases in the application of an old method to new data as this would be detected in the magnitude and direction of the resultant of all the composition difference vectors. That this is insignificant compared with the size of the errors indicates that the Garnier, Osguthorpe and Robson parameters have remained unbiased even when applied to double the number of structures from which they were derived.

The direction of the lines in the plot tend to run top-left to bottom-right. This indi-

306

Table 2. Matrix of predicted against observed structural classes of globular proteins (represented by three letter PDB code). The table is derived from *Figure 7* (see text also) and is based on the predicted composition of secondary structure. Proteins not in the diagonal cells (heavy border) are wrongly predicted. Most confusion is between the $\beta+\alpha$ and all-β classes

Observed class	*Predicted class*				
	$\alpha\alpha$	$\beta\alpha$	$\beta+\alpha$	$\beta\beta$	*None*
$\alpha\alpha$	MBN CPV C2C HMN 4	CYT UTG MHB 3			
$\beta\alpha$		SRX PGM FXN PFK TIM ABP LDH DFR ADK PGM ADH GPD B5C 13	RHD 1		
$\beta+\alpha$	HIP 1	LZM SNS 2	CAC SNT CPA SSI LYZ TMV ACT INS SBT RNS 10	PTI TLN CRN BP2 4	
$\beta\beta$			CNA PAB CHA PCY 4	REI NXB EST CRY FAB APP SOD 7	
None				FDX RXN	 2

cates that the composition error between β- and α-structure is correlated (with a coefficient of almost 0.5), implying that on average the right amount of structure is predicted but that the type is wrong. Extreme examples of this behaviour can be found in plastocyanin which is an all-β protein and has been predicted to have 34% α-helix. Similarly, haemoglobin, an all-α protein, is predicted as 20% β-structure. This effect results largely from the lack of discrimination that many hydrophobic residues have between α- and β-structure. The redundancies can be seen more clearly in the Chou and Fasman parameter sets (rather than in those of Garnier, Osguthorpe and Robson). In these, a few residues (such as glutamic acid) clearly discriminate between the two structure types, however, many are indifferent. The result is a reasonable distinction between the occurrence of structure and non-structure, but poor discrimination between structures.

Table 3. Correspondence of functional and structural classifications (47,48). Intracellular proteins remain in the reducing environment of the cytoplasm while the intercellular proteins are exported to the external oxidizing environment and, hence, usually contain disulphide bonds. The approximate agreement with the relationships shown is indicated in each cell of the table

	Enzyme	*Non-enzyme*
Intracellular	β/α	α
	79	56
Intercellular	β	$\alpha+\beta$
	66	52

(ii) *Plot partitions.* The plot can best be partitioned into structural classes (on the basis of their known structure) by the 15 and 35 percentiles. The results of this compartment-alization are shown in a table of predicted against observed structural class (*Table 2*) and if the assignment were perfect, entries would only occur in the diagonal cells. This is followed reasonably well except for a poor distinction between the all-β class and the $\beta+\alpha$ class. Many of the misclassified structures bind relatively large ligands (e.g. the high-potential iron protein that binds a large iron−sulphur cluster). Some of the all-α type proteins that bind haem (a large hydrophobic group) are often found to contain spurious β-predictions. Examination of these predictions in relation to the structure indicates that they are often associated with the segments of chain that form the haem binding pocket. Thus if, from biochemical evidence, a protein is known to bind a large ligand, then its predicted structures must be considered with caution.

4.5.2 *Amino acid composition plot*

Nishikawa and Ooi (47) demonstrated that if the residue composition of the protein is available, it is possible to predict the structural class of the protein. (When sequences are chemically determined this information is often known prior to the sequence. However, with the increased use of nucleic acid sequencing techniques the sequence is now available immediately.) They initially considered the residue composition of a large number of sequences that could be classified as enzyme/non-enzyme and intra-/inter-cellular. Each protein was located in the multi-dimensional space of amino acid frequency at a point defined by its composition. In this space the mean positions of the enzyme/non-enzyme defines an axis that was found (coincidentally) to be roughly at right angles to the corresponding intra-/intercellular axis. Together these define a co-ordinate frame in which the displacement due to each amino acid type can be found. Thus for a new protein composition, the vector components of each amino acid need only be added to identify the probable protein type.

The importance of the work to structure prediction is, however, the observation that the above classification corresponds significantly with the structural class of the protein as shown in *Table 3* (48). Why this useful structural corollary should exist is not obvious, however, it is possible that the correspondence relates to the evolutionary descent of the proteins.

4.5.3 *Prediction of domain boundaries*

The difficulty of structure prediction increases with the length of the sequence. Thus any method that can partition a long sequence into self-contained domains greatly increases the chance of a correct fold being predicted. Attempts to analyse sequence characteristics indicative of domain divisions have suffered from an ill-defined definition of what constitutes a domain.

This vital problem can, however, often be approached by the use of other clues. For example, predicted transmembrane regions serve to delimit domains or biochemical evidence, such as known enzymic cleavage sites and introns can suggest domain junctions. Further evidence from homologous sequences will be discussed later.

5. EMPIRICAL TERTIARY STRUCTURE PREDICTION

The methods discussed in the preceding sections have concentrated on the prediction of secondary structure. In the section on pattern recognition it was seen how, as larger structures are considered (e.g. super-secondary structure), more implicitly tertiary structural information is incorporated. This is done at the expense of the predictive methods becoming less general and it was seen how the identification of structural class becomes more important in this context. In this section I continue this trend towards the description of increasingly specific structures culminating in a brief description of structure building by direct sequence homology with a known structure.

5.1 Secondary structure docking

5.1.1 *Combinatorial approach*

(i) α/α *packing*. Analysis of protein structure has revealed general patterns of hydrophobicity associated with the packing together of secondary structures. These were often too unspecific to allow any conclusions to be drawn as to the nature of the interaction (e.g. a hydrophobic patch on the face of an α-helix might pack against another helix or a β-sheet). In 1978 Richmonds and Richards (40) analysed the packing of α-helices in one protein, sperm whale myoglobin, with the hope of finding more specific hydrophobic patterns that would help predict how helices pack together in general. Three classes of helix interactions were observed, each with a characteristic separation and interhelical angle and, importantly for future predictions, each packing class had a typical pattern of hydrophobic residues that packed together.

A simple pattern matching algorithm was then used to relocate possible hydrophobic packing patches along the sequence of myoglobin and from the results an attempt was made to reassemble the structure of myoglobin by bringing together, or docking, complementary patches. The predicted sites did not, of course, define a unique structure and each possible combination of patches was tried, resulting in 100 million possibilities! Fortunately, many of these predicted structures could be discounted on the grounds of two simple stereochemical criteria:

(i)   There must be a sufficient number of residues between sequential α-helices to span the connecting distance.

(ii) α-helices must not interpenetrate.

These filters led to a reduced list of only 20 structures (49).

Further application of known biochemical distance constraints subsequently reduced the list of possibilities to two structures, one of which was a close approximation to the native (50).

(ii) *β/β packing.* With the aim of extracting similar information, the packing of pairs of β-sheets was analysed in several all-β proteins (51−53). Typically, the sheets in these structures pack 10 Å apart at a 30° angle and the hydrophobic residues that mediate the packing trace anti-complementary diagonal patches on the surface of the sheets (this is a consequence of the standard twist of the β-sheet). The combinatorial approach was again used to generate all possible strand arrangements and these were filtered for those which could be legitimately connected. In addition, the toplogical observation that connections between β-strands are almost invariably right-handed was used to further reduce the number of possible strand arrangements. A large number of possible strand arrangements often remained but when these were rank-ordered on the quality of their anti-complementary hydrophobic patch and number of hydrogen bonds, the correct arrangement was generally in the top 300.

(iii) *β/α packing.* A similar approach was applied to the β/α class of protein structures (30,53). These have an additional advantage over the all-β class of protein as they often have parallel β-strands which reduces the possible combinations by half. Using this and other topological constraints, the number of possible folds was found to be more restricted than for the all-β class with typically the correct fold found in the best 100 structures and for one protein, in the top six.

5.1.2 *Limitations of the combinatorial approach*

The applications of the combinatorial approach outlined above indicate the possibilities for tertiary structure prediction if an exact definition of secondary structure is known. Without this the number of possible structures becomes much larger as the secondary structures can no longer provide rigid distance constraints on the end−end connections. The method must, therefore, be seen as a prediction 'bridge too far' as it is apparent now that secondary structure prediction is probably inherently incapable of providing the exact structure end definitions required by the combinatorial approach.

Despite this it is possible to see how a less compartmentalized union of the two approaches might interact to boot-strap towards a mutually consistent structure definition.

5.2 **Finger-print templates**

5.2.1 *Turn patterns*

The continued analysis of specific protein sub-structures has revealed many associated residue patterns and often the stereochemical reasons why particular residue types are preferred in certain locations. Recently, interest has concentrated on the turns between secondary structures. In the light of the limitations of the combinatorial approach, this is important as the identification of a turn of defined length restores the effectiveness of the end−end distance filter and thus restrains the relative adjacency of secondary structures in the bundle of rods.

(i) *β-β hairpin.* An analysis of the β-β hairpin by Sibanda and Thornton (54) indicated considerable structural homology within each type of turn found at this topological position. Most importantly, they also observed distinct amino acid preferences at particular positions in the turn. Initial tests of the predictive value of these patterns using

310

the general template matching program of Taylor (55,56) indicate that the identification of turns is greatly improved resulting in an improved overall secondary structure prediction (57).

5.2.2 *Nucleotide binding patterns*

Wierenga and Hol (58) observed and explained the conservation of a relatively specific pattern of residues (principally glycines) associated with the binding of a nucleotide moiety to the first $\beta\alpha\beta$ unit of the Rossmann fold (6). This pattern has obviously great predictive power and has been used to justify the prediction of a nucleotide binding site in the sequences of some oncogene products (58,59).

The unique location of this pattern as the first $\beta\alpha\beta$ in the domain is especially useful in extending predictions beyond a single $\beta\alpha\beta$ unit as it provides a foundation from which less certain predictions can be built (or in combinatorial terms, greatly reduces the possibilities). This combination of partial sequence homology and secondary structure prediction has been formalized and automated within the general template matching procedures of Taylor (55).

A template was designed to incorporate the features of Wierenga and Hol's nucleotide binding 'finger-print' and matched in conjunction with the general $\beta\alpha\beta$ template described in Section 4.3.3. By designing the finger-print template to score highly, its inclusion in the final selection of fitted templates that constitute the prediction can be ensured. This has the effect that the general $\beta\alpha\beta$ template fits must accommodate the presence of the secondary structure associated with the finger-print. As the finger-print is relatively specific there is usually no ambiguity in its location. Its effect is thus not only beneficial to the prediction of its own local secondary structure but also to the prediction of surrounding structure. Furthermore, by inference from other known structures it is almost certain that the $\beta\alpha\beta$ following the finger-print template will lie adjacent to it.

Since the identification of the Wierenga and Hol pattern, other subtle variants on the nucleotide binding loop have also been identified (60,61) and are associated with more specific types of binding (e.g. ATP, GTP, etc.).

Of particular interest is one of the patterns described by Walker *et al.* (61) that lies generally some 100 residues C-terminal to the Wierenga−Hol pattern. This is again associated with nucleotide binding and contains an α-β structure that lies adjacent in the sheet to the N-terminal strand of the Wierenga−Hol pattern. Identification of this pattern again greatly restricts the possibilities of the structures following Wierenga−Hol finger-prints. Indeed, it has been found for the cation-dependent ATPase sequences that the predicted topologies of the nucleotide binding domain are limited to only a few possibilities (62).

5.3 Structure prediction by sequence homology

In this section I consider ways of extending the finger-print template approach to a more general representation that allows a wide variety of patterns to be simultaneously used. The method forms a bridge from the finger-print templates discussed in the preceding section and the prediction of structure by direct inference from a closely homologous sequence of known structure.

The problem often arises that although no good overall homology exists between a new sequence and one of known structure, odd fragments can be found that, given the biochemical context, suggest a structural relationship. Because such fragmentary sequence homology is interspersed with only tentative homology, or none at all, and that large insertions or deletions might occur, the application of conventional alignment algorithms usually fails to detect a significant alignment. The problem, however, is obviously well suited to an approach by template matching techniques and in the following section I describe the application of the template matching algorithm of Taylor (56) to the problem. This method allows the structural and biochemical requirements of a sequence to be explicitly stated and automatically used to identify regions that correspond to predefined structures.

5.3.1 *Consensus template matching*

The method is an extension of a pattern recognition algorithm previously used for secondary structure prediction to identify likely super-secondary structures (22,41). The approach is based on matching templates with the sequence and finding the best combination of templates which satisfy predetermined rules. In the above application the templates consisted of patterns of secondary structure and were subject to the simple rule that β-structural elements must not be predicted in the same place as α-helical elements [Section 4.3.3(ii)]. The templates used for homology matching consist of patterns of residues and can be subject to more complex selection rules reflecting the greater restrictions on template occurrence. The method is, however, independent of the structure of the templates and it is possible to use the type of template described below with the secondary structure templates described previously. Such a combination is useful for very remote homologies where little conservation remains apart from hydrophobic patterns of residues and local preferences for a particular secondary structure.

The degree to which it is desirable to impose features on a sequence alignment depends on the certainty with which they are expected to occur. For example, in a search for immunoglobulin-like domains, a strong requirement will be that the sequence contains two widely separated cysteines suitable for forming the intra-domain disulphide bond. Similarly, if a particular enzyme is suspected, certain residue types may be required to constitute the active site but may be mutable within known physico-chemical limits (e.g. a required hydrogen bond donor). In addition to these specific requirements, which depend on the particular protein family, there is the general requirement that insertions and deletions are expected in regions which are exposed to solvent in the tertiary structure and especially in regions which are not part of a secondary structure.

The method requires that the features expected to be found are stated in the form of templates where each template has a one-to-one correspondence of positions with the region of sequence in which the features occur. Each position in the template can be assigned a property which can range from an absolutely conserved residue type to an unassigned gap. Generally, the positions are assigned to physico-chemical properties of the amino acids, such as hydrophobicity, but any combination of amino acids can be defined to match a required property or function.

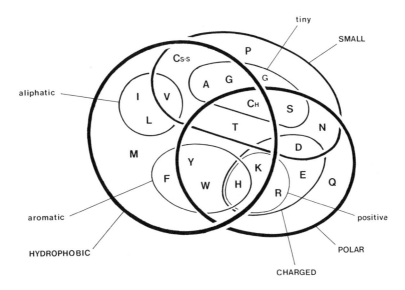

Figure 8. The Venn diagram shows the relationship of the 20 naturally occurring amino acids to a selection of physico-chemical properties which are important in the determination of protein tertiary structure. The diagram is dominated by properties relating to size and hydrophobicity. The amino acids are divided into two major sets, one containing all amino acids which contain a polar group (POLAR) and a set which exhibits a hydrophobic effect (HYDROPHOBIC). A third major set, SMALL, is defined by size and contains the nine smallest amino acids. Within this is an inner set of smaller residues, TINY, which have at most two side chain atoms. The location of Cys is ambiguous and two locations are indicated. Other sets include full-charge (referred to as CHARGED) which contains the subset POSITIVE (negative is defined by implication) and AROMATIC and ALIPHATIC. The latter set is not as general as the name implies and includes only those residues containing a branched aliphatic side chain. Because of its unique backbone properties, proline was excluded from the main body of the diagram.

5.3.2 *Classification of amino acid conservation*

The features associated with the templates will naturally be those that have been observed previously in other sequences that have known tertiary structures or can be directly related to a known structure via a good sequence homology. Thus, by aligning sequences with a sequence of known structure it is possible to observe the type of property that is conserved at a given position and use it subsequently in identifying the corresponding position in a new sequence. To do this systematically requires that the conservation is rigorously defined and preferably exhibits no bias towards any particular sub-group of sequences that might be included in the alignment.

(i) *Minimal property sets.* A general representation of a sequence alignment was achieved by representing each position by a property of amino acids (e.g. hydrophobicity, charge, size, etc.). The diversity of physico-chemical properties, however, leads to a complex set of relationships between the amino acids. The more important properties from the point of view of structure have been incorporated into a Venn diagram [*Figure 8*, after Dickerson and Geiss (63)] for ease of visualization. Unions and intersections of the

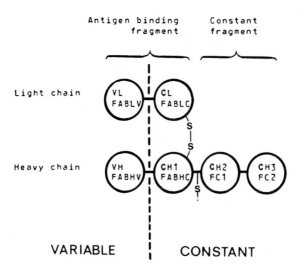

Figure 9. Schematic representation of half an immunoglobulin molecule. Each circle represents a globular domain consisting of two β-sheets 'tied' together by a disulphide bond (not shown). Inside each domain two different abbreviations for the domain type are shown. In these V indicates the variable type domain and H indicates heavy. (FAB, fragment, antigen-binding; FC, fragment, constant.) The complete molecule is a dimer linked through the unsatisfied disulphide bond (S-).

sets included in the Venn diagram were then used to describe the characteristics of a particular combination of residue types brought together by alignment (64). For example, for the alignment of residues SSTAVP (one letter code) the property set that includes these residues is 'SMALL' which also includes ASN and CYS. Thus by assigning a position in a template to a profile corresponding to 'SMALL', the template matches a more general occurrence of the pattern than is observed in the sequences considered. It was hoped that by matching general properties of amino acids the validity of the consensus template sequence could be extended beyond the limitations of the data which determined it.

(ii) *Properties conserved in aligned sequences.* As a test application, I considered the alignment of a large number of closely related immunoglobulin sequences (55). The general impression gained from this was that, as sequences were added to the alignment, the regions of conservation shrank to encompass only the secondary structures that formed the core of the tertiary structure. Within these regions just the buried hydrophobic positions retained any specificity.

5.3.3 *Domain recognition*

Following the approach outlined above, a set of templates was derived for each conserved region in several different immunoglobulin domains (see *Figure 9* for outline of domain structure). These were then matched to the sequences under the constraint that the correct order of the β-strands must be maintained. It was found that the inherent power of pattern recognition by template matching derives from the ability to concentrate on those features of the sequence which are conserved. This aspect is dominant when matching templates to sequences that contain patterns which are similar to the template patterns, but in the application of the templates to sequences increasingly remote from their source, the constraints imposed on the interactions of the templates become more important.

Despite a trend towards a general β-strand pattern of hydrophobic conservation alternating with no conservation, the individual strand types contained enough specificity to be useful in searches. The specificity resulted from deviations in the ideal alternating pattern of hydrophobicity, combined with a few absolutely conserved residue identities. The conserved cysteines played an important role by defining a region which must be spanned by the sequentially intervening strands.

Application to sequences which were not part of the data from which the templates were derived, revealed that the templates continued to behave in a coherent manner where the sequences were thought to be related to the immunoglobulins, but for other sequences, effectively random patterns were produced. Comparison with a more conventional alignment algorithm (65), however, indicated that the templates developed from one set of sequences are not sufficiently flexible to allow for the occurrence of reasonable insertions predicted by the conventional method at positions where no insertion had been observed in the immunoglobulins. The conventional method, in contrast, was too flexible and allowed insertions where, for structural reasons, they would be better avoided.

The ability to define interactions between templates is one of the most important aspects of the approach as it allows the definition of patterns of templates which can be used to constrain not only intra-domain secondary structure relationships, but also possible choices between alternative domain structures. For example, two sets of templates representing the variable and constant domains of the immunoglobulin structure (V and C respectively) were used so that templates from one set were fitted to the exclusion of the other. This can be represented as:

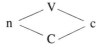

(NB: lower case 'n' and 'c' denote the chain termini).

The same template sets, when fitted in series, can, similarly, be represented as:

$$n-\ V\ -\ C\ -c$$

More specific sets of templates derived from the immunoglobulin heavy and light chains (suffixed H and L respectively) define a more complex network, as follows:

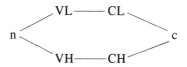

Further distinction of the sub-domains of the constant-heavy chain results in a template network which, as can be seen from comparison with *Figure 9*, closely reflects the intact immunoglobulin structure.

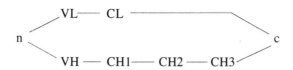

The ability to make the identification of these conserved patterns conditional on each other greatly increases the probability of successful recognition. If sufficient protein finger-prints are collected from as many families as possible they can be structured using the template formalism into a large tree of associated and incompatible patterns which can be used to identify the nature of an unknown sequence.

5.4 Structure building by direct sequence homology

If the sequence of unknown structure is closely ($>50\%$) homologous to a sequence of known structure then a sequence alignment of the two will predict a corresponding structural alignment (66). I will say nothing of the various methods that might be used to produce such an alignment as these are adequately treated in Chapter 13. However, in this section I will briefly review the techniques that are available to construct a molecular model and the confidence that might be placed in such a model. For a fuller account of the state of this rapidly changing field the reader is recommended to consult a recent review [e.g. (67)].

5.4.1 *Model building*

Until relatively recently the only software tools available for model building were those designed by protein crystallographers to build structure into their electron density maps. These generally ran on very expensive interactive graphics systems that allowed fragments of structure to be moved and rotated relative to each other. Perhaps the most commonly used and known of these systems is FRODO (68). Recently, however, many new systems have appeared, some of which also allow electron density to be displayed but most of which are dedicated to the construction of new protein structures by the modification of known structures. The interest of commercial companies in this area has led to the development of such systems being increasingly taken over by professional software companies who maintain and distribute the code.

(i) *Modelling loops.* An inherent problem in building one structure from another is how to model insertions and deletions. These generally occur on the protein structure surface and so are less troublesome than might be expected as in that position they do not greatly alter the packing of residues in the protein core. Two solutions to this problem are usually advanced.

(i)　　The first builds the connecting chain in a stereochemically allowed, but otherwise random, way and relies on final energy minimization to find the best conformation.

(ii)   The second relies on finding a connecting loop from another structure, of the same length as the new insertion, to fill the gap and is often referred to as 'spare-part' model building (69).

(ii) *Modelling side chains.* Most efficient model building systems will substitute the new side chains so that they have a maximum correspondence of atom positions with the residue they replaced. This is often sufficient to maintain adequate packing. In more compact regions such as the hydrophobic core, adjustments of neighbouring side chains are usually required but it is exceptional if the main chain conformation has to be altered.

Despite this, poor packing in the core is the main feature that distinguishes model built structures from crystal structures (70).

5.4.2 *Accuracy of models*

Having built a model by homology of one structure to another it must be carefully considered how valid is the extrapolation. This problem has been analysed by Chothia and Lesk (71), who considered pairs of homologous sequences both of which have known structures. By plotting the sequence homology of the pairs against their structural correspondence (as measured by a least squares superposition of the two structures) they obtained a plot from which the expected structural similarity of any two structures could be read for a given sequence homology. Combined with detailed analysis of the types of structural changes found between members of protein families, they concluded that between 20 and 50% residue identity the changes seen can be large with shifts in individual secondary structures as big as 7 Å. These, however, tend to be correlated in such a way that the active (or binding) site geometry is maintained (72). Above 50% homology the changes are slight and a model built structure can be almost as good as a moderate resolution crystal structure. Where the residue types are substituted the side chains tend to have a different orientation but otherwise stay the same if the residue type is unaltered.

6. CONCLUSIONS

Because of the inherent uncertainty in structure prediction, I have concentrated in this chapter more on the underlying theory of the methods rather than on how to run the programs. This, I believe, is important to avoid the situation where structure prediction programs are run as 'black boxes'.

For those who still retain some faith in structure prediction, the following section provides, in outline, a strategy to be followed when faced with a sequence of unknown structure. However, when following this route, it must always be borne in mind that the structure of the new sequence may be quite unlike any yet seen and the sequence should consequently not be forced to conform to a known structure if this implies inconsistencies.

6.1 **A practical approach**

In approaching a new sequence with the aim of predicting its structure it is recommended that the order in which techniques are described in this chapter is reversed for their

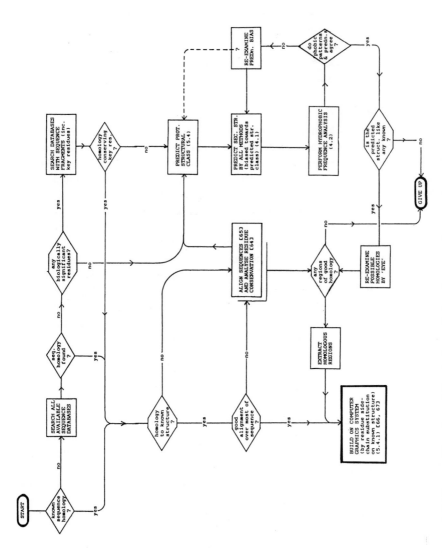

Figure 10. Flow diagram indicating possible paths to follow in the prediction of structure for a protein sequence. In each box where action is required, a reference is given either to a paper in the literature or a section of this chapter.

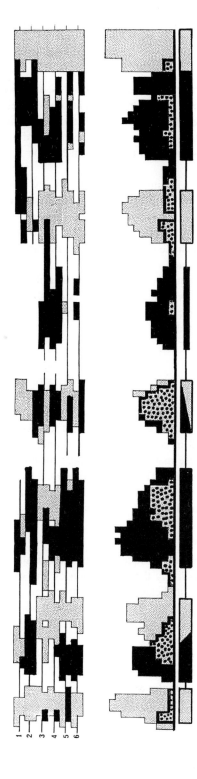

Figure 11. Six relatively homologous aligned sequences are shown. These are represented as a line with predicted secondary structures determined by two standard methods. One prediction (19) is shown above each line and the other (21) below. α-helix is indicated as solid, β-strands as shaded and coil or turn as a simple line. This representation allows the inhomogeneity of the different methods on the different sequences to be readily appreciated. From the sequence homology it can be expected that the structures will have the same basic pattern of secondary structure, thus some reconciliation must be achieved for the often conflicting predictions. A simple approach to this can be made by adding the predictions at each position and taking the most dominant as shown in the plot and 'consensus' predictions below. This, however, still leaves some ambiguous predictions.

application. The application of methods depends, of course, on what is known about the sequence and to help with this the flow diagram in *Figure 10* can be followed.

6.1.1 *Homology searches*

(i) *Overall homology.* Firstly the sequence should be matched against any available sequence databases and for this the Protein Identification Resource (73) is to be recommended for ease of use. It will save a lot of time pondering over dubious predictions if a sequence homology is found with a sequence of known structure enabling an accurate structure to be built (as described in Section 5.4.1).

(ii) *Fragmentary homology.* Failing the discovery of any overall homology, searches should also be made for any partial homologies such as with the patterns associated with various binding sites. Identification of a partial homology must be treated with caution as these can occur by chance (74). Thus, any available information of biochemical activity should influence both the choice of patterns that are searched for and the interpretation of any that are found.

If several homologous sequences are available then the regions that are conserved in these may be associated with activity and consequently be conserved also in other sequences. They thus provide a good probe to search the databases for other sequences that may be remotely related.

6.1.2 *Template matching*

If any region of conservation can be associated with a structure, then, using the general template matching method of Taylor (55,56), this can be used to constrain the prediction of other neighbouring structures. However, failing the identification of any specific structural homology clues, the finger-print templates associated with general turn types can be matched in conjunction with secondary structure patterns.

The interpretation of these matches will be more meaningful if the structural class of the protein can be identified. This can be performed by all the methods described in Section 4.5 but should still be considered in the context of known biological function and location (e.g. intra-/intercellular, enzyme/non-enzyme, membrane bound, etc.).

6.1.3 *Secondary structure prediction*

The increasing trend towards the sequencing of biologically related molecules raises a new problem, since it is likely that the sequences will exhibit a greater degree of mutual similarity than their predicted secondary structures. Thus, not only must a compromise be found between the results of different methods but a further compromise solution must be found to reconcile the different results of the same method on slightly different sequences (see *Figure 11*).

Using methods that segment the sequence (e.g. templates) this can be done in a combinatorial fashion to produce a 'best' compromise. However, simply adding profiles or predictions together is a simpler and equally effective way. When averaging different secondary structure prediction methods, it must be remembered that the average of several, possibly wrong, answers does not give a right answer.

7. THE AVAILABILITY OF PROGRAMS

Many of the protein secondary structure prediction methods mentioned in this chapter, including those of Chou and Fasman (19), Garnier, Osguthorpe and Robson (21) and Lim (37) are in the Protein Secondary Structure Prediction Suite from Leeds University. The programs, written in FORTRAN 77, are available for the DEC VAX minicomputer running the VMS operating system, and IBM and Amdahl computers using the VM/CMS operating system. Some conditions are imposed concerning limits to redistribution and usage of these programs. At the time of writing the suite is available for £50 to academic users and for £250 to commercial users, from E.Eliopoulos and A.J.Geddes, Department of Biophysics, University of Leeds, Leeds LS2 9JT, UK.

The author's structure prediction programs using pattern recognition techniqes (41,42) written in FORTRAN 77 and Pascal for a DEC VAX using the VMS operating system are available from the author at the Laboratory of Molecular Biology, Department of Crystallography, Birkbeck College, University of London, Malet Street, London WC1R 7HX, UK.

8. REFERENCES

1. Monod,J. (1972) *Chance and Necessity,* Vintage Books, New York.
2. Rees,A.R.C. and Sternberg,M.J.E. (1984) *From Cells to Atoms,* Blackwell Scientific, Oxford.
3. Schulz,G.E. and Schirmer,R.H. (1979) *Principles of Protein Structure,* Springer-Verlag, New York.
4. Richardson,J.S. (1981) *Adv. Protein Chem.,* **34,** 167.
5. Pauling,L., Corey,R.B. and Branson,H.R. (1951) *Proc. Natl. Acad. Sci. USA,* **37,** 205.
6. Rao,S.T. and Rossmann,M.G. (1973) *J. Mol. Biol.,* **76,** 241.
7. Sternberg,M.J.E. and Thornton,J.M. (1976) *J. Mol. Biol.,* **105,** 367.
8. Argos,P., Rossmann,M.G. and Johnson,J.E. (1977) *Biochim. Biophys. Acta,* **57,** 83.
9. Weber,P.C. and Salemme,P.R. (1980) *Nature,* **287,** 82.
10. Chothia,C., Levitt,M. and Richardson,D. (1977) *Proc. Natl. Acad. Sci. USA,* **74,** 4130.
11. Richardson,J.S. (1977) *Nature,* **286,** 495.
12. Richardson,J.S., Richardson,D.C., Thomas,K.A., Silverton,E.W. and Davies,D.R.J. (1976) *J. Mol. Biol.,* **102,** 221.
13. Levitt,M. and Chothia,C. (1976) *Nature,* **216,** 552.
14. Nemethy,G. and Scheraga,H.A. (1977) *Quant. Rev. Biophys.,* **10,** 239.
15. Robson,B. and Platt,E. (1986) *J. Mol. Biol.,* **188,** 259.
16. Levitt,M. (1976) *J. Mol. Biol.,* **104,** 59.
17. Robson,B. and Osguthorpe,J.D. (1979) *J. Mol. Biol.,* **132,** 19.
18. Cohen,F.E. and Sternberg,M.J.E. (1980) *J. Mol. Biol.,* **137,** 9.
19. Chou,P.Y. and Fasman,G.D. (1974) *Biochemistry,* **13,** 212.
20. Nishikawa,K. (1983) *Biochim. Biophys. Acta,* **748,** 285.
21. Garnier,J., Osguthorpe,J.D. and Robson,B. (1978) *J. Mol. Biol.,* **120,** 97.
22. Taylor,W.R. and Thornton,J.M. (1984) *J. Mol. Biol.,* **173,** 487.
23. Kabsch,W. and Sander,C., personal communication.
24. Kabsch,W. and Sander,C. (1984) *Proc. Natl. Acad. Sci. USA,* **81,** 1075.
25. Sander,C. and Schultz,G.E. (1979) *J. Mol. Evol.,* **13,** 245.
26. Ptitsyn,O.B. and Finkelstein,A. (1983) *Biopolymers,* **22,** 15−25.
27. Rose,G.D. (1978) *Nature,* **272,** 586.
28. Bussetta,B. and Hospital,M. (1982) *Biochim. Biophys. Acta,* **701,** 111.
29. Chothia,C., Levitt,M. and Richardson,D.C. (1981) *J. Mol. Biol.,* **138,** 321.
30. Cohen,F.E., Sternberg,M.J.E. and Taylor,W.R. (1982) *J. Mol. Biol.,* **156,** 821.
31. Schiffer,M. and Edmundson,A.B. (1968) *Biophys. J.,* **8,** 29.
32. Palau,J. and Puigdomenech,P. (1974) *J. Mol. Biol.,* **88,** 457.
33. Crick,F.H. (1953) *Acta Crystallogr.,* **6,** 689.
34. Eisenberg,D., Weiss,R.M. and Terwilliger,T. (1984) *Proc. Natl. Acad. Sci. USA,* **81,** 140.
35. Finer-Moore,J. and Stroud,R.M. (1984) *Proc. Natl. Acad. Sci. USA,* **81,** 155.
36. Eisenberg,D., Schwarz,E., Komaromy,M. and Wall,R. (1984) *J. Mol. Biol.,* **179,** 125.

37. Lim,V. (1974) *J. Mol. Biol.*, **88**, 873.
38. Rawlings,C.J. and Stockwell,P. (personal communication)
39. Cohen,F.E., Abarbanel,R.M., Kuntz,I.D. and Fletterick,R.J. (1983) *Biochemistry*, **22**, 4894.
40. Richmonds,T.J. and Richards,F.M. (1979) *J. Mol. Biol.*, **119**, 537.
41. Taylor,W.R. and Thornton,J.M. (1983) *Nature*, **301**, 540.
42. Taylor,W.R. (1984) *J. Mol. Biol.*, **173**, 512.
43. Matthews,B.W. (1977) *Biochim. Biophys. Acta*, **405**, 442.
44. Levitt,M. and Greer,J. (1977) *J. Mol. Biol.*, **114**, 181.
45. Kabsch,W. and Sander,C. (1983) *Biopolymers*, **22**, 2577.
46. Kabsch,W. and Sander,C. (1983) *FEBS Lett.*, **155**, 179.
47. Nishikawa,K. and Ooi,T. (1982) *J. Biochem.*, **91**, 1821.
48. Nishikawa,K., Kubota,Y. and Ooi,T. (1983) *J. Biochem.*, **94**, 997.
49. Cohen,F.E., Richmonds,T.J. and Richards,F.M. (1979) *J. Mol. Biol.*, **132**, 275.
50. Cohen,F.E. and Sternberg,M.J.E. (1980) *J. Mol. Biol.*, **138**, 321.
51. Cohen,F.E., Sternberg,M.J.E. and Taylor,W.R. (1980) *Nature*, **285**, 378.
52. Cohen,F.E., Sternberg,M.J.E. and Taylor,W.R. (1980) In *Protein Folding*. Jainikie,R. (ed.), Elsevier, Amsterdam, p.131.
53. Cohen,F.E., Sternberg,M.J.E. and Taylor,W.R. (1981) *J. Mol. Biol.*, **148**, 253.
54. Sibanda,B.L. and Thornton,J.M. (1985) *Nature*, **316**, 170.
55. Taylor,W.R. (1986) *J. Mol. Biol.*, **188**, 233.
56. Taylor,W.R. (1986) *J. Mol. Biol.*, **188**, 254.
57. Thornton,J.M., Sibanda,B.L. and Taylor,W.R. (1985) In *Methodology Surveys in Biochemistry and Analysis*. Reid,E., Cook,G.M.W. and Morre,D.J. (eds), Plenum Press, New York, p.23.
58. Wierenga,R.K. and Hol,W.G.J. (1983) *Nature*, **302**, 842.
59. Sternberg,M.J.E. and Taylor,W.R. (1984) *FEBS Lett.*, **175**, 387.
60. Moller,W. and Amons,R. (1985) *FEBS Lett.*, **186**, 1.
61. Walker,J.E., Saraste,M., Runswick,W.J. and Gay,N.J. (1985) *EMBO J.*, **1**, 945.
62. Taylor,W.R. and Green,N.M. (1986) *Biochemistry*, in press.
63. Dickerson,R.E. and Geis,I. (1969) *The Structure and Action of Proteins*, Harper and Row, New York.
64. Taylor,W.R. (1986) *J. Theor. Biol.*, **119**, 206.
65. Needleman,S.B. and Wunsch,C.D. (1970) *J. Mol. Biol.*, **48**, 443.
66. Blundell,T.L., Sibanda,L. and Pearl,L. (1985) *Nature*, **304**, 273.
67. Blundell,T.L. and Sternberg,M.J.E. (1985) *Trends Biotech.*, **3**, 228.
68. Jones,A.T. (1978) *J. Appl. Cryst.*, **11**, 268.
69. Greer,J. (1981) *J. Mol. Biol.*, **153**, 1027.
70. Novotny,J., Bruccoleri,R. and Karplus,M. (1984) *J. Mol. Biol.*, **177**, 787.
71. Chothia,C. and Lesk,A.M. (1986) *Proc. R. Soc. Lond. A*, **317**, 345.
72. Lesk,A.M. and Chothia,C. (1980) *J. Mol. Biol.*, **136**, 225.
73. Barker,W.C., Hunt,L.T., Orcutt,B.C., George,D.G., Yeh,L.S., Chen,H.R., Blomquist,M.C., Johnson, G.C., Seibel-Ross,E.I., Hong,M.K. and Ledley,R.S. (1984) *Protein Identification Resource, Nat. Biomed. Res. Fed.* Washington DC.
74. Argos,P. and Leberman,R. (1985) *Eur. J. Biochem.*, **152**, 651.

CHAPTER 13

Molecular sequence comparison and alignment

J.F.COLLINS AND A.F.W.COULSON

1. INTRODUCTION

Related problems of detecting similarity between biosequences arise in different biological contexts; the computational techniques available can generally be applied across the whole range of these problems.

In any particular case the biologist may need to know the answers to three questions.

(i)   Are the computational tools already available to him adequate to solve the problem?

(ii)   If not, will any other published program (available as a transportable package, or accessible by way of an electronic network) do the job?

(iii)   Finally, if he has to write (or get written) his own program, what algorithm should it use?

It is important that he understand the algorithms being used, as well as specific features of their implementation, so that

(i)   he can make programs work as well as possible and

(ii)   he can check that the program (however derived) is really doing what he thinks it is doing.

In this chapter we concentrate on the properties of the fundamental algorithms (rather than giving specific instructions for particular programs).

The problems of finding similarities in biosequences are similar to those of finding the occurrence of particular (perhaps mis-spelt) words in long texts, or of identifying signals in the presence of noise. Effective computational techniques can be adopted from these other domains. Major difficulties in doing so are introduced by the fact that genuinely similar biological sequences or sub-sequences may not have the same length. Good alignments can often only be achieved by postulating deletions in one strand or insertions in the other. Since these two possibilities are conceptually equivalent, it is convenient to refer to both by the term *indel* (1).

Similarity searches require a solution to one of three types of problem.

Problem I. Given two finite sequences, what pattern of indels makes the most plausibly similar alignment between them? Most commonly the sequences are those of complete genes (or gene products) and the similarity is examined for evidence of evolutionary divergence (see Chapter 14).

Problem II. Which sub-sequence(s) of an indefinitely long sequence show(s) the greatest similarity to a short query sequence (taking account of all possible indels)? A simple example of this problem is finding a restriction site in a DNA sequence. A harder form of Problem II is finding prokaryotic promoter sequences by attempting to match a consensus sequence at all positions. In this case the permitted degree of mismatch to the consensus varies at different positions in the sequence; indels (of a limited range of lengths) can only be postulated at certain positions. The database search problem ('does anything already published resemble this newly determined sequence?') might be treated in this way but is better regarded as an example of the next problem type.

Problem III. Which pair(s) of sub-sequences, each drawn from indefinitely long (collections of) sequences, show(s) the most plausible similarities taking into account all possible indels? A newly determined sequence or partial sequence is best compared with a collection of published sequences in this way because it has often been found that significant similarity is limited to only a part of a single gene. Further examples of problem III are those of comparing a sequence with itself to detect repeated sequences, with its reverse to detect palindromes or (for DNA and RNA) with its reverse complement to detect potential 'hairpin' or 'stem-loop' structures (Chapter 11).

We will use the term *alignment* to refer to a particular arrangement of a pair of sub-sequences, perhaps containing indels. For example, the alignment

$$\text{-A-B-X-Y-C-D-E-}$$
$$\text{-a-b-----c-d-e-}$$

indicates that the five corresponding upper- and lower-case letters are to be compared with each other, and that two residues are to be regarded as having been inserted in one sequence (or deleted from the other). The similarity of an alignment is expressed by summing a score for each pairwise comparison of residues (and for each comparison of a residue with a deletion). The simplest scheme gives a score of unity to each match, and zero to every mismatch or deletion. As an alternative to similarity, one can score the distance between two sub-sequences in a particular alignment. The simplest distance scoring scheme scores zero for each match, and unity for each mismatch or single residue deletion. The summed distance is the minimum number of individual changes required to transform one sub-sequence into the other, and can therefore be regarded as a measure of the evolutionary distance between them.

Practical scoring schemes vary widely. A class of scheme which is both theoretically and practically important is the metric. Three constraints must apply if a scheme is to conform to the definition of a metric.

(i)   The distance between two sequences is zero if the sequences are identical.

(ii)   The distance between Sequence A and Sequence B must be the same as the distance between Sequence B and Sequence A. Few, if any, biologically valuable distance scoring schemes will not conform to these constraints.

(iii)   The final condition is that the sum of the distances between Sequence A and Sequence B, and between Sequence B and Sequence C must be greater than or equal to the distance between Sequence A and Sequence C. (The three conditions must be met for all A, B and C).

Scoring schemes for which the third (triangle inequality) condition is not met are sometimes used. The theoretical importance of metrics is that it has been proved that the algorithms discussed in Section 3 will certainly find the best alignment possible with any given scoring scheme, but only if the scheme is a metric. If the triangle inequality is not met it cannot be guaranteed that the algorithms will find the best alignments. The conditions under which a deliberate breach of the triangle inequality rule is desirable are so restricted that it seems unlikely that the effect on the algorithms will be serious. However, some programs which allow the user to vary parameters at will over a wide range do not prevent a choice of values which will inadvertently break the rule, and the user should check carefully to see that this is not happening.

A similarity search has two phases: a first of finding similarities, and a second of assessing their significance. At one extreme both problems are very simple. A run of 12 matched amino acid residues (or 30 bases) is easily found, and cannot possibly have occurred by chance. Such a pair of sub-sequences must be functionally very closely related, and can be assumed to have a common evolutionary origin. Generally the biologist will still be interested in sub-sequences showing a much lower degree of similarity, since he knows that function can be preserved despite extensive substitution and insertion or deletion of residues. One important judgement to make about different computational methods is how well they perform in finding low degrees of similarity. The analogy with signal processing is particularly appropriate. One can think of a set of functionally similar sub-sequences as examples of a degraded signal which one is trying to identify in the presence of noise. Assessing the performance of different methods requires specific models for the properties of the 'degraded signal' and for the background noise. In this chapter we shall assume that the noise (i.e. the 'non-similar' sub-sequences) can be represented as random sequences of characters with the appropriate frequency distribution. This is justified because although short-range ordering is detectable in both nucleic acid and protein sequences, these effects are not large. For example, it would be hard to assess whether a particular sequence of 200 amino acid residues was a genuine protein sequence, or had been derived from a known sequence by random shuffling, and even harder if the sequence had been shuffled with some constraint preserving nearest neighbour frequencies.

The main model we shall use for the properties of signals (functionally related sub-sequences) is that proposed by Dayhoff (2) for the evolution of proteins. This choice is appropriate because the model is more highly developed than any other, and because the biologist's purpose in comparing protein sequences is very frequently the detection of putative evolutionary relationships. In addition, many of the conclusions we reach about the detectability of these signals may be extended to other kinds of sequence similarity. However, the reader should be aware of the extent to which our conclusions are based on the properties of particular models: if he has good reason to believe that the properties of the signals in which he is interested are significantly different from those of the models, he should modify the analysis accordingly. Relationships representing cases of convergent evolution would certainly need a different model for their detection.

Another way of expressing this is to point out that we avoid giving an unequivocal definition of similarity. However, any running program contains an implicit definition of both similarity and non-similarity. Before a needle can be found in a haystack, the searcher has to know how to tell the difference between the needle and hay. The scientific responsibility of deciding whether the definitions used by a program are appropriate to his problem ought to remain with the biologist, and he should not abdicate it in favour of the program writer.

Three computational approaches have been widely used for biosequence similarity searches; the next sections of this chapter describe each approach in detail and compare them with each other. The final section deals with the problem of searching a large database for similarities to a single query sequence.

2. THE 'DOT-PLOT'

Consider a rectangular array (a match matrix) whose columns are labelled with successive residues of one (protein or nucleic acid) sequence $(A_1 . . . A_n)$, and whose rows are labelled with successive residues of a second sequence $(B_1 . . . B_m)$ with which the first is to be compared.

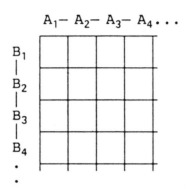

The cell (j,i) can be used to represent the result of the comparison of the ith residue of the first sequence with the jth residue of the second. All possible pairwise comparisons are represented by elements of the array, and all possible sub-sequence alignments are represented by paths in the array. If there are no indels, the paths are diagonals from upper left to lower right. Note that there is no agreed convention for the direction of increased sequence numbers in drawing dot-plots and related diagrams.

2.1 The simple dot-plot

The simplest kind of dot-plot is created by putting a mark (a 'dot') in each cell (j,i) for which A_i matches B_j. Matching sub-strings appear as diagonal lines of dots. Mismatches produce gaps in the line, and indels, when appropriate, make it possible to pass from one diagonal line of dots to another by a horizontal or vertical step. A horizontal step represents a deletion in the B sequence, or an insertion in the A sequence. A vertical step represents an insertion in the B sequence or a deletion in the A sequence. A sub-sequence of either sequence which is repeated in the other gives rise to a horizontally or vertically aligned pair of diagonal runs of dots, and repeats in both to a rectangular array of four diagonal runs. In a comparison of a sequence with itself, the main diagonal is fully occupied, the array is symmetrical about the main diagonal, and all features off the main diagonal represent repeats. For this reason, self-comparison dot-plots frequently display only one triangular region of the match matrix, bounded by the main diagonal. These properties of the simple dot-plot are illustrated by *Figures 1, 2* and *3*.

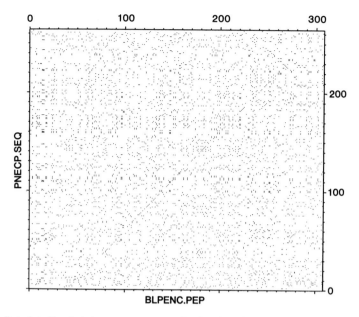

Figure 1. Dot-plot of two beta-lactamase sequences showing clear diagonal runs of matches indicating close relatedness. Each dot represents one residue match. BLPENC.PEP is the beta-lactamase from *Bacillus licheniformis*; PNECP.SEQ is the beta-lactamase from the PBR322 plasmid.

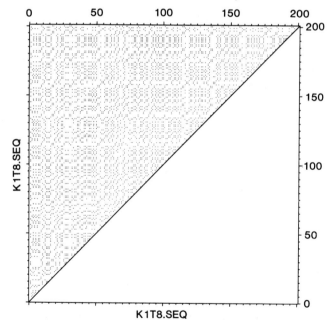

Figure 2. Dot-plot of a short DNA sequence from *Plasmodium falciparum*, in a self-comparison. Each dot represents one base match. The diagonal features visible parallel to the main diagonal occur in a regular pattern, indicating that this fragment is highly repetitious, with a basic repeat distance of 21 bases.

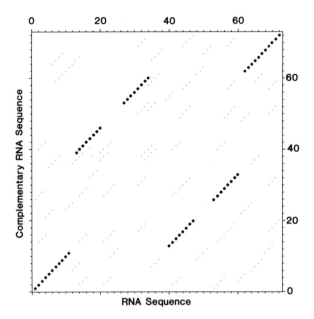

Figure 3. Filtered dot-plot (minimum 3 matches in a row), showing patterns associated with secondary structural features in an RNA sequence. The RNA sequence is compared with its inverse complement; the features associated with a Y-shaped structure are emphasized. Such patterns can be detected by eye, but estimation of the optimum folded structure needs additional thermodynamic criteria to distinguish the most stable structural features among many complementary regions which may overlap along the sequence.

Each residue of a given type will give rise to a dot wherever it is compared with a residue of the same type in the other sequence. Hence, if a_i and b_i are the numbers of residues of the ith type in the two sequences, the total number of dots on the plot is

$$\sum_i a_i.b_i \qquad\qquad \text{Equation 1}$$

The maximum possible signal is the length of the shorter sequence, and the rest of the dots make up the background noise. As the total size of the match matrix is increased regions of signal, which will generally be smaller than the maximum and will be broken up by mismatches and indels, become harder to perceive.

This effect is enhanced because patterns of no particular significance also appear in larger dot-plots. If there are q types of residue (q=20 for proteins and q=4 for nucleic acids), then the dot-plot can be considered to be made up row-by-row by the selection of one of only q different rows according to successive residues of the B sequence. Each of the q different rows corresponds to the distribution of residues of a given type along the A sequence. Repetitions (arising by chance) of short patterns of residues then give rise to symmetrical patterns in the dot plot which the eye perceives very readily, and which may mask diagonal signals. This effect is more serious with nucleic acid sequences because of the small alphabet (*Figure 5*). It is also beginning to be visible in the protein comparison of *Figure 4*. However, it is still the general background noise which obscures the signal in this case, a repeat in the protein sequence (see below).

For weaker signals to be detected in large dot-plots, some kind of filtering is evidently needed to enhance the signal-to-noise ratio.

328

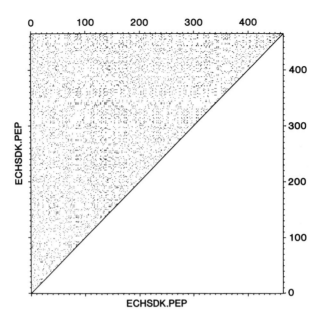

Figure 4. Self-comparison of a host-specificity protein from the restriction-modification system of *Escherichia coli* K12 (24). Each dot represents one residue match.

2.2 Filtering by translation

A dot-plot constructed from protein sequences will have reduced 'noise' compared to one constructed for the corresponding nucleic acid sequences because of the increased alphabet and the reduced total size of the comparison. Under all normal circumstances the 'signal' will be maintained or increased since the function of translated genes is expressed at the protein level.

This argument suggests that it will almost always be more effective to prepare dot-plots from translated sequences when the mode of translation is known or can be guessed. *Figure 5* shows the nucleic acid dot-plot corresponding to part of *Figure 1*; the reduced visibility of the signal in *Figure 5* is striking.

2.3 Threshold filtering

2.3.1 *Definitions*

A filter is designed to increase the signal-to-noise ratio by systematically removing dots from the plot in a way which is more likely to remove dots arising from accidental matches than from genuine similarity. The effectiveness of a particular filtering procedure therefore depends on the properties both of the signal and of the noise.

A flexible and efficient method of filtering is to put a dot only where the local concentration of matches exceeds a set threshold. A window of length w is a diagonal path of this length in the match matrix. The set of all such paths represents all possible pairwise comparisons of sub-sequences of length w, drawn one from each of the compared sequences. If the stringency is set to s, a dot is placed at the centre of a given window only if at least s out of the w residue-pairs match. The simple dot-plot is a special case with $s = w = 1$.

329

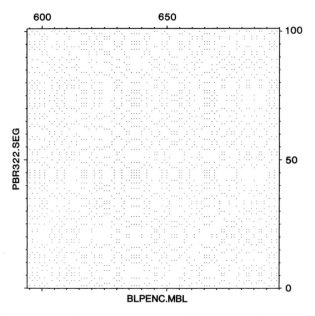

Figure 5. Dot-plot using part of the gene sequences coding for the two beta-lactamases used in *Figure 1*. Each dot represents one base match. It is not easy to detect the signal region corresponding to the matching residues in the proteins, because of the high noise level; other patterning is also apparent but accidental in nature.

2.3.2 *Effect of filtering on 'noise'*

The total number of dots in a simple dot-plot is given by Equation 1, and these can be regarded as distributed at random except in regions of significant similarity. The expected distribution of matches (the number of windows of length w containing at least s matches for all values of s and w) can be calculated in terms of the probability p that there is a match at a given position in the match matrix.

If the frequencies of residues of type i in sequences A and B are F_{ai} and f_{bi} respectively, the probability p of a match occurring by chance at any position in a match matrix is

$$\sum_i f_{ai} f_{bi}$$

If the frequencies are uniform over all types of residue in both sequences, p = 0.25 for nucleic acid sequences and p = 0.05 for protein sequences. Naturally occurring proteins have a highly non-uniform frequency distribution of amino acids. We have calculated p for a number of collections of protein sequences, and find that its value is always close to 0.058. We shall therefore use p = 0.06 as a representative value for proteins, and p = 0.25 for nucleic acids.

We have calculated the expected number of dots arising by chance for a range of values of window length and stringency. The probability Q(s,w) of at least s matches in a window of length w is given by

$$\sum_{i=s}^{w} P(i,w)$$

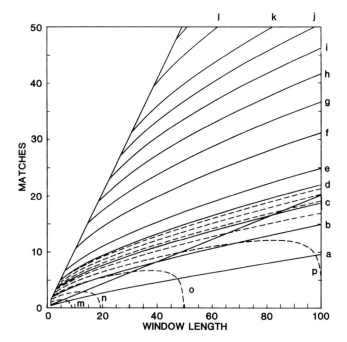

Figure 6. Probability contours for finding a given number or more matches in a window of variable size, for protein comparisons, taking the probability of a match $= 0.06$. Probabilities on the contours are: a: 10^{-1}; b: 10^{-3}; c: 10^{-5}; d: 10^{-7}; e: 10^{-9}; f: 10^{-14}; g: 10^{-19}; h: 10^{-24}; i: 10^{-29}; j: 10^{-34}; k: 10^{-44}.
The dotted contours represent the one-event expectation for comparison of the $N \times N$ type, where $N = 10$ (m), 20 (n), 50 (o), 100 (p), and also for $N = 200, 500, 1000$ and 2000 (unmarked).
The properties of a degraded signal, retaining only 20% identity with a second sequence, are indicated by the line from the origin to 20 matches in a window of length 100.

where $P(s,w)$ is the probability of exactly s matches in a window of length w. $P(s,w)$ is most easily calculated iteratively, using the relations

$$P(s,w) = p.P(s-1,w-1) + (1-p). P(s,w-1)$$
$$P(i,j) = 0$$
$$i>j$$

and the starting points

$$P(0,1) = 1-p; P(1,1) = p$$

Figures 6 and *7* show contour diagrams of $-\log_{10}Q(s,w)$ for $p = 0.06$ and $p = 0.25$ respectively.

These diagrams have a general utility in allowing an estimate to be made very rapidly of the probability that a given similarity (however discovered) could have arisen by chance.

However, the main purpose of this calculation was to find the expected effect of a given type of filter on the appearance of the background of a dot-plot. A window of length w can be placed $(n-w+1).(m-w+1)$ times on a plot representing the comparison of sequences of length n and m. The expected number of dots remaining after

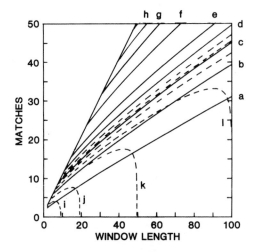

Figure 7. Plot of the probability that at least a given number of matches will be found in a variable length window, taking the probability of a match = 0.25, for nucleic acid comparisons. The contours represent different filter combinations which have the same probability of success in a random series of trials. Contour probabilities:

a: 10^{-1}; b: 10^{-3}; c: 10^{-5}; d: 10^{-7}; e: 10^{-9}; f: 10^{-14}; g: 10^{-19}; h: 10^{-24}.

In addition, the dotted contours indicate the events likely to occur once in a comparison of N by N bases, of random sequences, where N = 10 (i), 20 (j), 50 (k), 100 (l), and also for N = 200, 500 and 1000 (unmarked).

the application of a filter (s,w) is the product of this number and the value of Q(s,w) which can be estimated from *Figures 6* or *7*. The dotted contours in *Figure 6* show the combinations of s and w which are expected to reduce the background to one dot for comparisons with n = m = 10, 20, 50, 100, 200, 500, 1000 and 2000. The 'edge-effects' are not large, except for window lengths which are close to the length of the longest diagonal, and these contours show that the background of random dots can be reduced to any arbitrary level by filters with a range of values of s and w.

The important question for designing a filter is whether a 'signal' is more likely to pass a filter at the left or the right end of these contour lines: a short window and relatively high stringency, or a long window with relatively lower stringency.

2.3.3 Effect of filtering on signals

If a signal does not pass the threshold test of a given filter, it will be lost from the dot-plot. We could deal with this in a probabilistic way as well, but there is a more straightforward approach. The 'worst case' for detecting similarity in a pair of sub-sequences arises when the mismatches are uniformly distributed. If there is a region of genuine similarity containing s matches in total length w, it is certain to pass a (s,w) filter, and some parts of it will also pass any shorter filter with the same ratio of stringency to window length. It will pass a more stringent filter only if the matches happen to be bunched together. Filters of a given fractional stringency are represented in plots like *Figures 6* and *7* by straight lines passing through the origin, such as the 20% stringency line shown in *Figure 6*. A long signal containing 20% matches will certainly remain on a plot filtered by any combination of window length and stringency on or below this line, but may not for any point above it. Since these lines are always

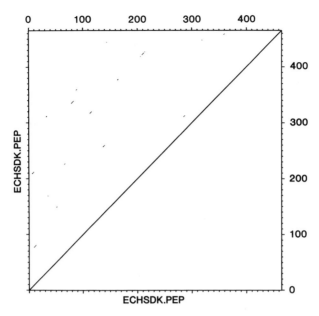

Figure 8. Self-comparison of the host-specificity protein ECHSDK.PEP, using a (5,10) filter. See *Figures 4, 9* and *10*.

steeper than the local contours, the random background which conceals the signal will be lower, the longer the window length. However, if the window length is longer than the signal, the fractional stringency will fall, and the signal will be less easily detected.

We conclude that the highest ratio of signal to noise will be achieved with a window length equal to the true length of the similar sub-sequences.

2.3.4 *Conclusions*

Any comparison of two sequences should start with an unfiltered dot-plot. If diagonal features are obvious, the problem is solved. If the background is so high that significant similarities may be concealed, plots should be drawn with increasingly stringent filters until the background is sufficiently reduced. *Figures 6* and *7* may be used to estimate a suitable starting point for a given size of comparison. This window length should be as long as the potential signal which is sought. A further advantage of long window lengths is that integral steps in the stringency allow a finer adjustment in the background than is possible with short windows.

The length of expected signals is for the biologist to decide. Conserved elements in proteins are often found to be entire domains, which are of the order of 100 amino acid residues or 300 bases in length. Indels, if present, reduce the effective length of the signal. If there is no information about the expected length of signals, then the analysis must be repeated with a range of window lengths. An effective tactic is to superimpose the plots; features which appear at several window lengths are more likely to be significant.

The general utility of long windows can be demonstrated with *Figures 8* and *9*, which represent the comparison in *Figure 4*, filtered with (5,10) and (17,100) respectively.

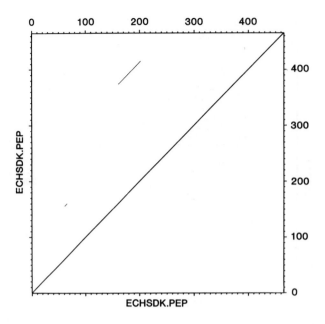

Figure 9. Self-comparison of the host-specificity protein ECHSDK.PEP, using the filter (17,100). See *Figures 4, 8* and *10*.

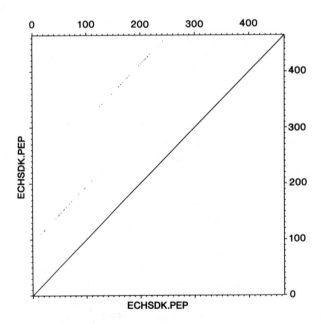

Figure 10. Self-comparison of host-specificity protein ECHSDK.PEP, using a filter of 17 matches in a window of 100 residues, using the COMPARE/ALL option in the UWGCG package. Two regions of potential duplication are visible. Compare *Figure 9*.

These filters reduce the noise to about the same extent in each case, but the signal is only apparent when the long window is used. Long windows introduce a potentially confusing artefact, since the noise tends to cluster into diagonal lines (and any real signal which is shorter than the window is smeared out). This effect is neatly removed in the UWGCG package of programs (3) by the COMPARE/ALL option. When a window passes the filter threshold, a dot is placed not at the centre of the window, but at the position of each contributing match; such a plot is shown as *Figure 10*.

Several of the widely used analysis packages do not allow dot-plots to be constructed in the ways that this argument suggests are most effective. Thus DIAGON (ref. 4) only allows filters of the (n,n) type, and MicroGenie (Beckman-RIIC, Ltd) limits the maximum window length to 50 residues and the minimum stringency to 50%. Such high stringencies will rarely be effective in detecting low degrees of similarity in proteins. Even the limit of 100 in the window length of the UWGCG package may be too low for nucleic acid comparisons.

2.4 Reduced alphabet

2.4.1 *Introduction*

The biologist does not generally regard all pairs of non-identical amino acids as being equally dissimilar. In weakly homologous sub-sequences, lysine and arginine or leucine and isoleucine are often regarded as functionally equivalent. Signals might therefore be enhanced by, for example, rewriting all the arginines in a sequence as lysines before performing the comparison. Such a grouping of residues together is called alphabet reduction, since the number of different symbols is made smaller. For this reason, the number of random matches in a given comparison is also increased. Whether signals are made more or less detectable by alphabet reduction depends on the relative size of the two effects.

2.4.2 *Effect of alphabet reduction on noise*

On a probability plot such as *Figure 6*, the threshold for detectability of a signal is represented by the point of intersection of the vertical for a particular window length with one of the probability contours (which one depends on the total size of the comparison, and on the effective threshold chosen by the user). When two letters of the alphabet are merged, the effect is to increase p, and therefore to increase the stringency of the filter which will be needed to reduce the background noise to any given level. *Figure 11* shows the most important probability contours as a function of p; the effect of increasing p on increasing the background noise is represented by the slopes of these contours.

If f_i and f_j are the frequencies of residue types i and j respectively, the increase in p on making i and j equivalent is 2 $f_i f_j$. At the 20% level of matches, the change in stringency for unit change in p is about 1.7.

2.4.3 *Effect of alphabet reduction on signal*

For a specific model of a low level of signal, we will use that developed by Dayhoff (2) which represents protein sequences which are assumed to have separated by divergent

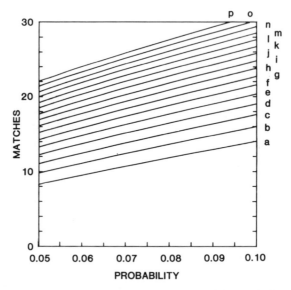

Figure 11. Effect of changing the probability of matches, through alphabet reduction, on the stringency of filters designed for a window of length 100. The improbability contours allow the selection of equivalent pairs of filters at different matching probabilities. Contours (at 0.5 unit intervals) range from (a) 10^{-1}, to (p) $10^{-8.5}$.

evolution so that only 20% of residues match. This corresponds to 250 accepted point mutations per 100 residues. (The model will be described in greater detail in section 2.5.) *Table 1* is the array of numbers, M_{ij} which gives the probability that a residue of type i has been substituted by residue type j after this evolutionary interval. If residue types i and j are treated as equivalent, the expected change in the proportion of matches is $f_i.M_{ji} + f_j.M_{ij}$. If this number is larger than the change in threshold produced by the corresponding increase in p, the signal will have become more easily detectable; if smaller, the effect of increased background noise will more than outweigh the average increase in signal.

Any potential alphabet reduction can be assessed in this way. For example, the frequencies of alanine and isoleucine residues in Dayhoff's set (*Table 2*) are 0.087 and 0.037 respectively; the change in p on making them equivalent is therefore 0.0064. This corresponds to an increase in threshold stringency of about 0.01; the expected increase in signal is 0.0056, so this pairing should be rejected.

The residue pairings which are most likely to lead to an enhanced signal to noise ratio are those with the largest value of the parameter

$$(f_i M_{ji} + f_j.M_{ij})/2f_i.f_j$$

Values of this parameter have already been given by Dayhoff as the 'odds matrix', *Table 3*. The critical value at which noise reduction takes over from noise enhancement is about 2.4. Five pairings only are found to produce an expected increase in detectability of signals: Phe/Tyr, Val/Ile, Leu/Met, Glu/Asp and Lys/Arg. It should be emphasized that the enhanced signal to noise ratio arises when a significantly large number of comparisons is attempted.

For the class of signal considered here, it seems that drastic reduction of the alphabet

Table 1. Mutation probability matrix for an evolutionary distance of 250 PAM's (2). Values × 100.

Final Residue	Initial Residue																			
	Ala	Arg	Asn	Asp	Cys	Gln	Glu	Gly	His	Ile	Leu	Lys	Met	Phe	Pro	Ser	Thr	Trp	Tyr	Val
Ala	13	6	9	9	5	8	9	12	6	8	6	7	7	4	11	11	11	2	4	9
Arg	3	17	4	3	2	5	3	2	6	3	2	9	4	1	4	4	3	7	2	2
Asn	4	4	6	7	2	5	6	4	6	3	2	5	3	2	4	5	4	2	3	3
Asp	5	4	8	11	1	7	10	5	6	3	2	5	3	1	4	5	5	1	2	3
Cys	2	1	1	1	52	1	1	2	2	2	1	1	1	1	2	3	2	1	4	2
Gln	3	5	5	6	1	10	7	3	7	2	3	5	3	1	4	3	3	1	2	3
Glu	5	4	7	11	1	9	12	5	6	3	2	5	3	1	4	5	5	1	2	3
Gly	12	5	10	10	4	7	9	27	5	5	4	6	5	3	8	11	9	2	3	7
His	2	5	5	4	2	7	4	2	15	2	2	3	2	2	3	3	2	2	3	2
Ile	3	2	2	2	2	2	2	2	2	10	6	2	6	5	2	3	4	1	3	9
Leu	6	4	4	3	2	6	4	3	5	15	34	4	20	13	5	4	6	6	7	13
Lys	6	18	10	8	2	10	8	5	8	5	4	24	9	2	6	8	8	4	3	5
Met	1	1	1	1	0	1	1	1	1	2	3	2	6	2	1	1	1	1	1	2
Phe	2	1	2	1	1	1	1	1	3	5	6	1	4	32	2	2	2	4	20	3
Pro	7	5	5	4	3	5	4	5	5	3	3	4	3	2	20	6	5	1	2	4
Ser	9	6	8	7	7	6	7	9	6	5	4	7	5	3	9	10	9	4	4	6
Thr	8	5	6	6	4	5	5	6	4	6	4	6	5	3	6	8	11	2	3	6
Trp	0	2	0	0	0	0	0	0	1	0	1	0	0	1	0	1	0	55	1	0
Tyr	1	1	2	1	3	1	1	1	3	2	2	1	2	15	1	2	2	3	31	2
Val	7	4	4	4	4	4	4	5	4	15	10	4	10	5	5	5	7	2	4	17

Table 2. Relative amino acid frequencies in accepted point mutation data (2).

Gly	0.089	Arg	0.041
Ala	0.087	Asn	0.040
Leu	0.085	Phe	0.040
Lys	0.081	Gln	0.038
Ser	0.070	Ile	0.037
Val	0.065	His	0.034
Thr	0.058	Cys	0.033
Pro	0.051	Tyr	0.030
Glu	0.050	Met	0.015
Asp	0.047	Trp	0.010

is likely to be counterproductive in making weak signals detectable. This argument applies strictly only to signals at the 20% level of matches remaining (and it also depends, of course, on the detailed appropriateness of Dayhoff's overall model). It would be possible to apply this analysis to larger or smaller evolutionary distances, but it seems unlikely that there are common circumstances in which it will be useful to make a more drastic simplification.

2.5 Similarity scoring

2.5.1 Introduction

So far, we have only considered the result of a pairwise comparison of two residues to have two possible results ('match' or 'mismatch'). However, it is usual for biologists to recognize intermediate grades of similarity between these two extremes. In principle, the similarity score should reflect the relative probability, given a specific proposal for the relationship between the two sequences, that a particular pair of residues are found in alignment. For example, if two sequences are being compared for evidence that they are related by divergent evolution, the similarity score for each pairwise comparison should reflect the odds that the corresponding interchange of residues has occurred when two sequences have diverged to the appropriate extent. If the similarity score is actually the logarithm of these odds, the sum of the pairwise values for a particular window will be a measure of the probability that the corresponding alignment represents a similarity of the kind being sought.

It should be clear that appropriate similarity scores can only be calculated from a specific model of similarity. For example, in the case of divergent evolution, quite different sets of values are required for different evolutionary distances. Unfortunately rather few models of similarity have so far been developed to the point of providing well-defined sets of similarity scores. The user must either develop his own, or more likely use a set which may be less than ideal in his particular case, but which is expected to have a general utility. The same problem arises in assessing the expected increase in detectability of signals by these methods. The analysis can only be carried out with reference to a particular scoring scheme. We use here the scheme due to Dayhoff (2) since it is very widely used. The analysis should be easy to repeat for any other scoring scheme.

The Dayhoff scheme arises from an attempt to predict the pattern of substitutions in proteins at large evolutionary distances from the pattern observed at short distances.

Table 3. Log of odds matrix, × 100, that in the alignment of two sequences which have evolved to a distance of 250 PAM's, any pair of residues represent the same residue in the common ancestral sequence (2).

	Ala	Arg	Asn	Asp	Cys	Gln	Glu	Gly	His	Ile	Leu	Lys	Met	Phe	Pro	Ser	Thr	Trp	Tyr	Val
Ala	18																			
Arg	-15	61																		
Asn	2	0	20																	
Asp	3	-13	21	39																
Cys	-20	-36	-36	-51	119															
Gln	-4	13	8	16	-54	40														
Glu	3	-11	14	34	-53	25	38													
Gly	13	-26	3	6	-34	-12	2	48												
His	-14	16	16	7	-34	29	7	-21	65											
Ile	-5	-20	-18	-24	-23	-20	-20	-26	-24	45										
Leu	-19	-30	-29	-40	-60	-18	-34	-41	-21	24	59									
Lys	-12	34	10	1	-54	7	-1	-17	0	-19	-29	47								
Met	-11	-4	-17	-26	-52	-10	-21	-28	-21	22	37	4	64							
Phe	-35	-45	-35	-56	-43	-47	-54	-48	-18	10	18	-53	2	91						
Pro	11	-2	-5	-10	-28	2	-6	-5	-2	-20	-25	-11	-16	-46	59					
Ser	11	-3	7	3	0	-5	0	11	-8	-8	-28	-2	-21	-32	9	16				
Thr	12	-9	4	-1	-22	-8	-4	0	-13	1	-14	0	-16	-31	3	13	26			
Trp	-58	22	-42	-68	-78	-48	-70	-70	-28	-51	-18	-35	-6	4	-56	-25	-52	173		
Tyr	-35	-42	-21	-43	3	-40	-43	-52	-1	-9	-9	-44	-42	-70	-49	-28	-27	-2	101	
Val	2	-25	-17	-21	-19	-19	-18	-14	-22	37	19	-24	18	-12	-12	-10	3	-62	-25	43
	Ala	Arg	Asn	Asp	Cys	Gln	Glu	Gly	His	Ile	Leu	Lys	Met	Phe	Pro	Ser	Thr	Trp	Tyr	Val

Table 4. Relationship between observed differences and estimated evolutionary distance, according to the Dayhoff model (2).

Observed % difference	*Evolutionary distance in PAM's*
1	1
5	5
10	11
15	17
20	23
30	38
40	56
50	80
60	112
70	159
80	246

A large collection was prepared of pairs of sequences of proteins (either real or inferred ancestral sequences) which differed at no more than 15% of residues. It was assumed that at this level of difference there were no successive changes at any one site, so that it was possible to derive values for the probability that a residue of any given type should mutate to any other given type (or should not mutate at all) during an evolutionary interval in which one point mutation was accepted per 100 residues (an evolutionary distance of 1 PAM). The pattern of accepted point mutations expected at longer evolutionary distances was then derived by multiplication of this matrix of probability values by itself the appropriate number of times (n times for a distance of n PAMs) (*Table 1*). At longer distances there is a greater and greater chance that multiple mutations will have occurred at any given site, and *Table 4* shows the values derived by Dayhoff connecting the overall proportion of observed mismatches with the corresponding evolutionary distance in PAMs. The appropriate similarity score for a particular pairwise comparison at any given evolutionary distance is the logarithm of the odds that a particular pair of amino acid residues has arisen by mutation rather than by chance, and these odds can be derived by dividing the values in the mutation probability matrix by the overall frequencies of the amino acid residues in the original data set. *Table 3* shows these similarity scores appropriate to an evolutionary distance of 250 PAMs (equivalent to 20% matches remaining).

Wilbur (5) has pointed out that the ratio of the frequencies (in Dayhoff's data) of mutations which require at least two base changes to those that require only one are not consistent with the underlying model, which supposes that the probability of interconversion of a pair of amino acid types is independent of the position of the residue concerned in a sequence. However, no scheme based on a more sophisticated model has been presented, and the use of the 250 PAM odds matrix as a measure of distant similarity has been justified empirically (see below).

2.5.2 *Effect on background noise*

From what has been said, it should be clear that the ideal similarity score depends on the expected evolutionary distance in the alignments being sought. However, Dayhoff showed that similarity scores based on an evolutionary distance of 250 PAMs (corresponding to 80% mismatches) are quite close to the optimum similarity scores over a wide range of evolutionary distances, from about 50 to several hundred PAMs. Dayhoff also compared this scoring scheme with several radically different alternatives (based for example on the genetic code) and concluded that none was better for detecting distant relationships. With reservations, the same conclusion was reached by Feng *et al*. (6).

We shall therefore continue the analysis only for this case; it is generally agreed that the difficulty in detecting similar sub-sequences rises sharply as the overall proportion of exact matches falls below 20%. This scoring scheme was introduced for use in dot-plots by McLachlan (7) who also introduced two additional refinements. The first was to use a triangular weighting scheme so that the pairwise similarity scores, before being summed, were multiplied by a constant dependent on the position in the window; comparisons at the centre of the window carried the most weight. The second refinement was to use a probability calculation to determine the appropriate threshold for a plot. The 'double matching probability' is calculated for the expected distribution of scores when sub-sequences of the appropriate window length are drawn at random from unlimited pools of residues having the same composition as the sequences being compared. The threshold for display of scores can then be chosen as, for example, a certain number of standard deviations away from the mean.

The weighting scheme introduces a new type of parameter which has to be chosen in designing a dot-plot and it seems to be an unnecessary complication. Intuitively, it seems unlikely (since the filtering is based on the improbability of a particular set of matches) that the weighting scheme would make any difference to the detectability of signals. Closer analysis shows that this is true for the case when the window is much shorter than the signal, but not when the window is comparable to the signal in length. Under these circumstances, a signal whose shape matches the triangular weighting distribution (i.e. one in which the similar residues are clustered in the middle of the alignment) is more easily detected than average. However, this is at the expense of a reduced detectability for other patterns in the signal (including the uniform distribution). Unless there is some reason to believe that the signal being sought has a shape and size corresponding to the weighting scheme used, triangular weighting will therefore be counterproductive.

The weighting scheme was introduced by McLachlan to improve the definition of the ends of regions of similarity, and it does have this effect. However, if the detection of the ends of the signal is important, it would be better to use a weighting scheme which effectively differentiates the signal by subtracting the sum of the scores for one half of the window from the sum for the other half. In what follows, we assume that the score for a window is just the sum of the scores for each pairwise comparison, without weighting. *Figure 12* shows the probability contours for the expected distribution of scores when proteins of average composition are compared using the similarity scores of *Table 3* with window lengths from 1 to 100. This plot can be used, like *Figure 6*, to estimate the number of similar regions which will be found in a given comparison for a particular choice of window length and threshold, and also for estimating the significance of regions of similarity.

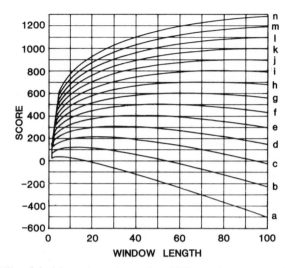

Figure 12. Probability of obtaining a given score, or better, when random sequences (with residue frequency given by *Table 2*) are compared using the Dayhoff similarity scores of *Table 3*, for window lengths from $1-100$. The contours show decreasing probabilities, from 10^{-1} (a) to 10^{-14} (n), in unit steps. There is a negative average expectation for pairs of residues chosen at random but in the ratios given in *Table 2*, of -8.2456.

2.5.3 *Effect on signal*

The scoring scheme is optimal for proteins which have diverged according to the Dayhoff model to the extent of 250 PAMs. The similarity score expected for such a pair of sequences can be calculated from the frequency of amino acid residues, the probability of their interconversion at this evolutionary distance and the appropriate score from *Table 3*. This value is 10.8 per unit window length. A line of this slope can therefore be compared with the 20% line of *Figure 6*, and the relative detectability of signals by the two methods of exact matching and similarity scores, can be shown by comparing the probability values through which these two lines pass for each window length. *Figure 13* shows this comparison, which may be summarized by saying that use of this similarity score reduces by about half the distance over which a signal must persist before it will be detected.

2.5.4 *Conclusions*

For the particular case considered here, use of a similarity scoring scheme produces a significant increase in the detectability of signals. This improvement is the largest possible, because signals were considered which conformed exactly to the model used to establish the scoring scheme. In real cases the improvement will be less marked, and it is possible to imagine circumstances under which any particular scoring scheme will be counterproductive. In principle, it would be possible to tailor the scoring scheme to the similarity being sought. For example, the Dayhoff scheme could be used to derive scoring schemes for a range of evolutionary distances, and these could be used in turn. In practice few, if any, users have thought this worthwhile. If the extent of similarity is not known in advance, it is clearly wise to examine dot-plots with both exact matching (to detect strong signals) and similarity scoring (for more remote sub-sequences).

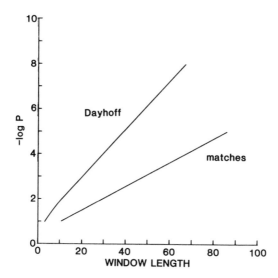

Figure 13. Behaviour of Dayhoff scoring and identity scoring methods, in the case of two sequences which have diverged to the 250 PAM extent, or 20% residual identity. The negative log. of the probability of obtaining the expected signal in windows of different lengths by chance is plotted against the window length. The Dayhoff scheme allows the signal to be detected in much shorter runs of residues, and is to be preferred where indels may have reduced the signal length significantly.

3. EXHAUSTIVE ALIGNMENT ALGORITHMS

3.1 **Introduction and definitions**

The previous sections have been concerned with the match matrix, containing the results of all possible character comparisons between two sequences, and the visualization of regions of significant similarity. We now examine a series of algorithms which make an exhaustive examination of a score matrix, which is derived from the match matrix. These methods originated as distance scoring methods (8) and owe much to Sellers (9,10). Similarity scoring methods have been developed by Needleman and Wunsch (11) and extended by Smith and Waterman (12).

The score matrix has dimensions $(n+1, m+1)$ where n and m are the lengths of the two sequences. The additional row and column compared to the match matrix contain dummy values initialized in the algorithms. Cells in the score matrix represent states associated with possible partial alignments, and all alignments between the two sequences can be represented by paths through the score matrix, using three types of transition between cells:

(i) diagonal moves, from node $(j-1, i-1)$ to node (j,i); that is, in the sense of the sequences. These correspond to matches or mismatches in the alignment, where characters A_i and B_j are aligned.

(ii) horizontal moves, from node $(j, i-1)$ to node (j,i); these correspond to indels, with an unmatched character A_i in the alignment.

(iii) vertical moves, from node $(j-1, i)$ to node (j,i); these correspond to an unmatched character B_j in the alignment.

To allow all possible alignments to be represented, there must be transitions associated

343

with initial indels in either sequence, and the dummy first row and column provide for these. The score at each cell is found by considering the score at the three cells which can act as precursors to (j,i), and generating new possible scores at (j,i) by adding penalties, for transitions which are indels or mismatches, and bonuses for transitions associated with matches. These potential scores are compared, and the most favourable path extended by allocating to (j,i) the score associated with that path. The score is therefore a measure of the quality of the alignment to that point. By extending this process till the cell (n,m) has its score determined, the quality of the best alignment is known, and by a process of backtracking through the score matrix, the alignment can be determined by the choices preferred at each stage. Starting with the score at cell (0,0) as zero, the score at any cell is given by the weighted sum of the matches, mismatches and indels on the optimal path leading to that node.

$$S = m.M - x.X - g.G \qquad \text{Equation 2}$$

where

   S = quality or score
   M = number of matches on path
   X = number of mismatches on path
   G = number of indels on path
   m = positive weighting of a match
   x = penalty for mismatch
   g = penalty for indel

Once the initial value at (0,0) has been set, the value at (1,0) and (0,1) can be set as indel penalties. The score at (j,i) depends on scores at (j−1,i), (j,i−1) and (j−1,i−1), and the result of the character comparison A_i and B_j. Scores are usually calculated row by row throughout the matrix. Backtracking from the last cell (n,m) to (0,0) to discover the optimal quality path can be seen as a re-examination of the choices which led to the path reaching cell (n,m).

Implementation of this algorithm is constrained by the available computer storage. Since much of the matrix can be shown to be outside the extreme limits for the optimal path, time can be saved by never setting scores into those cells. With some ingenuity, space can also be saved by mapping the score cells needed onto a smaller array. However, some of the flexibility in running this algorithm will be lost in these circumstances, and a straightforward implementation seems preferable, if the resources available are adequate.

Another possible method of saving space is to store the information about the paths entering cells, rather than the scores themselves. This needs a smaller matrix of byte or character variables, and it is only necessary to retain scores in the last row calculated, while the scores for the next row are determined. Thus, only storage for two rows of the score matrix is needed, while the path information is accumulated in the path matrix. Backtracking is quicker, and there are no overheads in re-examining character comparisons.

Backtracking encounters another problem which is often hidden from the user. Scores from different paths can be identical, and a cell may be set to a quality score consistent with two or more possible alignments. If the score matrix-only method is used, a rule has to be applied when backtracking to determine which of the branching paths to follow, giving the order of preference of diagonal, horizontal and vertical moves. This rule is built in, and the user may be unaware of it, though it may be possible to discover it. If the path matrix method is used, the decision of which path to follow may be made when assigning the path cells, or, by using coding which recognizes the multiple path possibilities, deferred till the backtracking stage; the rule to be used must still form part of the program. Ideally, the user should be given explicit information about the rule(s) employed, and their control, if this facility is available.

3.2 Scoring schemes

3.2.1 *Distance scores*

By setting match weighting to zero, the quality of a path depends only on the errors in the alignment. Varying the relative weighting of mismatches and indels may result in different optimal paths or families of paths being discovered.

When mismatches and indels are both given unit weighting, the quality of a path is exactly the number of single character changes (and insertions or deletions) needed to convert one into the other. This is often referred to as the evolutionary distance between two sequences, but it must be remembered that it is a minimum interconversion cost. In the biological context, it does not imply that an evolutionary process generated the sequences being compared from a common ancestral sequence (see Chapter 14).

3.2.2 *Similarity scores*

(i) *Scoring based on matches.* By scoring matches positively, paths with many matches tend to be favoured. Even though distance scoring and match scoring methods both produce alignments in which a match is always preferred to any other possibility, these methods can still find different paths or families of paths.

In the extreme case, where mismatches and indels are zero weighted, the optimal paths have the maximum number of matches possible for the sequences being compared. These paths may form an alignment of interest, but when the sequences are distantly related, the free choice of indels available to maximize the number of matches often gives alignments containing many scattered indels, which do not lend themselves to easy interpretation in terms of biological events. Hence, both types of error are usually given a non-zero weighting, to reduce the number of indels to an acceptable proportion. It should be noted that if the ratio of mismatch to indel weighting exceeds two, the triangle inequality fails, and no mismatches will ever be reported in optimal alignments. Instead, adjacent indels on opposite strands will be found. Each choice of parameters implies a prior model of the mechanism by which changes may have occurred in sequences (cf. Chapter 14). Since the relationship between sequences is in effect being prejudged, any one set of alignments should not be taken as necessarily more significant than other sets of alignments found using different parameters. The interpretation of alignments should be made pre-eminently on the basis of those regions that show high similarity, and less heavily on regions of disturbed and unstable alignment. Alignments which depend critically on the parameters used are less satisfactory than those (elements of) alignments which are more independent of the assumed model.

(ii) *Similarity rating of all pairs of residues.* If the Dayhoff or some other scheme, which rates all possible comparisons of amino acid residues by a variable measure representing similarity, is used to provide the scores for all paths to be extended diagonally, then the quality of an alignment includes a quantitative measure of the closeness of all aligned residues, and is thus sensitive to much more of the information in the sequences. This method is more likely to produce biologically interesting alignments, as conservative substitutions score very highly.

The need for triangle inequality should still be considered, but it is possible to judge some mismatches so unlikely, that two adjacent indels on opposite strands might represent a more probable alignment.

3.2.3 *Non-uniform indel penalties*

When a fixed indel penalty is used, alignments of poorly related sequences often contain regions with multiple small and disconnected indels. However, alignments with indels either fewer in number, or showing more tendency to occur in groups are often preferred, on the grounds that isolated insertion or deletion events are less likely than occasional events leading to the gain or loss of several residues at once. To model this process, the indel penalty is set at two levels; the first, when a path which last moved diagonally is extended by an indel, is relatively high. The second level penalty is used when the last transition in the path was an indel, and a lower penalty is applied to its extension. Where the user can choose the weightings, care is needed. If the extension penalty is less than half a mismatch penalty, it is not safe to assume that the triangle inequality still holds, and the algorithms cannot be guaranteed to find the best paths. This can be avoided by setting an upper limit to the length of indels, but there can be no biological justification for this until the degree of relationship of the sequences has been broadly established.

3.3 **The total alignment algorithm (Type I problem)**

The process described above for the generation of the score matrix is concerned with the optimal path through the matrix from (0,0) to (n,m); this can then be converted into the optimal alignment of the sequences. We give here some short examples (*Figure 14*) to demonstrate the differences in alignment produced when differing parameters are used. It is important to bear in mind that most implementations only trace back a single alignment, using rules to decide which path to follow in cases of branching. This is expedient, but there is a case for tracking twice, using complementary rules at branch points, to examine extreme alignments; comparison of the extreme alignments will show regions that are constant, and other regions which vary in the two alignments. Interpretation can be made in a more informed manner if these distinctions can be drawn in the displayed alignment.

3.4 **The best location algorithm (Type II problem)**

The algorithm for the alignment of a short sequence at the best location within a longer sequence is similar to the algorithm for the Type I total alignment problem. It differs in that every location in the longer string is regarded as a potential starting place for the desired alignment. The quality of a path is handled in a similar way to the Type

```
match=0; mismatch=-1; indel=0; score=-18
Maximum no. of matches, or 'Longest common subsequence'
    ***   **** * ** * **** * **
CSNLSTCV---LS-AYWKDLNNYHRFSGMGFGPETP
CSNLSTCVLGKLSQELHK-LQTYPR-TNTGSG--TP


Backtracking priority rules reversed for choice between indels
in different strands
    ***   **** * ** * **** * **
CSNLSTCV---LSAYW-KDLNNYHRFSGMGFGPETP
CSNLSTCVLGKLSQELHK-LQTYPRTNT-GSG--TP


match =0; mismatch=-1; indel=-1; score = -18
Minimum evolutionary distance
    ***  ***** ** * **** * **
CSNLSTCV---LSAYWKDLNNYHRFSGMGFGPETP
CSNLSTCVLGKLSQELHKLQTYPR-TNTGSG--TP

match =0; mismatch=-1; indel=-2; score = -19

    ****** *********** **
CSNLSTCVLSAYWKDLNNYHRFSGMGFGPETP
CSNLSTCVLGKLSQELHKLQTYPRTNTGSGTP


Similarity scoring methods (Table 3).   * indicates a negative scoring
element; . indicates a neutral scoring element; all the others score
positively.

Indel=0; score = 1023.0
Maximum possible similarity score
    * ** ***       *** ** ***  * * **
CSNLSTCVLSAYWK---DLNN----YH-RF--SGMGFGPETP
CSNLSTCVL-G--KLSQELHKLQTY-PR-TNTG-S-G--TP

Indel=-20.; score = 779.0
    ***       *** * *  *.*
CSNLSTCVLSAYWKDLNN----YHRFSGMGFGPETP
CSNLSTCVLGKLSQELHKLQTYPR-TN-T-GSGTP

Indel=-30.; indel extension penalty=-3.; score =799.0
    ***    *** * *  * * **
CSNLSTCVLSAYWKDLNN----YHRFSGMGFGPETP
CSNLSTCVLGKLSQELHKLQTYPR-TNTGSG--TP
```

Figure 14. Results from total alignment algorithm, under a variety of conditions. Sequences are bovine calcitonin (upper sequence), and salmon calcitonin (lower sequence). ∗ indicates mismatch or error.

I problem, and the same type of matrix for score or path history is constructed. The implementation differs in that all unmatched characters in the long (horizontal) sequence attract no penalties if they precede the first character from the short sequence. The dummy first row is therefore set entirely to zero, and the Type I process used to complete the path matrix. The last row in the matrix contains all the possible exit points for the desired alignment, and the backtracking is applied at those cells with the best quality rating. A number of alignments, rather than a single one, may be found, if they have the same rating, and it is possible that they represent separate locations at which good alignments are possible. Paths from some starting points will not persist to the bottom row (being cut off by better quality paths), so that the set of scores for the paths reaching the last row does not necessarily contain information about any paths other than the best. For specific searches, e.g. the search for short consensus sequences, or for control or restriction sites, this algorithm has advantages over the Type I algorithm. *Figure 15* illustrates some of these points.

```
Identity based alignments:

match=0; mismatch= -1; indel= -1; score = -17.
Minimum evolutionary distance weightings.
            *       * ** **   * *** **** * * *
EGSSLDSPRSKRCGNLSTCMLGTYTQDL-N--KF--HTF-PQTSIGVGAPGKKRDMAKDLETNHHPYFG
          CSNLSTCVLSAYWKDLNNYHRFSGMGFGPETP

match=0; mismatch= -1; indel= -2; score = -17.
Increased weighting of indels reveals unstable nature of previous alignment,
to the right of residues DL, without loss of quality.
            *  * ** **   ** * ***** ***
EGSSLDSPRSKRCGNLSTCMLGTYTQDLNKFHTFPQTSIGVGAPGKKRDMAKDLETNHHPYFG
          CSNLSTCVLSAYWKDLNNYHRFSGMGFGPETP


Using similarity scoring (Table 3) for all pairs of residues:

Indel= 0; score = 1257
Maximum possible similarity score.

EGSSLDSPRSKRCGNLSTCMLGTYT-QDLNKFHT-FPQTSIGVG-APGKKRDMAKDLETNHHPYFG
          CSNLSTCVLSAY-WKDLNNYH-RF---S-GMGFGP----------ET---P

Indel= -10.; score = 1064
Even a small indel penalty produces more compact alignments.

EGSSLDSPRSKRCGNLSTCMLGTYT-QDLNKFHTFPQTSIGVG--APGKKRDMAKDLETNHHPYFG
          CSNLSTCVLSAY-WKDLNNYHRF--SGMGFGPETP

Indel = -20.; score = 1004
Doubling the indel penalty removes all but two indels, forced by a
particularly unfavourable mismatch.

EGSSLDSPRSKRCGNLSTCMLGTYT-QDLNKFHTFPQTSIGVGAPGKKRDMAKDLETNHHPYFG
          CSNLSTCVLSAY-WKDLNNYHRFSGMGFGPETP
```

Figure 15. Detection of the best location for a short test sequence (bovine calcitonin) within a longer stretch of a calcitonin precursor from rat, using the exhaustive algorithm for the Type II problem. The upper sequence is the precursor sequence; the lower sequence is bovine calcitonin. The parameters used are listed in each case.

3.5 The best local similarity algorithm (Type III problem)

This algorithm is another variant of the general scheme for assessing alignments exhaustively in a score matrix. In the basic algorithm, the scoring scheme is chosen to produce alignments of average score zero, in a comparison of random sequences. Every cell in the matrix is treated as a potential starting point for a region of significant alignment, and no indel penalties attach to overhanging regions of either sequence; the first row and column of the score matrix are set to zero. At any cell where the best path score falls below zero, the current path is judged to be below expectation. and it is not extended, but the score is reset to zero, and the cell is treated as a potential starting point for a new region of local similarity. At the completion of the operation, the regions of greatest local similarity can be identified by backtracking from the maxima in the matrix. The scoring scheme for protein comparisons is most sensitive when the graded similarity schemes are used. It is particularly useful since partial alignments can be tried when only fragmentary sequence data are available, or when it is important to detect structural or functional analogies. The local alignment algorithm is successful in drawing attention to significant sequence analogies, which the total alignment algorithm may fail to find if the neighbouring region of sequence is poorly related.

Scoring schemes in which alignments of a pair of random sequences will probably

```
Using identities (marked with bar):
match=5.; mismatch=-1.0; indel=-3.0; minimum threshold = 10.0

CSNLSTCV----LS-AYWKDLNNYHRFSGMGFGPETP      Score 56
|||||||     ||     |  | |    | |   ||
CSNLSTCVLGKLSQELHK-LQTYPR-TNTGSG--TP

C-VLS   Score 11    LS    Score 10
|  ||               ||
CSNLS               LS
```

```
Using similarity scoring (Table 3), and minimum threshold of 100.0;
positive pairings marked with bars, neutral pairings with dots:
Indel= -50. (The average value of a match is 48.)

CSNLSTCVLSAYWKDLNNYHRFSGMGFGPETP      Score 753
||||||||||   ||||| | ||  | |||||
CSNLSTCVLGKLSQELHKLQTYPRTNTGSGTP

LSAY    Score 167    CV-LSAY    Score 149
| ||                |  ||||
LQTY                CSNLSTC

CV-LGK    Score 129    MG-FGPE    Score 104
|  ||.                 || ||||
CSNLST                 LGKLSQE
```

Figure 16. Best matching regions reported by the Type III algorithm, using bovine calcitonin and salmon calcitonin. Bovine calcitonin = CSNLSTCVLSAYWKDLNNYHRFSGMGFGPETP (upper sequence) Salmon calcitonin = CSNLSTCVLGKLSQELHKLQTYPRTNTGSGTP (lower sequence) Reversing the sequences produced the same major alignment in each case (cf. *Figure 17*). Similarity-based alignments seem more plausible in this comparison.

have negative scores can be deliberately chosen to limit the regions found to those with more significant degrees of similarity; these are shorter, and clearer, than the rather more extended paths that are often found when the average alignment scores zero. *Figure 16* illustrates some of these points.

3.6 Conclusions

These exhaustive algorithms are the essential tools of molecular sequence comparisons. They do not suffer the drawbacks of methods developed when computer memory or speed was limited; the results can be treated as more authoritative than those obtained by *ad hoc* filtering processes. The Type I algorithm deals with whole sequence comparison; this is more applicable to proteins then nucleic acid sequences, unless the nucleic acid sequences represent complete molecules or homologous genes. The Type II algorithm is under used, and could be valuable in pattern detection studies, particularly when the 'consensus' is well defined. The Type III algorithm is particularly effective in locating highly related regions of sequences that are only distantly related overall. Though dot-plots give the most complete over-view of sequence comparisons, they can be difficult to interpret without the support that an exhaustive algorithm can give to proposed alignments.

The significance of total or partial alignments is difficult to assess on an absolute basis. Though some idea of the likelihood of a particular region of an alignment can be gained (e.g. from *Figures 6* and *7*), individual comparisons are made with residues in their natural frequencies of occurrence, which may be very atypical, and with indels available to improve further the optimal path. The statistical properties of the distributions of the scores found when searching by the exhaustive algorithms have been described by Goad and Kanehisa (13) for nucleic acids, and they included randomized sequences to provide an empirical baseline set of measurements. Smith, Waterman and Burks (14) have described the distributions of similar sets of results for nucleic acids as non-normal in their properties, and have given empirical methods for calculation of significance, based on a theoretical analysis of the properties of the distributions. It is important to have such estimates available, but in the case of protein comparisons there are sound reasons for believing that the conserved properties of structural domains in proteins will produce regions of local homology that may be statistically of low significance, yet will have considerable biological importance.

In any case, neither nucleic acid nor protein sequences can be treated as random sets of residues or bases; the use of random sequences for establishing the expectation of any searching method may take account of variation in composition, but local subsequences may have evolved with structural or functional constraints. When the distribution of results cannot readily be found, the behaviour of suitable randomized sequences (e.g. with some regard to the composition and nearest neighbour frequencies) may provide the best measure of significance, even at the expense of the additional computing resources required.

Methods for multiple sequence comparisons have attracted some attention (15,16) but this problem is considerably more demanding in space and computational complexity, and is not yet commonly available to molecular biologists.

Protein comparisons have occupied most of the discussion given above; the problem of nucleic acid comparisons is in general capable of less refinement, because of the smaller alphabet involved, and the smaller degree of interpretation that can be applied to shorter sub-sequences in the whole. Folding regions are the most striking exception to this, and while some programs detect such regions, they are usually constrained to short displacements; perhaps the exhaustive search algorithms will be more informative in the long run (cf. Chapter 11).

4. SIMILARITY BUILDING ALGORITHMS

4.1 Introduction

A third approach to the computerized comparison of biosequences is a family of methods derived from that introduced by Korn and Queen (17), and despite serious shortcomings, these methods are still available. These algorithms, like those discussed in the previous section, systematically search the match matrix from all possible starting positions for paths which include a high degree of similarity. Similarity, however, is restricted to exact matches. The main differences from the exhaustive algorithms are that only a limited set of methods of continuing the path are considered, and that these possible extensions are considered in a fixed order. The first allowable method of continuing the path is followed, even if a method further down the list would eventually lead to a better alignment. A path terminates when there is no allowable way of continuing it, and also if the overall proportion of matches falls below some threshold. In many versions of this algorithm, the *a priori* improbability of the alignment is estimated, and only those for which this value is greater than a set threshold are presented to the user.

Different implementations of this general scheme differ principally in the range of possible path extensions considered. Path extensions may be one or more diagonal steps, indels, or steps in which unequal numbers of residues are passed over in each strand (the latter can be regarded as combinations of diagonal steps and indels, but the order of single horizontal, vertical or diagonal steps is not defined). Diagonal steps producing a match between the sequences are always preferred; otherwise there is a defined order in which steps involving mismatches are considered. The version of the algorithm used in the SEQ program of Bionet (IntelliGenetics Inc.) will be used as a specific example. In this case the algorithm is partly controlled by the user by setting a parameter called LOOPLEN (abbreviated to L in what follows). The type and order of extensions considered when there is no match on the diagonal step can be described in pseudo-code:

L can be set in the range 0 to 3; if L = 0, no mismatching extensions are considered. Otherwise, the order is

    for j = 1 to L,
      extend with j mismatches
then
    for j = 1 to L,
      extend with a deletion of j in one strand, then
      extend with a deletion of j in the other strand
then
    for j = 1 to L−1,
    for k = j + 1 to L.
      extend with j residues passed over in one strand, k in the other, then
      extend with k residues passed over in one strand, j in the other.

The acceptability of a step depends on a second user-settable parameter, AFTERMISS (abbreviated here to A). An alignment is continued by the first step considered which is followed by A out of three diagonal match steps.

These algorithms are difficult to program because of the complex logic required to avoid pursuing a path which is contained in one already discovered, and because the probability calculation has to take account of the fact that only a subset of the paths containing any particular proportion of matches is discoverable by the algorithm.

Provided the program is correct, these difficulties need not concern the user. More serious are the theoretical drawbacks that the methods will not always find the best alignments, that the user cannot be sure that everything which he would consider significant will be presented to him, and that there is no simple way of epitomizing the types of similar alignment that the method will find or fail to find.

Two compensating virtues have been claimed for the method. The first is that the alignments presented are scored, and can be ranked in order of their *a priori* improbability. This improbability can be identified with their probable significance.

A second virtue is that these methods may be faster than the exhaustive algorithms. However, the practical significance of differences in speed between programs is that, as the problems considered become larger, the time required grows until it is no longer practicable to use the program; a difference in speed is only important if a large range of problems is accessible to one program and not to the other. Waterman (18) has pointed out that (for particular versions of the two types of algorithm) the total running time

Example A.
```
                      **** *****
      1      CSNLSTCVLSAYWKDLNNYH       20
      1      CSNLSTCVL----GKLSQELH      17
   % = 55.000       P( 20, 11) = .406E-04      E = .036


                  *** **** * ** * *** * **
      4      LSTCVLSAYWKDLNNYHR-FSGMGFGP       29
     12      LS-QEL----HK-LQTYPRINIGSG-TP      32
   % = 37.037       P( 27, 10) = .388E-04      E = .035



     31      TP       32
     31      TP       32
   % = 100.000      P(  2,  2) = .581E-01      E = 52.283
```

Example B. Calcitonin sequences reversed.
```
              **
      1      PTEPG     5
      1      PTGSG     5
   % = 60.000       P(  5,  3) = .518E-01      E = 46.625


              **** ***** ** **** **
      9      GSFRHYNNLDKWYASLVCTSL-NS       31
      3      GS--GIN--TRPYTQL--KHLEQS       20
   % = 29.167       P( 24,  7) = .357E-04      E = .032


              **** **
     23      SLVCTSLNSC       32
     20      SL--KGL-VC       26
   % = 40.000       P( 10,  4) = .183E-01      E = 16.471


              **
     23      SL-VC     26
     28      SLNSC     32
   % = 60.000       P(  5,  3) = .518E-01      E = 46.625



     24      LVCTSLNSC       32
     24      LVCTSLNSC       32
   % = 100.000      P(  9,  9) = .234E-07      E = .000
```

Figure 17. Comparison using the Korn−Queen algorithm (SEARCH option of program PEP in Bionet), of (A) calcitonin sequences from cattle (upper sequence) and salmon (lower sequence), and (B) the same calcitonin sequences, reversed. The parameters in each case were set as follows: Minmatch = 2, Loopout = 3, Aftermismatch = 1, Expect = 100.0 and Percentmatch = 10. Differing alignments are reported in the two cases, and the estimated significance of the alignments including the nine consecutive matches differ by a factor of 1.7×10^3. The longest alignments in each case are not related, and neither can be simply related to the nine-match alignments to produce the accepted overall alignment between the two short sequences. The uppermost alignment (20,11) can be improved to (16,11) or (21,12) by minor adjustments by eye.

for both grows as the square of the length of the sequences being compared, and this implies that there is no great difference in the range of sizes of problems which can be handled by the two methods.

4.2 Practical considerations

The key to understanding the behaviour of this algorithm is the parameter AFTER-MISS. The danger in using an algorithm which does not 'look ahead' and choose the 'best' way of continuing a path is that, in following a signal, the program will be deflected from the alignment sought because in some region where the signal is weak, adventitious noise will allow an erroneous path extension. The only check that the path chosen is correct is that the number of matches which follow the correct step is likely to be higher than those following a by-way. This discrimination will be weaker at lower values of AFTERMISS (and will also be weaker for nucleic acid than for protein comparisons; it might be expected that a more effective algorithm for protein comparisons would examine a longer region ahead of a mismatch step). So the best assurance that the algorithm will follow the true signal is to set AFTERMISS at the high end of its range. This in turn implies that the overall proportion of matches in the alignment must be high.

We conclude that this type of program will work most reliably in detecting signals which contain very high local concentrations of matches. The circumstances under which the method might compete in effectiveness with the dot-plot are that the signal should consist of short highly-matched regions broken up with a large number of short mismatched regions or indels. Theoretically, there appear to be no circumstances under which Korn−Queen algorithms are to be preferred to the exhaustive local algorithms of Section 3.5, since, with an appropriately chosen scoring scheme, the alignments found by the latter should be a superset of those found by the Korn−Queen method. If the exhaustive algorithm, in the implementation available, is prohibitively slow, the Korn−Queen method may have to be used. The output from the program should be carefully examined, since it will often be possible to improve the alignments by eye. *Figure 17* gives an example of the application of this type of program to two short sequences, to illustrate some of these problems.

We have seen that in all methods of comparison of protein sequences, the detectability of weak signals is greatly enhanced by the use of similarity scoring in place of the simple match/no match scoring scheme. This facility is not available in widely-used versions of the Korn−Queen type of algorithm.

5. DATABASE SEARCHING

5.1 **Introduction**

Databases are a source of new molecular insights into the origin, role and evolution of proteins and nucleic acids (see Chapter 4). The information in known sequences, and in their relationship to each other has been very revealing to molecular biologists, but it has required considerable dedication to make the extensive searches needed to uncover these discoveries. Now the accumulated knowledge of protein and nucleic acid sequences has reached substantial proportions (~ 1 million amino acid residues in protein sequences, and 6 million bases in nucleic acid sequences), it is both potentially more rewarding and more difficult to undertake these searches. There are other important reasons to search databases; for example, to aid in the early identification of novel (and possibly fragmentary) sequence data. Gene sequences are routinely tested for open reading frames, and the sequences of putative proteins included in the databases when their existence is strongly suspected but their properties are unknown. Eventually, their identification by analogy with known sequences will be possible, and database searches including the putative protein sequences serve a valuable role. Database search techniques could also be used as an alternative to a consensus search as a way to identify functional short sub-sequences. A consensus sequence is an attempt to represent by means of ambiguities, variable length gaps and mixed sequences what it is that, for example, prokaryotic promoter sequences have in common. It is conceptually difficult both to derive a consensus and to define closeness to it, and these difficulties are reflected in a relatively poor performance of programs for the purpose. An effective alternative which has not been widely used is to employ high-speed database searching techniques to compare every known example of the functional sub-sequence with the sequence in which the site is being sought.

Ideally, a database search would be performed by using any of the methods described in Sections 2 and 3 to make pairwise comparisons of each query sequence with each member of the database. Of course, nobody is ever going to examine several thousand dot-plots by eye. The problem of scanning such plots is not dissimilar from that of analysing spark-chamber photographs automatically. This is a routine operation in high-energy physics research, but there appear to have been no attempts to apply similar methods to molecular biology.

There is no such problem in scanning the output of programs performing the exhaustive algorithms of Section 3, since the alignments can be ranked in order of the similarity scores derived from the parameters used in these methods. However, a single pairwise comparison of moderate-length sequences may take several seconds to a minute on a conventional multi-access system, and when this is multiplied several thousand times, complete database searches become impracticably expensive on such machines.

Such searches can be carried out on so-called supercomputers. At present, few molecular biologists have routine access to machines of such power, but if there is a recognized need for these resources, that at least provides the first step in acquiring them. In any case the continuing rapid technical development of computer hardware means that the power of accessible machines is always increasing and before long adequately powerful machines may be routinely available. For these reasons we describe in Section 5.3 the experience of ourselves and others in using supercomputers to perform exhaustive database searches.

In the absence of a powerful enough machine to perform an exhaustive search, the biologist must find some way to limit the number of pairwise comparisons which have to be made by the methods of Sections 2 and 3. The most important way to do this is to find some non-exhaustive, approximate method which can rapidly compare the query sequence to each of the database sequences in turn, and which it is hoped will single out all those comparisons which will yield significant similarities when examined by the exhaustive methods. Filters of this kind are described in Section 5.2.

Finally, the sequences searched may be limited to a subset of the whole database defined by attributes other than the sequence. For example, the search might be carried out only over sequences derived from a particular range of species, or only over those known to express a particular funtion. The principle DNA and protein databases either order entries into natural groups, or use semi-systematic names for entries (or both), and on many computer systems there are system utilities (such as text editors, or methods for handling ambiguously named files) which allow these features to be used to select a useful subset of the database. More generally, there may be utilities (such as SEARCH in VAX/VMS) which allow the whole of every entry to be searched for the occurrence of keywords such as species name or functional attribute. In addition, many centres either use general purpose database management systems or have developed special purpose systems for retrieving sequences based on non-sequence attributes. The programs 'PSQ' and 'NAQ' (NBRF, Washington) are particularly worth mentioning (see Chapter 5), because they are effective, transportable (on VAX/VMS systems) and inexpensive. The methods described in this paragraph are particularly valuable if they reduce the number of sequences to be examined to such a small number that the remaining pairwise comparisons can be made by non-approximate methods. If this is not the case, the kind of filtering in Section 5.2 has to be used anyway, and since it is possible to apply these methods to the whole database, there is little to be gained by not doing so.

5.2 **High speed filters**

All regions that can be aligned closely give rise in the match matrix, to locally high concentrations of matches along a single diagonal or on a narrow band of diagonals. Wilbur and Lipman (19) devized a fast method to obtain the total number of occurrences of single matches (or doublets, triplets, etc) on each diagonal. This score is summed for a group of adjacent diagonals, in order to take account of possible indels, and the value compared to the mean and standard deviation of all values in order to identify the most statistically significant regions of the total comparison. The highest value of this score for each pairwise comparison was used as an overall measure of similarity between the sequences, and the $(20-30)$ comparisons with the highest values were selected for more careful comparison. In a refinement (ref.20) the five best diagonals in each pairwise comparison were rescored with the Dayhoff similarity scores, and the highest of these used as the similarity score between the two sequences.

These programs gain their speed principally by the method which they use to locate short runs of matches (words or patterns). The first stage is the construction, from the query sequence, of a list of all the singlets (or doublets, triplets etc) which occur in it with a pointer to their position. Each successive word in the database is then checked against this list using a standard rapid search method, and a tally kept of the matches on each diagonal. One can speed up an individual search by inverting this process, and generating a wordlist (a dictionary) from the database, and checking successive words from the query sequence against it, since the database only changes occasionally, while each query sequence is novel. This is a practical and effective method on any computer system which has a large enough real or virtual memory. There is of course an overhead in the initial preparation and maintenance of the dictionary, and a continuing overhead of storage requirement for the dictionary which is now larger than the original database.

This approach, in the 'Prelate' system (Department of Molecular Biology, University of Edinburgh) is implemented in an extremely fast short pattern location scheme, which is extended to query sequences by treating them as a set of (overlapping) patterns to be found. The fastest result when searching such a dictionary structure, is obtained when a pattern is absent. A disadvantage is that while the scheme is ideal for searches based on identities, it requires additional work if residue similarity is to be sought. Because of its speed, the use of this approach to investigate the occurrence of short thematic patterns is a promising area for development. The dictionary approach has been extended to sub-strings which are not in their original order nor formed from adjacent characters, allowing disconnected or partially matching sub-strings to be detected with the same speed and efficiency as simple sub-strings in the query sequence.

Dictionaries have been used in various systems of sequence analysis, but one aspect has not perhaps been given due weight. The use of sub-strings in sequence comparisons

```
Identifier Hits Description
** TCBO      8 Calcitonin - Bovine and sheep
   TCPG      6 Calcitonin - Pig
   TCON      2 Calcitonin 1 - Salmon
   TCEE      2 Calcitonin - Eel
   TCON2     2 Calcitonins 2 and 3 - Salmon

Other sequences contained a maximum of 1 pattern match.
```

Figure 18. Fast dictionary classification of bovine calcitonin, using 'Prelate' system with a pattern length of four residues, and offsetting the pattern by four residues each time. The dictionary was prepared from the NBRF Protein Database, Version 4 (2676 sequences). Top 5 results shown.

is well known; but the use of preprocessed fixed length strings has to be preferred in database searching applications, as it has fewer space overheads. The length of the fixed string should be chosen neither too short (when many references will be found, but no single one will necessarily be of interest) nor too long (when any reference will be significant, but most patterns will never be found in the database). Naturally occurring sequences of length five have an approximately 50% chance of being unique in the database (based on NBRF Release 5.0; 694 014 amino acid residues); so many pentapeptides never occur in the database that this probability will only change slowly as the database expands. Sequences of length three are matched relatively frequently, and the overheads of processing these short sub-string matches are high.

We have found that tetrapeptides, with characters that are adjacent, or spread over a short neighbourhood, form the basis of an extremely fast protein screening program, which fulfils the requirements of a database searching program successfully (see *Figure 18*).

The design of the re-ordered sub-strings on which to base additional dictionaries can exploit the short range ordering in proteins which arise from the common structural features such as helices and sheets, where specific types of residues tend to recur at characteristic spacings.

The most important use of database searches is for the recognition of functional similarities which were not expected from the known properties of the new sequence. When a new globin or serine protease sequence becomes available, it will be carefully compared with all other members of the same class; a database search is mainly needed when the class (in this sense) of a sequence is unknown, or to demonstrate a similarity between classes. Most interest attaches to similarities which are so strong that there can be no doubt about a functional and/or evolutionary relationship. All the methods described in this section will locate similarities of this kind [for example, the similarities of angiogenin to ribonucleases (21); the tyrosine kinases and oncogenes (22)], and the choice between them is therefore only a question of local availability.

5.3 Application of the exhaustive algorithms

It might appear that the filtering which can be achieved by the approximate methods should be adequate for most purposes. However, certain tasks would still be difficult by the methods described so far, e.g. pattern detection over more extended regions. Exhaustive algorithms, of Types II and III, will always be safer, ensuring that the constraints of the approximate methods have not lost significant results.

Two groups have implemented exhaustive search algorithms on supercomputers. Smith, Waterman and Burks (14) report results obtained with a FORTRAN program

```
Results:
1    Bovine calcitonin                                        Score 96    Quality 100.00

          1    CSNLSTCVLSAYWKDLNNYHRFSGMGFGPETP    32
          1    CSNLSTCVLSAYWKDLNNYHRFSGMGFGPETP    32

2    Calcitonin - pig                                         Score 84    Quality 87.50
                             **    *
          1    CSNLSTCVLSAYWRNLNNFHRFSGMGFGPETP    32
          1    CSNLSTCVLSAYWKDLNNYHRFSGMGFGPETP    32

3    Calcitonin presursor - rat                               Score 30    Quality 45.46
                    *    * ** **    ** *
         85    CGNLSTCMLGTYTQDLNKFHTF    106
          1    CSNLSTCVLSAYWKDLNNYHRF     22

4    Calcitonins 2 & 3 - salmon                               Score 28    Quality 58.33
                             *****
          1    CSNLSTCVLGKLSQDL    16
          1    CSNLSTCVLSAYWKDL    16

5    Calcitonin - eel                                         Score 27    Quality 100.00

          1    CSNLSTCVL    9
          1    CSNLSTCVL    9

6    Calcitonin 1 - salmon                                    Score 27    Quality 100.00

          1    CSNLSTCVL    9
          1    CSNLSTCVL    9

7    Calcitonin - human                                       Score 26    Quality 39.39
                 *    * ** ** * ** *
          1    CGNLSTCMLGTYTQDFNKFHTF    22
          1    CSNLSTCVLSAYWKDLNNYHRF    22

8    gag-fps polyprotein - avian sarcoma virus prcii  Score 20    Quality 83.33
                        *
        149    SAMGFGPE    156
         23    SGMGFGPE     30

9    ig kappa chain v-i region - human ag                     Score 20    Quality 83.33
                                *
         61    RFSGSGFG    68
         21    RFSGMGFG    28

10   gag-fps polyprotein - fujinami sarcoma virus            Score 20    Quality 83.33
                        *
        357    SAMGFGPE    364
         23    SGMGFGPE     30
```

Figure 19. Exhaustive database search for the best local homologies, using bovine calcitonin as the query sequence, against the NBRF Version 5.0 database using the ICL DAP (22). Parameters were: Match, $+3$; Mismatch, -1; Indel, -3; Threshold 15. 119 results were collected at or above the threshold value, in 48 sec. The best result, predictably, was bovine calcitonin; the next 9 are also shown.

running on a CRAY-1. This machine is representative of the most widely distributed type of supercomputer, the vector processor. The characteristic architectural feature of these machines in their use of a small number (in the case of the CRAY-1, one) of complex multistage processors in which several operations, identical except for the content of the data, are active (using different stages) at the same time. Such a device is often called a pipeline, but an assembly line provides a better analogy.

A FORTRAN program performing the local similarity algorithm of Section 3.5, and applied to DNA sequences with a simple scoring scheme, made 240 pairwise comparisons per minute when the average length of each sequence was 800 nucleotides.

Automatic vectorization does not necessarily lead to efficient implementation of these algorithms. The use of a parallel processor array, such as the 4096-processor ICL DAP, offers significant improvements over simple vectorization, because the basic connectivity of the machine is exploited when the sequences are mapped onto the processor array, and while the intermediate results are calculated. We have written implementations of the Types II and III algorithms for the DAP, specifically for the task of database searching (23). The Type III program can scan for up to 4096 regions of interest (above a preset, but adjustable, level of scoring) and return them in order of merit and aligned for display, taking approximately $1-2$ sec/residue in the query sequence (*Figure 19*). This speed appears to be broadly comparable to that achieved on the CRAY-1, though the DAP costs only about 10% of the price of the vector machine. These particular machines are both obsolete, but both have descendants of related architecture.

The significant point is that exhaustive database searches are practicable using machines which were designed 20 years ago. Since the power of machines continues to grow faster than the sequence databases, it is likely that approximate methods will soon be replaced even for routine database searches by exhaustive algorithms.

The problem of sequence comparison is one biological task that is certainly well suited to processor array architecture machines, and these computers will form a valuable resource as database size and query complexity continue to increase.

6. REFERENCES

1. Sankoff,D. and Kruskal,J.B. (eds) (1983) *Time Warps, String Edits and Macromolecules*, Addison-Wesley, Reading, Massachusetts.
2. Dayhoff,M.O., Schwartz,R.M. and Orcutt,B.C. (1978) in *Atlas of Protein Sequence and Structure*, Volume 5 Supplement 3, Dayhoff,M.O. (ed.), NBRF, Washington, p. 345.
3. Devereux,J., Haeberli,P. and Smithies,O. (1984) *Nucleic Acids Res.*, **12**, 387.
4. Staden,R. (1982) *Nucleic Acids Res.*, **10**, 2951.
5. Wilbur,M.J. (1985) *Mol. Biol. Evol.*, **2**(5), 434.
6. Feng,D.F., Johnson,M.S. and Doolittle,R.F. (1985) *J. Mol. Evol.*, **21**, 112.
7. McLachlan,A.D. (1971) *J. Mol. Biol.*, **61**, 409.
8. Ulam,S.M. (1972) in *Applications of Number Theory to Numerical Analysis*, Zaremba,S.K. (ed.), Academic Press, New York p.1.
9. Sellers,P.H. (1974) *SIAM J. Appl. Math.*, **26**, 787.
10. Sellers,P.H. (1980) *J. Algorithms*, **1**, 359.
11. Needleman,S.B. and Wunsch,C.D. (1970) *J. Mol. Biol.*, **48**, 443.
12. Smith,T.F. and Waterman,M.S. (1981) *J. Mol. Biol.*, **147**, 195.
13. Goad,W.B. and Kanehisa,M.I. (1982) *Nucleic Acids Res.*, **10**, 247.
14. Smith,T.F., Waterman,M.S. and Burks,C. (1985) *Nucleic Acids Res.*, **13**, 645.
15. Sankoff,D. and Cedergren,R.J. (1983) in *Time Warps, String Edits and Macromolecules*, Sankoff,D. and Kruskal,J.B. (eds), Addison-Wesley, Reading, Massachusetts, p. 253.
16. Fredman,M.L. (1984) *Bull. Math. Biol.*, **46**, 553.
17. Korn,J.L., Queen,C.L. and Wegman,M.N. (1977) *Proc. Natl. Acad. Sci. USA*, **74**, 4401.
18. Waterman,M.S. (1984) *Bull. Math. Biol.*, **46**, 473.
19. Wilbur,W.J. and Lipman,D.J. (1983) *Proc. Natl. Acad. Sci. USA*, **80**, 726.
20. Lipman,D.J. and Pearson,W.R. (1985) *Science*, **227**, 1435.
21. Strydom,D.J., Fett,J.W., Lobb,R.R., Alderman,E.M., Bethune,J.L., Riordan,J.F. and Vallee,B.L. (1985) *Biochemistry*, **24**, 5486.
22. Lorincz,A.T. and Reed,S.I. (1984) *Nature*, **307**, 183.
23. Lyall,A., Hill,C., Collins,J.F. and Coulson,A.F.W. (1986) in *Parallel Computing '85*, Feilmeier,M., Jouber,G. and Schendel,U. (eds), North-Holland, Holland, p. 235.
24. Gough,J.A. and Murray,N.E. (1983) *J. Mol. Biol.*, **166**, 1.

CHAPTER 14

Inference of evolutionary relationships

M.J.BISHOP, A.E.FRIDAY and E.A.THOMPSON

1. INTRODUCTION

1.1 Assessing sequence homology

When two or more nucleic acid or protein sequences are compared it is very obvious whether they are the same or different. It is not so obvious how to assess the similarities between them. This requires a *statistic* (which is a measure of the similarity) and a *model* which determines the probabilities for the values of the chosen statistic. Statistics which have been employed include the single-matching and the double-matching measures described by McLachlan (1) (see Chapter 13), and the largest exact homology of Karlin *et al.* (2). Models are of two fundamentally different types; there are descriptive models which assign probabilities on the basis of *descriptive assumptions* about the characteristics of a sequence (independence of and equiprobability of transitions between the four types of base, for example), and there are biological models which attempt to describe the effects of *biological processes* upon a sequence. There are two components to the biological assessment of similarities between sequences: a historical component and a functional component. Sequences may be similar because they have evolved from a common ancestral sequence and this is the historical component. Sequences may also be similar because they have a similar role in the cell (and may be convergent), and this is the functional component. Molecules may have a similar role in the cell (for example, perform similar enzymatic functions) but differ greatly in sequence. The properties they hold in common may be due to a three-dimensional conformation not obviously reflected in the primary covalent linkages. Because we cannot obtain adequate historical samples of biological macromolecules, making statements about how the similarity of present day sequences depends on their history is a matter for inference under a statistical model of evolutionary change at the molecular level. The functional component of sequence similarity is open to experimental study by site-directed mutagenesis (3). This powerful tool enables us to probe the function of DNA, RNA or protein molecules. In this way we can, at least in theory, experimentally determine the functional constraints on the evolution of a particular piece of DNA. The function of RNA and protein molecules may often be correlated with the intramolecular folding and intermolecular interactions in three dimensions. A number of biophysical techniques including X-ray crystallography (4), electron microscopy (5) and nuclear magnetic resonance spectroscopy (6) are available to study these interactions. An important goal is the ability to predict the nature of such interactions from a knowledge of primary structure alone. An approach to this is through molecular modelling using computer simulation (7) or interactive graphics (8). Knowing the functional constraints, we are then in a position to consider the historical component of sequence similarity (9).

1.2 Inferring an evolutionary relationship

One evolutionary problem that can be formulated is the following. Given the sequences of similar pieces of DNA from a number of organisms, or given the sequences of members of a multigene family from within a single genome, how do we infer the pathways of genetic descent which relate the sequences? To do this we require a model for the topology of the pathways of genetic transmission and a model for the processes of genetic change involved. This in turn implies that the processes of genetic change at the sequence level are open to experimental study, which indeed they are. We also, ideally, need to have a knowledge of the functional constraints as determined by experiment.

1.3 Inferring the nature of an evolutionary process

Another evolutionary problem may be formulated as follows. Let us suppose that we have in our possession an accurate knowledge of the pathways of genetic transmission relating a set of sequences. This knowledge includes the times (in the past) of branching in the genetic graph. (In practice, we do not possess an accurate knowledge of the history of genetic transmission except on a very short time scale.) We have also determined the major functional constraints by experimental methods. We are then in a position to infer the nature of the evolutionary process by which the sequences have changed.

1.4 Inferring the functional constraints

A third problem is not concerned with evolutionary inference, but uses knowledge of the pathways of genetic transmission and the processes of genetic change to infer the nature of the functional constraints acting on a biological macromolecule. The experimental data might consist of a large number of sequences of a protein, such as myoglobin, from a variety of organisms. Let us suppose that the evolutionary tree, with times, is known. Let us also suppose that the nature of the processes of genetic sequence change is known. Functional constraints on the evolution of the protein are then revealed by lack of fit of the sequence data to the model.

1.5 The status of evolutionary inferences

In the literature of sequence comparison we find examples of work

(i)   attempting to reconstruct phylogeny (for example, 10),
(ii)   attempting to reconstruct the nature of the processes of genetic change (for example, 11),
(iii) attempting to measure functional constraints on a molecule (for example, 12), and
(iv)   attempting to do all three from the *same* data.

The latter category must be mistaken. *It is not possible to infer phylogeny, the processes of genetic change, and the functional constraints from the same data.* In our view the practice of evolutionary inference is in its infancy. We possess, as yet, neither the results of extended studies of functional constraints by site-directed mutagenesis nor adequate results of experimental studies of the nature of processes of change at the molecular sequence level in natural systems. We are aware that there are notable biases in genomic composition imposed, for example, by bias in codon usage. There are notable differences in the DNA repair systems of organisms as closely related as primates and rodents. If, given the knowledge about functional constraints and the other sources of bias, we adopted a model of evolutionary change which could reasonably be accepted on experimental grounds and found that for the majority of genes for a set of organisms the best estimates of phylogeny were almost always the same, then we might accept this common estimate and attempt to interpret the nature of the unusual features of the processes in any exceptions. (Although we might find that there exists no estimate which can be accepted; the data may not contain historical information.) Studies are likely to be rewarding for recently diverged groups such as the hominoids and the sibling species of *Drosophila*. Over long time periods the phylogenetic infor-mation contained in sequences disappears. Though major patterns may emerge, on cur-rent evidence it seems unlikely that the details of the radiation of the phyla or the origin of the tetrapods will be revealed from sequence data. This view may, however, be over-pessimistic, and certainly our present knowledge is a pale shadow of that which we shall possess in years to come. The enterprise is scarcely practical on present data, yet it is one in which many seem to wish to indulge. Much of the literature of molecular evolution is confused as to what constitute the data which have been observed, what constitutes the model (the assumptions which have been made) and how to evaluate the relative merits of the competing hypotheses which are being considered. Outlining how to set about this is a practical matter which warrants the inclusion of a chapter in this book.

2. PATHWAYS OF GENETIC TRANSMISSION

2.1 The evolutionary tree model

Genealogical relationships define the path which the genetic material has followed (13) and, in the absence of direct observation, the only measure of genealogical relation-ships is the extent of observable genetic differences between individuals. Genealogical reconstruction, therefore, involves finding an appropriate time-dependent model of genetic change. Evolutionary inference differs from genealogical inference in that the time scales and the genetic differences are much greater in the former. Different models of genetic change are appropriate to the different sorts of data but many of the prin-ciples of reconstruction remain the same.

Two individuals are genealogically related if one is the ancestor of the other, or if they share a common ancestor (13). The relationships between individuals may be depicted by a network of parent—offspring connections. Since the genetic events oc-cur in time the network has an origin and a direction and can be drawn on a single time scale (*Figure 1*). Extrapolation from the genealogical network to the evolutionary tree involves an increase in the amount of time elapsed. A tree is a graph with no clos-ed loops in which one node, the root, has no predecessor and every other node has one predecessor. We use the term *tree pattern* for the shape of such a graph. An evolu-tionary tree is a tree pattern drawn against a single time scale. The evolutionary tree

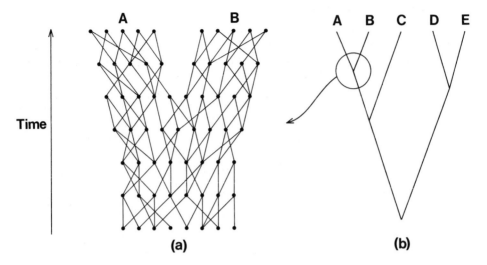

Figure 1. A diagram (after Hennig) of a genealogy showing (**a**) the pathways of transmission of genetic information between individual organisms (**b**) part of a phylogenetic tree drawn to show how it may be interpreted in terms of the pathways of genetic transmission between individual out-breeding organisms.

is the simplest description of the pathways of transmission of the genetic material, and its use can be justified only if it adequately describes these pathways (exceptions are discussed in Section 2.2).

2.1.1 *The case in which the evolutionary tree is specified by other evidence*

In this case we need only concern ourselves with estimates of the times of the branch points on a single tree pattern. The tree may have been estimated from other molecular datasets, in which case the times are relative, or from palaeontological evidence, in which case the times are sidereal and obtained by the methods of geochronology. For a tree based on molecular data it is sufficient to know the sidereal time of one branch point to convert the remaining relative branching times to sidereal branching times. In many groups of organisms no palaeontological evidence is available, and then the relative dates estimated from molecular data must stand on their own.

2.1.2 *The set of evolutionary hypotheses is the set of trees*

At present there are no known analytical solutions which enable one to proceed directly to the best tree (for a given criterion) for a particular set of data. This fact is of considerable practical importance, because the set of possible trees is the set of competing evolutionary hypotheses which are under consideration. Felsenstein (14) considers the problem of the number of tree patterns, reviewing previous work and providing a simple recurrence relation which enables calculations to be made of the number of (rooted) tree patterns, of either bifurcating or multifurcating structure, for various numbers of tips. Whilst for four tips there is a modest number of 15 (bifurcating) or 26 (all) tree patterns, for 10 tips there are approaching 35 million (bifurcating) and 280 million (all). The numbers rapidly become such that it is not computationally feasible to evaluate even the subset of all plausible trees for more than about 10 tips.

2.2 **More complex topologies**

It is easy to find examples of cases of genetic transmission violating the tree model. This has the effect that branches are bridged to form closed loops and so the genealogical graph is no longer a tree. At the individual level the tree model is violated by matings between related individuals, and at the population level by hybridization between populations of different species. At the molecular level there are processes which allow genetic exchange between rather than within lineages (infectious elements, Section 3.1) and therefore may be called 'horizontal transmission'. Within genomes it is known that a variety of rearrangements can take place. The order of gene duplication events and divergence of genes within a multigene family can be represented by a tree model only if processes of interaction such as gene conversion or multiple unequal crossing-over have not occurred. If they have occurred then they will have given rise to horizontal transmission within the gene phylogeny and lead to closed loops in the graph representing the pathways of descent of the genes. There is a practical point about inference here. Exceptions found to the validity of a tree model do not necessarily invalidate use of this description in other situations. Evolutionary inference on a graph representing paths of genetic transmission of arbitrary complexity is possible in principle. In practice, lack of constraints on the problem and the inadequacy of the data may make the problem insoluble.

3. PROCESSES OF GENETIC SEQUENCE CHANGE

The results of changes to DNA sequences may be described in terms of substitution, insertion or deletion of one or more bases or by inversions of two or more bases. Such descriptions may be made without reference to the processes of change and without knowledge of the steps which did in fact occur. Processes of change may be studied experimentally and the individual steps may be observed in favourable cases. This is one of the key advances towards the understanding of molecular evolution. Most of the experimental work has been done in the bacterium *Escherichia coli* (15), and unfortunately very little is known about eukaryotic systems.

3.1 **Processes of genetic sequence change as observables**

It would be wrong to think of the DNA molecules in a cell as highly stable, and the cell as passively accepting those few alterations to the DNA which are thrust upon it. The true situation is quite the reverse. There are complex enzymatic pathways which ensure the stability of the DNA in the cell by the mechanisms of DNA repair, and complex cellular mechanisms which ensure variability by reassortment of Mendelian factors, generation of recombinant chromosomes and movement of mobile elements. As these cellular systems depend upon proteins which are themselves evolving, the precise mode of their action may vary in different species. As well as these intragenomic activities there is also a component of variation imposed by infectious processes from outside the genome. A serious study of the evolution of a group of organisms should begin with a survey of knowledge on these subjects, which unfortunately, is likely to reveal that for many groups very little is known at present. All we can do here is to give isolated examples of experiments which provide the sort of knowledge which is needed.

3.1.1 *Replication*

The fidelity of DNA replication is remarkably high, but not perfect. From a consideration of the energies of the interactions involved in base pairing we would expect an error frequency of 10^{-4} to 10^{-5} per base pair replicated. The error frequency measured in *E. coli* or phage T4 is in the range 10^{-8} to 10^{-10}. The bacterial polymerase enzymes are thought to examine the base pair after the new base has been added and to remove the most recently added base if incorrect. This is known as *proof reading* and depends on a $3'-5'$ exonucleolytic activity of the DNA polymerase. Knowledge of DNA replication has been summarized by Kornberg (16,17).

3.1.2 *Repair*

Faithful replication is not enough to safeguard DNA through generations of organisms. There are a number of enzymatic systems which protect DNA against the consequences of damage and these form a *repair system* (18). Some parts of the repair system may be shared with the recombination system (Section 3.1.3). The repair system of *E. coli* is comparatively well characterized. The repair systems of mammals are poorly known at present but there appear to be differences between, for example, primates and rodents. Better knowledge of repair systems will in future lead to refinement of evolutionary models.

3.1.3 *Recombination*

Recombination is a major mechanism of evolutionary change which permits the separation of mutant genes and their testing for viability in new assortments. The molecular mechanisms of recombination are complex and have been described by Dressler and Potter (19). The frequency of recombination is not necessarily constant throughout the genome but may depend on factors such as chromosome structure. Individual recombination events studied in the fungus *Ascomycetes* reveal the existence of *gene conversion*: the strand of DNA representing one allele is corrected into the sequence of the other allele by the repair system. Genes in multiple copies may suffer conversion of part of one copy and sequences have been observed for which such a mechanism of change is a plausible explanation of differences.

3.1.4 *Mobile elements*

In both prokaryotes and eukaryotes there exist sequences which can move from one site to another and these sequences are called mobile or transposable elements (20). The mechanism of the transposition involves a manipulation of the DNA within which the transposable element exists, rather than the production of a free piece of DNA. Models of evolution involving mobile elements (21) are more complex than the simple model described in Section 5 which deals only with substitutions and single insertions and deletions.

3.1.5 *Infectious elements*

Viruses contain nucleic acid molecules which can have an existence both within and outside a genome. Elements which move in and out of genomes may introduce new genetic material. Recombinant DNA molecules are an essential part of genetic engineering techniques and we may suspect that recombinant DNA molecules have played a major part in evolution. Obtaining the evidence that this is the case is more difficult. One important example concerns the viral oncogenes and the realization that some oncogenes evolved from precursors in genomes of infected cells. This led to the identification of some of the cellular genes whose alteration can lead to neoplasia (22).

3.2 The need to consider generation time

It is often assumed that the mutation rate in DNA is uniform in sidereal time irrespective of the organisms concerned because the efficiencies of DNA repair mechanisms are similar and errors accumulate during the regular maintenance of DNA. It may be, however, that most of the errors arise during DNA replication, in which case the number of DNA replications from zygote to gamete divided by the generation time should be used to scale the rates. While these points have been discussed at length (23), there is, unfortunately, very little experimental evidence which may be brought to bear on the resolution of the problem.

3.3 Processes of genetic sequence change inferred from comparative studies

Rather than erecting a model for the processes of genetic sequence change and then attempting to infer phylogenetic relationships between genes or organisms, it is possible to assume that a phylogeny is known (from some other evidence) and that the purpose of comparative studies on sequences is to infer the nature of the processes by which they have changed. Comparative studies of sequences from closely related organisms may suggest to us that previously unrecognized processes of genetic change may be involved. This approach has been reviewed in detail by Dover (24). It has the merit of suggesting new mechanisms of change, but the onus of demonstrating that the mechanisms do indeed operate rests upon the experimentalist.

4. METHODS OF INFERENCE FOR EVOLUTIONARY PROBLEMS

4.1 Methods employing heuristic approaches

Very often when molecular biologists are comparing genes which they have sequenced they align the sequences by eye and produce a blow-by-blow account of the sequence changes which they suppose to have occurred in evolution. This is the narrative approach and suffers from the fact that there is no objective way of assessing the relative merits of alternative scenarios. The heuristic approach is to set up some reasonable intuitive criterion of sequence similarity and to proceed by finding a best-fit of reconstructed sequence changes onto a tree shape joining the sequences. The score associated with the tree shape is recorded and then a search is made for tree shapes with a better score until the best is found. Many approaches of this type have been devised (often under the heading of 'parsimony methods') and we shall not attempt to review them here but refer instead to the recent reviews of Felsenstein (25,26) and Sankoff and Kruskal (27). A heuristic approach may be satisfying to its inventors and may in particular cases even produce answers which are correct in the sense that there is a 1:1 mapping from the best-fit tree shape onto the graph representing the pathways of genetic transmission which were in fact followed. We consider that statistical inferences of phylogenies under time-dependent probabilistic models of genetic sequence change are more in the spirit of mainstream scientific investigation (28,29). Statistical inference focuses attention on the validity of the model and away from the appeal of the various intuitive methods. Paradoxically, some advocates of heuristic approaches reject the statistical approach because they claim that the model used is wrong (for example, that genetic sequence change is not stochastically uniform) and then suggest that the solution is to use parsimony methods. A more reasonable response would be to provide a statistical model which is in accord with the processes of change as they believe them to be occurring.

An early example of the heuristic approach was the work of Dayhoff (30), often referred to in subsequent literature as the empirical approach. Dayhoff selected 71 groups of closely related proteins, aligned them, reconstructed phylogenetic trees by a parsimony method and estimated ancestral sequences at the nodes of the trees. This enabled her to estimate the number of times each one of 20 amino acids had been converted into one of the other 19 in evolution. Dayhoff produced a mutation probability matrix from which it is possible to compute a measure of evolutionary distance given the observed amino acid differences between two protein sequences. Holmquist (31) has discussed the properties of Dayhoff's measure and compared it with estimates of evolutionary distance derived from several other approaches including his own probabilistic model (see Section 4.2).

There has been an unfortunate propensity for students of molecular evolution to mix the heuristic, empirical and stochastic approaches without regard to established principles of inference. This has brought confusion to a subject with more than enough inherent complexity.

4.2 Methods employing probabilistic models of genetic sequence change

The statistical approach to evolutionary problems posed by genetic sequence data stems from the suggestion of Zuckerkandl and Pauling (32) that amino acid changes in protein sequences might occur according to a Poisson process and that the accumulation of fixed mutations would therefore be exponential with time. Kimura (33), Neyman (34) and Holmquist (35) made early contributions to the field. Recent accounts of progress are given by Felsenstein (28), Kimura (29), Bishop and Friday (36), Li *et al.*, (37) and Hasegawa *et al.* (10, 38 – 40).

The essence of the approach is that a stochastic model of genetic sequence change is developed: this evolutionary model describes genetic changes along the pathways of genetic transmission (evolutionary tree). Historical estimates of the relationships of the present day sequences which constitute the data may then be made in terms of parameters estimated on the basis of the model. The fact that the models currently in use are not very realistic does not invalidate the statistical approach in general, although ideally, uncertainty about the model should be incorporated into confidence statements about our inferences. As our knowledge of the biological processes is improved the models can be refined.

4.3 A method of statistical inference

Inference requires the *Model* (M), which is that part of the description which may be regarded as given; the alternative *Hypotheses* (H), which attribute particular values to the parameters of the model; and the *Data* (D), which are the observed outcomes. There are a number of methods of inference but we prefer and shall use here likelihood inference.

The *likelihood principle* (41) advocates that hypotheses be compared on the basis of their relative likelihoods. The likelihood L of a hypothesis H_1 on data D under the model M is:

$$L(H_1) = P(D|H_1,M) \qquad \text{Equation 1}$$

where P is the probability of observing the data given the hypothesis and the model. Similarly, for a hypothesis H_2 on the *same* data

$$L(H_2) = P(D|H_2,M) \qquad \text{Equation 2}$$

All the information in the data D about the relative merits of the two hypotheses is contained in the *likelihood ratio*:

$$L(H_1)/L(H_2) = P(D|H_1,M)/P(D|H_2,M) \qquad \text{Equation 3}$$

or, equivalently, in the *support difference*, defined as the natural logarithm of the likelihood ratio. The likelihood principle may seem rather abstract at this stage, but we will show later how it may be applied in practical examples of inference.

5. A SIMPLE MODEL OF DNA SEQUENCE DIVERGENCE

Consider two sequences B and C which are assumed to have diverged from a common ancestor A which existed at some time t in the past (*Figure 2*). Let $\pi(A)$ denote the overall probability of observing a sequence A, and $P_t(A \to B)$ denote the transition probability from A to B over time t. Then the probability of observing sequences B and C is

$$\Sigma_A \ \pi(A) \ P_t(A \to B) \ P_t(A \to C) \qquad \text{Equation 4}$$

the sum over all possible ancestral sequences, A, of the products of the probability of observing that ancestor, the probability that the ancestor was converted to B and the probability that the ancestor was converted to C, both in time t, assuming independent evolution in the two lines of descent. This probability of observed data is then the likelihood for parameters of the model providing the probabilities π and P_t. Now

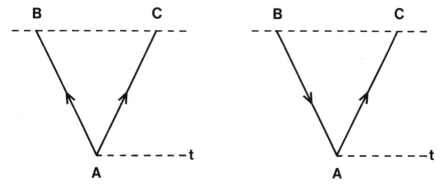

Figure 2. The diagram shows (left) the transitions between the DNA sequences of a common ancestor, A and those of two descendents B, C in time t. We may estimate the time $\tau = 2t$ for the transition of the sequence B to the sequence C (right) because of the Markov property of reversibility.

suppose that, presented only with sequences A and B, there is no feature of them which can be used to distinguish which is the ancestor and which the descendent. For sequences of genes that are not too divergent, this seems a reasonable assumption. We may then assume an underlying reversible stochastic process (42), and hence, rather than concerning ourselves with hypothetical ancestors A, we need consider only transition from either one of our two current sequences (B and C) to the other, over time period τ = 2t (for proof see ref. 43).

5.1 The exponential failure model

We suppose that the disturbances which result in failure accurately to repair DNA sequences arise according to a Poisson process. Failure to correct during time τ is caused only if at least one disturbance occurs. T, the time until a copying error, is a continuous random variable:

$$F(\tau) = P(T \leq \tau) = 1 - P(T > \tau) \qquad \text{Equation 5}$$

Now, $T > \tau$ only if there is no disturbance during time τ, and this is only the case if $X_\tau = 0$, hence:

$$F(\tau) = 1 - P(X_\tau = 0) = 1 - e^{-r\tau} \qquad \text{Equation 6}$$

$F(\tau)$ represents the cumulative distribution of an exponential failure law.

Under the model, the nucleic acid sequence may be viewed as a family of independent, identically distributed random variables (one for each site) which have the Markov property: conditional independence of the next transition and the past transition, given the present state. This system of nucleotide substitutions (*Figure 3*) forms a single,

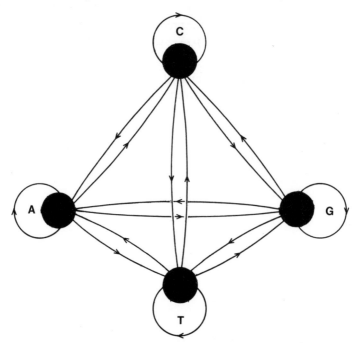

Figure 3. The Markov chain representing the 16 possible transformations of the four bases (A,C,G,T) of DNA. This system of nucleotide transitions forms a single, closed, communicating class of recurrent states.

closed, communicating class of recurrent states. A given site, which may be in any one of the states A, C, G or T, may remain in its previous state over a length of time t. On failure, there are 16 possible substitutions which have to be considered. For simplicity we may assume that all base substitutions are equiprobable (though this is not essential to either the model or the method). When a base is replaced, its replacement is one of A, C, G or T with equal probabilities of 1/4. When bases at a comparable site are identical, this may mean that no replacement has occurred, or that two or more events have culminated in restoration of the original identity of the base at that site by back mutation. From Equation 6, after time τ a base will have become any one of the other three bases with probability:

$$q = (1-e^{-r\tau})/4 \qquad \text{Equation 7}$$

and in the absence of other kinds of events will have remained the same with probability:

$$R = (1-3q) = (1+3e^{-r\tau})/4 \qquad \text{Equation 8}$$

5.2 Incorporating substitution, deletion and insertion

A realistic model for evolution of DNA sequences should incorporate substitution, deletion and insertion of bases. For simplicity, we again assume all base substitutions to be equiprobable, and for stationarity the (per base) rates of insertion and deletion must be equal. As a first approximation to a model, we consider each base separately (43). We suppose that substitutions occur at rate r, and that deletion of a base and insertion of a base (between any two existing bases) are events occurring at rate s. Then after time τ a base will have become any one of the three other bases with probability $q(r,\tau)$ as given by Equation 7 above, and will have suffered deletion or an adjacent insertion with probability $p(s,\tau)$. We introduce w where $p = q/w$ and consider the estimation of q (or equivalently $r\tau$) and of w, rather than of s or p; hence the form of p does not matter. Under the model, w will also depend upon τ. Independent insertions of several bases at any location are allowed under the model — with probabilities that are functions of powers of p. The model does not yet, however, accommodate the case of simultaneous insertion or deletion of a block of adjacent bases.

These considerations imply a transition matrix (*Table 1*) from bases of B to bases of C (or vice versa) over a time period $\tau = 2t$. (Note that 'transition' is used here in its general sense, *not* as versus 'transversion'.) The term $p/8$ for insertions, given in *Table 1*, is explained as follows. An insertion may be of any one of the four base types, and occurs between two bases. The total probability of all possible transitions of a complete set of bases must sum to one. Therefore it is necessary to consider an insertion as occurring with probability 1/2 before a given base and probability 1/2 after

Table 1. Transition matrix from bases of sequence B to bases of sequence C over time period $2t$. $(R = 1-3q-2p)$.

	A	C	G	T	insert[a]	delete
A	R	q	q	q	$p/8$	p
C	q	R	q	q	$p/8$	p
G	q	q	R	q	$p/8$	p
T	q	q	q	R	$p/8$	p

[a]This probability applies to each of the eight possible insert events described in the text.

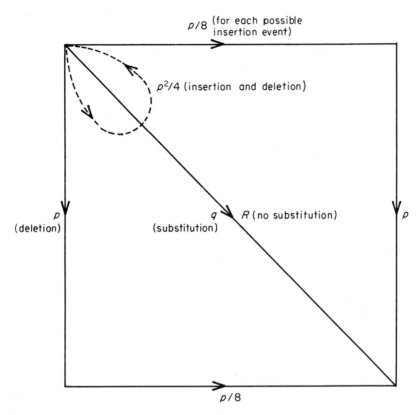

Figure 4. The basic cell showing the possible transformations of a single base, and the probabilities of each possible event.

it. Also insertions will occur with the same probability after the base preceding the given base and before the base following it. Hence any observed insertion must be assigned probability $p/8$, so that the total between two bases is:

(number of possible bases) × (number of locations between bases) × $(p/8)$ =
$$4 \times 2 \times (p/8) = p$$
The reasoning behind this is that it is computationally necessary to assign events as per base. Thus an insertion between two bases of the sequence must be (artificially) ascribed to one or the other. It is not necessary to distinguish the two possibilities; the point is to ensure that the probability of insertion is not counted twice in summing over bases.

Note that each row total in *Table 1* is one. An alternative representation of the content of *Table 1* is shown in *Figure 4*, where the node (n,m) denotes a location distance n bases along one sequence and m along the other. The diagonal transition (labelled R or q) denotes the substitution of the next base in the first sequence by the next base in the other. If these next bases in each sequence are the same there has been no substitution. A horizontal transition in *Figure 4* denotes insertion of a base into the second sequence, and a vertical transition denotes loss of a base.

Under the transition matrix a given base of a sequence B may:

(i)   disappear, with probability p;
(ii)   remain of length 1 via substitution or deletion+insertion, with probability $R + 3q + 2 \times 4 \times p \times p/8 = 1 - 2p + p^2$
(iii) become of length 2, with probability $(p/8)(8R + 24q + 16 \times 3 \times p \times p/8) = p - 2p^2 + 3p^3/4 = p - 2p^2$
(iv)   become of length 3, with probability approximately $3p^2/4$

The probabilities for transitions to greater lengths may be computed in similar fashion, but involve only powers of p of three or greater. When all the above probabilities are summed they come to less than 1. The missing probability, of order $p^2/4$, may be ascribed to the independent loss of a base in both sequences, or to gain + loss at a site in transition from one sequence to the other. Since there is no resultant base in either sequence, such an event may be represented by a loop from node (n,m) to itself (*Figure 4*). However, such events cannot be substantiated on the basis of the two sequences alone, and so we omit them from the computation of the likelihood. With this approximation, we now have a transition matrix which defines a probability distribution for evolution of a single, given base. The matrix also defines a probability distribution over all possible evolutionary events of a sequence, because each base of the sequence is treated independently in its contribution to the overall probability.

6. PAIRWISE ESTIMATES OF EVOLUTIONARY PARAMETERS UNDER A SIMPLE MODEL

6.1 Computing the likelihood

In order to compare alternative values for the parameters of the model, we require the likelihood; that is, the probability of the data under the model. In effect, this is the transition probability $P(B \rightarrow C)$, because only this transition depends on the evolutionary parameters. Consider two DNA sequences laid out as shown in *Figure 5*, along the sides of a lattice. Let $B^{(n)}$ denote the B sequence up to the nth base, and similarly $C^{(m)}$ the first m bases of sequence C. Then $P(B \rightarrow C) = P[B^{(N)} \rightarrow C^{(M)}]$, the probability of transition of the total sequences of length N and M, respectively, can be computed inductively as follows.

Let $L_{n,m} = P[B^{(n)} \rightarrow C^{(m)}]$, and consider the addition of an extra 'cell' to the sequences under consideration (*Figure 4*). The final base of each sequence may correspond to a substitution (denoted by a diagonal transition in the cell) or to a base being present in one sequence but not the other (corresponding to the vertical and horizontal transitions implied by insertion and deletion). Further, given that it is a substitution event that is under consideration, a substitution *has* occurred if the bases are different in the two sequences, and *has not* if they are the same. Thus

$$L_{n,m} = R\,(1-d)\,L_{n-1,m-1} + q\,d\,L_{n-1,m-1} + p\,L_{n-1,m} + (p/8)\,L_{n,m-1} \qquad \text{Equation 9}$$

where $d = 1$ if the (n,m) bases are different and $d = 0$ if they are the same. From Equation 9, $L_{n,m}$ may be computed for all points of the lattice, starting from $L_{0,0} = 1$, to obtain eventually $L_{N,M}$. This likelihood may be evaluated for different values of q

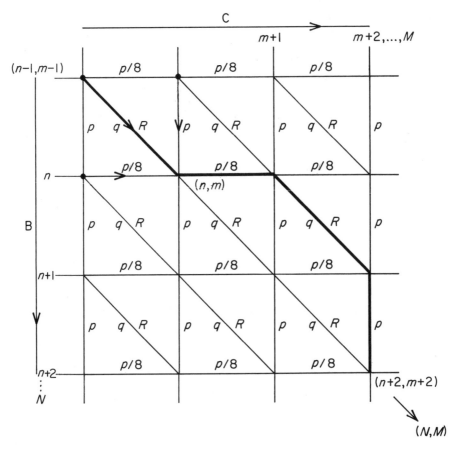

Figure 5. The combination of cells, providing for the transformation of a whole sequence. A block corresponding to three bases of each sequence is shown, together with the labelling of nodes and a possible route corresponding to a particular sequence of events.

and p (or of q and w), to obtain maximum likelihood estimates. Note that
$$P(B \to C) \ne P(C \to B)$$
since the sequences will normally have different equilibrium probabilities particularly if they are of different lengths. Thus if the sequences are interchanged the likelihood values $L_{n,m}$ will be changed in a non-trivial way. However, the same maximum likelihood estimates of parameters are found: the reversibility of the process (43) ensures this.

In principle, when making these estimates no constraints on the range of parameter values are necessary. The parameters w and q are identifiable. If the two sequences are identical, the maximum likelihood estimate is $p = q = 0$. If they differ widely, the maximum likelihood estimate may be that they do not overlap at all, and the estimate of w becomes very small (insertions are frequent, substitutions are irrelevant). To avoid such aberrations in practice it is convenient to limit the range of w in searching the likelihood surface. However, these constraints do not affect the estimates when sequences of similar genes in related species are considered.

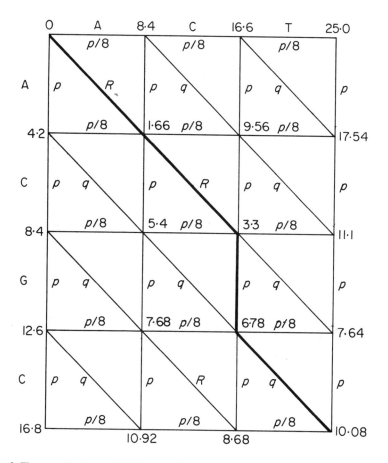

Figure 6. The example discussed in the text, of possible transformations from sequence ACGC to ACT. For this small example the likelihood values $L_{n,m}$ can be evaluated by hand to check computer results.

Note that each 'southeasterly' route through the lattice of *Figure 5*, from (0,0) to (N,M), defines a series of insertion, substitution and deletion events. The intermediate likelihoods $L_{n,m}$, evaluated at the maximum likelihood estimates of q and w, give the relative posterior probabilities for each particular sequence of transitions from (0,0) to (n,m). By searching for a route of nodes (n,m) with high $L_{n,m}$, we have an interpretation of the precise sequence of events involved in the transition between the two sequences. In particular, back-tracking from (N,M) to (0,0) by moving to the previous node of highest $L_{n,m}$ enables us to write out an alignment whose corresponding implied evolutionary events provide the best interpretation of the data. This may not correspond to the best alignment given by a parsimony method.

6.2 Worked example

As an example of the ideas of Section 6.1 consider a comparison of the two base sequences ACT and ACGC (*Figure 6*). The most plausible explanation is that one of C or G has been substituted by T, but there are other possibilities. For example, the T

373

Table 2. Maximum likelihood estimates of q and w for tRNA genes for aspartic acid from the mitochondrial genomes of man, mouse and ox.

Sequences	$-2 \ln L$	r_T	q	w	p
Man/Man	0.0	0.0	0.0	$>10^3$	0.0
Man/Permutation	173.0	0.91	0.15	6.2	0.024
Man/Mouse	85.8	0.16	0.04	1.9	0.021
Man/Ox	121.6	0.39	0.08	10.1	0.008
Mouse/Ox	134.8	0.42	0.09	4.4	0.020

'Permutation' is a random permutation of the sequence for man with the same base composition.

Table 3. Alignments corresponding to routes of highest likelihood values of the tRNA genes for aspartic acid from the mitochondrial genomes of man, mouse and ox.

Man/Permutation

```
AAGGTATTAGAAAAACCATTTCATAACTTTGTCAAAGTTAAATTATAGGCTAAAT.CCTATATATCTTA..
 ::     : : :: ::: ::: : :     :    : ::: :  :::::        :: : : :  : ::: ::
AAATAGTCAAAACAACAATTACTTTGTT.G.TTAAATTCTAATTACTTTGAAA.TTCATGGACATCATAGA
```

Man/Mouse

```
AAGGTATTAGAAAAACCA.TTTCATAACTTTGTCAAAGTTAAATTATAGGCTAA.A.TCCTATATATCTTA
::: :::::: :::: :: :: ::::::::::::::::::::::::::::   :: : ::  ::::::::::
AAGATATTAGTAAAATCAATTACATAACTTTGTCAAAGTTAAATTATAGATCAATAATCT.ATATATCTTA
```

Man/Ox

```
AAGGTATTAGAAAAACCATTTCATAACTTTGTCAAAGTTAAATTATAGGCTAAAT.CCTATATATCTTA
:::: :::: ::::: ::  :::: :::::::::::::: ::: : :  ::: ::: :: : :: :
GAGGTGTTAGTAAAACA.TTATATAATTTTGTCAAAGTTAAGTTACAAGTGAAAGTCCTGTACACCTCA
```

Mouse/Ox

```
AAGATATTAGTAAAATCAATTACATAACTTTGTCAAAGTTAAATTATAGATCAATAATCT.ATATATCTTA
 :: : :::::::::::  : : : :::: ::::::::::::::: ::: :  : :: ::    :: : :: :
GAGGTGTTAGTAAAAC.ATT.ATATAATTTTGTCAAAGTTAAGTTACAAGTGAAA.GTCCTGTACACCTCA
```

'Permutation' is a random permutation of the sequence for man with the same base composition.

could have suffered deletion, and the GC be an insertion. All the possible series of events contribute to the overall likelihood, which thus has many terms; every route through the lattice makes its contribution. When the likelihood is evaluated and maximum likelihood estimates made we obtain $r_T = 0.56$ (hence $q = 0.11$), $w = 0.88$ (hence $p = 0.12$), and hence $R = 0.44$. For these parameter values, the resulting values of $L_{n,m}$ at all nodes are shown in *Figure 4*. (The given values are in fact $-2 \ln L$.) The largest values of L (smallest of $-2 \ln L$) are approximately down the diagonal, showing that indeed a single substitution and a single deletion provide the most plausible interpretation of the transition events between the two sequences.

Table 4. Interpretation of the number of events for the alignments of *Table 3* of the tRNA genes for aspartic acid from the mitochondrial genomes of man, mouse and ox.

Sequences	Matches	Substitutions	Deletions	Insertions
Man/Man	68	0	0	0
Man/Permutation	36	29	3	3
Man/Mouse	59	8	3	1
Man/Ox	50	17	1	1
Mouse/Ox	48	19	1	3

'Permutation' is a random permutation of the sequence for man with the same base composition.

6.3 Illustration of the method

To illustrate the maximum likelihood method we used the transfer RNA (tRNA) genes for aspartic acid from the mitochondrial genomes of *Homo sapiens* (man) (44), *Mus musculus* (mouse) (45) and *Bos taurus* (ox) (46,47). We also generated a sequence which was a random permutation of the *Homo* tRNA gene (with the same base composition).

The results of maximum likelihood estimation are shown in *Table 2*. The maximum likelihood is expressed in terms of $-2 \ln L$ and estimates of q and w are given. *Table 3* presents alignments corresponding to routes of highest likelihood values and *Table 4* gives the interpretation of these alignments in terms of numbers of substitution, insertion and deletion events. Note that when identical man/man sequences are compared there is zero estimated divergence time, and no evidence for insertion or deletion. All other comparisons give finite non-zero estimates, the least similar pair of sequences being of man and a random permutation thereof. Also, in every case the estimate of w is greater than one, indicating that deletion (probability q) is less frequent than any particular substitution (probability p). With the exception of the man/ox comparison, estimates of p are remarkably similar. The estimation of q and w is illustrated in *Figure 7*.

6.4 Limitations of the model

We have shown how a model for the evolution of sequences may be developed, inferences made and parameters estimated on the basis of the model. The model is, of course, only a first approximation, and on the basis only of pairs of sequences inferences are necessarily limited (an extension to joint estimation under a more restricted model is described in Section 7). Nonetheless, two major parameters have been estimated: the 'time', $\tau/2$, of divergence (in terms of substitution rate, r, of bases), and the relative rate, w, of substitutions versus insertions and deletions. These are both identifiable parameters of our model and can be jointly and severally estimated. The model assumes that the same values of r and of w apply to both sequences; that is, *stochastically* the two sequences have evolved from a common ancestor under the same process. Note also that evolutionary stability requires that w is dependent on τ. Our estimate thus relates only to the two sequences concerned, and perhaps to similar sequences evolving over similar periods of time.

A major restriction of the current model is that every base is assumed to evolve independently; this is clearly not the case. (Another restriction is that the substitution

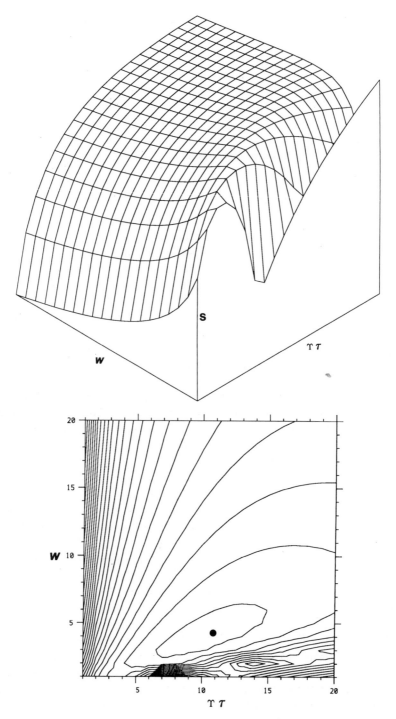

Figure 7. Estimation of the parameters r_T and w based on the sequences ACT and ACGC. The perspective view and contour diagram of the support surface show a maximum at $r_T = 0.56$, $w = 0.88$.

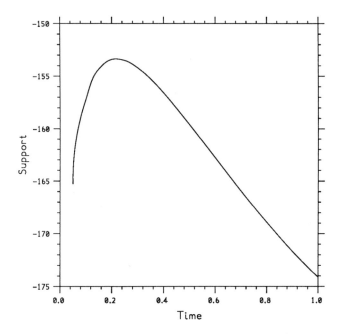

Figure 8. A support curve showing support values plotted against corresponding times t. The example shown is derived from the comparison of the nucleotide sequences of the tRNAs for aspartic acid from the mitochondrial genome of *Mus musculus* (mouse) and *Bos taurus* (ox). The curve shows a maximum at $t = 0.31$.

probability q is the same for all substitutions, but this is not essential to either the model or the method; different relevant probabilities q could clearly be used.) The restriction of considering only single bases is also not essential, although it is more cumbersome to circumvent. *Figure 5* shows the computation of the likelihood, each value depending on the values at the three preceding (northwesterly) nodes. In all there are 63 routes from node $(n-1, m-1)$ to node $(n+2, m+2)$. Under a more general model each of the sets of events corresponding to each route would have a probability, no longer necessarily determined by the single-cell events. By computing the likelihood at each node using not just the three routes from adjacent nodes, but summing over the 63 possible routes, one could take into account models in which three cells evolved jointly. Although cumbersome, this is feasible in principle. A first step might be to allow insertion and deletion of several bases. Rather than the deletion of three bases [the vertical route from (n, m) to $(n+3, m)$] having probability of order p^3, it could be assigned a probability of order p.

6.5 In the absence of deletion/insertion an analytical solution is obtained

If we ignore the occurrence of deletion and insertion then the probabilities of a base changing or remaining unchanged in time τ are given by Equations 7 and 8. If the DNA molecules each contain n positions and differ at d specified nucleic acid sites the associated likelihood is:

$$L = [(1+3e^{-r\tau})/4)]^{(n-d)} [1-e^{-r\tau}/4]^d \qquad \text{Equation 10}$$

The log likelihood S is:

$$S = (n-d) \ln(1+3e^{-r\tau})/4 + d \ln(1-e^{-r\tau})/4 \qquad \text{Equation 11}$$

with derivative with respect to τ

$$-3r(n-d)e^{-r\tau}/(1+3e^{-r\tau}) + rde^{-r\tau}/(1-e^{-r\tau}) \qquad \text{Equation 12}$$

which has a stationary point, the maximum, when:

$$e^{-r\tau} = 1-4d/3n \qquad \text{Equation 13}$$

A relative divergence time, rt (where $t = \tau/2$), in terms of nucleotide substitution, is therefore given by the expression:

$$rt = -1/2 \ln (1-4d/3n), \quad d/n < 3/4 \qquad \text{Equation 14}$$

Figure 8 shows a support (log likelihood) curve for comparison of the nucleotide sequences of the tRNAs for aspartic acid from the mitochondrial genomes of mouse and ox. Support values are plotted against corresponding values of time *t*.

6.6 Pairwise divergence times lead to a heuristic phylogenetic tree

When more than two sequences are being compared, we may estimate a divergence 'time' for each pair (as described in Section 6.5). (The estimate is of *rt* but may be scaled to time as described in Section 2.1.1.) The pairwise divergence time matrix can then be analysed by the unweighted average-link method of cluster analysis (48) to give an estimate of a phylogenetic tree, assuming the same stochastically uniform rate of evolution in all lineages. The method works by first clustering the sequences which are separated by the smallest times in the matrix. These groups are then combined and new values are computed for the times of divergence between the combined group and the other groups by averaging the appropriate times. This procedure of combining and averaging is followed until all groups are clustered together as one group. The newly-

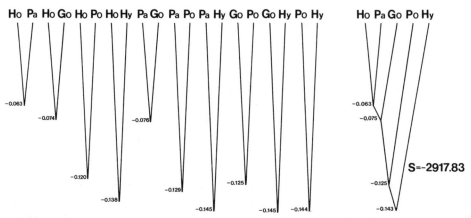

Figure 9. The diagram shows (left) the pairwise estimates of divergence time for each pair of the hominoid primates: *Homo sapiens* (man), Ho; *Pan troglodytes* (chimpanzee), Pa; *Gorilla gorilla* (gorilla), Go; *Pongo pygmaeus* (orang-utan), Po; and *Hylobates lar* (gibbon), Hy. The sequence data are the *Hind*III restriction fragments from the mitochondrial genomes (which include three tRNA genes and parts of two proteins) which were sequenced by Brown *et al.* (54), using their alignments. On the right is shown the combined tree derived by unweighted pair-group clustering of the pairwise estimates of divergence times.

derived, averaged times of divergence of the sequences may be more or less than the original pairwise time estimates. The fit to the original estimates gives some measure of overall confidence in the results (cf. 49). Pairwise estimation does not take full account of the information in the data, and joint estimation (Section 7) is preferable. However, pairwise estimation is computationally tractable whereas joint estimation may not be feasible for more than a few sequences. The pairwise estimate is a useful starting point in the search for the best joint estimate of phylogeny. The pairwise estimates of divergence times of partial mitochondrial genome sequences for some hominoid primates are shown in *Figure 9*, together with the tree estimate derived from clustering of these pairwise divergence times.

7. JOINT ESTIMATES OF PHYLOGENY UNDER A SIMPLE MODEL

Joint estimation simultaneously considers the information contained in a set of sequence data. The problem is to find both the tree pattern and the times that together result in the highest likelihood for the data in question under the model of evolutionary change. This problem is approached first by finding a method to evaluate the likelihood of a given tree pattern with particular times assigned to the branching points; then by finding a method for locating those times which maximize the likelihood for the given tree pattern. Finally, we need to employ efficient means to examine all or a range of possible tree patterns in the search for that pattern with those times which together result in an overall maximum likelihood.

Felsenstein (50−52) has described an algorithm which enables the likelihood to be calculated given a tree pattern and a set of times for the branching points. This pruning algorithm ensures that the nodes of a tree are inspected in an order such that the descendent node of a segment is inspected before the ancestral node of that segment. The two or more (in a polychotomous pattern) descendent tips of a node are stripped from the tree pattern at each step.

For this procedure the extant states of the corresponding positions in the DNA sequences are used to initialize the likelihoods for each site: the probability of occurrence of the base present is set to 1.0, and the probabilities of occurrence of the other three bases at the site are set to 0.

In order to find those values of the times for the branching points which result in the maximum likelihood, the trial times may be altered in turn until no increase in likelihood is achieved (within a reasonable level of accuracy). This is a simple hill-climbing algorithm of the type described by Felsenstein (50). Other algorithms have been proposed, for example, by Thompson (53) for gene frequency data and by Felsenstein (52) for DNA sequence data.

The problem of finding for certain the tree pattern which is associated with the global maximum likelihood has not been solved using means other than examining all possible tree patterns with their associated maximum likelihood time estimates. In Section 2.1.2 it was emphasized that this tactic rapidly becomes computationally intractable for more than quite small numbers of sequences. One way of dealing with the problem is to examine a subset of possible tree patterns in a systematic fashion such that relatively widely differing patterns are used as starting points for progressive rearrangement. At each cycle of pattern rearrangement further trial changes are made to the pattern of

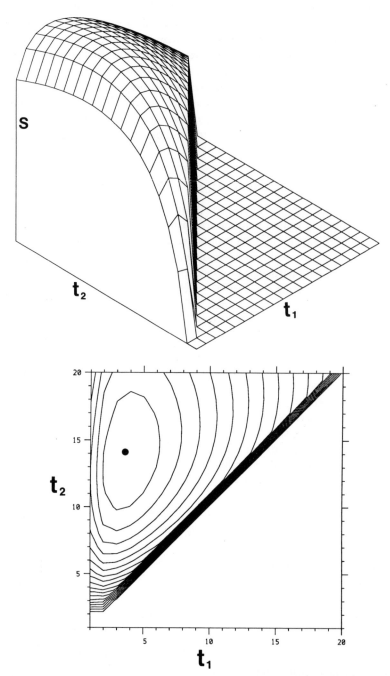

Figure 10. Joint estimation of the two times of branching of the mitochondrial tryptophan tRNA sequences of *Homo sapiens* (man), *Mus musculus* (mouse) and *Bos taurus* (ox). The tree of greatest support has the more recent divergence (time t_1) between man and ox, and an earlier divergence (time t_2) of mouse. The perspective view and contour diagram of the support surface are positioned on one side of the diagonal because time t_1 cannot be greater than time t_2 for this tree pattern.

highest likelihood found from the previous cycle. An algorithm proposed by Felsenstein (52) involves adding tips to the tree in sequence, with the new tip being offered in turn to each of the existing branches of the tree. The new topology with the highest likelihood is accepted and cycles of local rearrangement of branches are also carried out in an attempt to find a pattern of higher likelihood.

Bishop and Friday (36) used a different search strategy which starts from the 'big-bang tree', defined by Thompson (53) as that tree in which all lineages originate from a single point in the past and have evolved independently. The big-bang pattern is expanded by elevating each lineage in turn as a branch to each other lineage, and retaining the pattern of highest likelihood. After one cycle of expansion a single new branch point will have been added to the tree, and a new cycle of expansion is begun. Under this strategy no branching point is destroyed by further rearrangement once it has been created.

It should be emphasized that the methods described above for selecting tree patterns to be evaluated lack theoretical justification, but have been found to perform well in those cases where the results have been compared with the results obtained by evaluating all possible tree patterns. Necessarily, cases of complete evaluation are confined to examples of rather few sequences, and we do not yet know how well the search algorithms perform for larger numbers of sequences. A graphical illustration of joint estimation for the case of three species and two times is given in *Figure 10*.

We illustrate the procedure of joint estimation using the mitochondrial DNA sequences of five hominoids: human, chimpanzee, gorilla, orang-utan and gibbon. The sequences were determined by Brown *et al.* (54). The maximum likelihood tree estimated under the simple model proposed above was given by Bishop and Friday (36), and, together with a variety of other tree patterns and their associated likelihoods, by Bishop and Friday (55). Hasegawa *et al.* (10) have carried out maximum likelihood estimation of evolutionary trees on the same data under a different model.

The analysis of Bishop and Friday began by carrying out expansion of the big-bang tree pattern in which all five lineages originate from a single point in the past. An initial guess is made at the time ago, in relative terms, of the origin, and the program iteratively alters the value of this time until the maximum likelihood time for the big-bang pattern is obtained. As a convention times ago are expressed as negative numbers. The maximum likelihood big-bang result is of interest because, following Thompson (53), we may use it as a reference for the increases in likelihood obtained by evaluating more informative (i.e. more highly branching) tree patterns. The program then attempts to expand the big-bang pattern using the algorithm described. At each cycle of addition of a node to the tree (i.e. introduction of a new dichotomous branch point), that pattern incorporating the extra node which increases the likelihood most is retained (and printed). The next cycle of expansion then begins with the maximum likelihood tree so far, and the cycles continue until the tree is fully dichotomously branching.

For the five hominoid sequences the results of the expansion procedure are shown in *Figure 11*. The tree pattern of highest likelihood is shown for each of three cycles of addition of an extra node.

The pairwise estimate of both pattern and divergence times (see *Figure 9*) was also used as a starting point for submission to the program. Since the pairwise estimate was already fully dichotomously branching, no further rearrangement of the pattern was

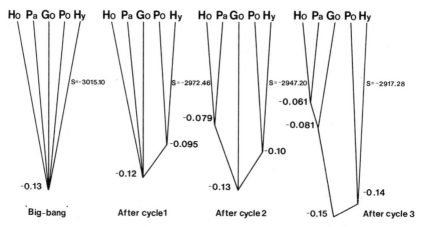

Figure 11. Expansion from the big-bang pattern for the same data used for *Figure 9*. The three cycles of expansion are shown, after which a completely dichotomous pattern has been achieved. The expansion procedure is described in the text. It is emphasized that once dichotomies have been created by the expansion algorithm they are not broken in subsequent cycles. The associated support values (log likelihood), *S*, are shown to the right of each tree. Abbreviations for species are as for *Figure 9*.

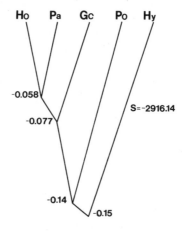

Figure 12. The maximum likelihood estimate of the pattern and times of divergence for the five hominoid species of *Figures 9* and *11*. Abbreviations for species are as for *Figure 9*. This pattern and these times were obtained by refinement of the pairwise estimate of *Figure 9*. The tree pattern derived by expansion (*Figure 11*) has slightly less support than the pattern shown here. In this case, therefore, the pairwise estimate proved to be close to the best found. This is often the case in practice, and the use of the pairwise estimate as a starting point for the joint estimation procedure is recommended.

carried out by the program. However, the estimates of divergence time were iteratively refined and slightly different times gave a somewhat greater support (*Figure 12*). Those times and the pattern as estimated by the pairwise procedure therefore became the maximum likelihood solution thus far.

In addition to these two starting points, a large variety of other patterns was evaluated. Here, as in most such exercises, many of these other patterns were in accord with published suggestions about the evolutionary history of the species made on the basis of data from comparative anatomical, palaeontological and other molecular studies. Despite an exhaustive search no estimate of higher likelihood was found other than that derived from refinement of the times on the pattern estimated by the pairwise procedure. In this case, therefore, the computationally much less expensive pairwise procedure gave a close solution to that determined by the joint procedure. It should also be emphasized that the solution reached by expansion from the big-bang pattern was comparable in terms of its support to the unrefined pairwise solution. In practice we have found, in many cases, that both the pairwise estimate and the solution reached by expansion are close to the estimate of highest likelihood reached by more exhaustive and computationally expensive investigation.

For five species there are 105 different rooted trees which are dichotomously branching and 236 different rooted trees if polychotomies are admitted. In studying these five hominoids we looked explicitly at about one-third of all possible patterns, and many more were evaluated during the running of the programs but were not printed out because their associated likelihoods eliminated them from further consideration. An additional indication of which patterns are likely to be worth consideraiton for minor rearrangement comes from cases in which times for the two branch points at either end of a branch converge on one another, and eventually become indistinguishable to the limits of accuracy. This we refer to as a 'collapse', and it may be taken to indicate that the pattern as submitted is unstable and should be rearranged by reordering the branch points which have collapsed. The likelihood associated with a collapsed pattern is without useful meaning until the pattern has been resubmitted explicitly in its new form.

Finally, it is worth pointing out that the algorithm used necessitates calculation of the posterior probability of finding each base at each site at each branch point of a tree under a given tree pattern. This information can be written out and gives an indication of the possible nature of the ancestral sequence at each node. However, this complete information may be very voluminous, and it will usually suffice to inspect the state of the most probable base in the reconstructed sequence.

8. AVAILABILITY OF PROGRAMS

Programs written in FORTRAN77 implementing the methods of inference of evolutionary relationships described in this chapter may be obtained by sending a 1/2 inch magnetic tape to M.J.Bishop. In the UK the programs may be used on the University of Cambridge Computing Service IBM 3081 mainframe which is accessible via JANET or PSS (see Chapter 5).

A phylogeny inference package called PHYLIP is available from Joe Felsenstein, Department of Genetics SK-50, University of Seattle, Seattle, Washington 98195, USA. The programs (written in Pascal) may be obtained by sending a 1/2 inch magnetic tape to J.Felsenstein. The programs are also available on diskettes in a variety of formats. Executable code for the IBM PC may be obtained from George D.F.('Buz') Wilson, Scripps Institution of Oceanography, La Jolla, California 92093, USA. Contact Buz Wilson in advance to find out how many diskettes are required.

9. REFERENCES

1. McLachlan,A.D. (1971) *J. Mol. Biol.*, **61**, 409.
2. Karlin,S., Ghandour,G., Foulser,D.E. and Korn,L.J. (1984) *Mol. Biol. Evol.*, **1**, 357.

3. Ackers,G.K. and Smith,F.R. (1985) *Annu. Rev. Biochem.*, **54**, 597.
4. Blundell,T.L. and Johnson,L.W. (1976) *Protein Crystallography*. Academic Press, London.
5. Aebi,U., Fowler,W.E. and Smith,P.R. (1982) *Ultramicroscopy*, **8**, 191.
6. Kaptein,E., Zuiderweg,R., Scheek,R., Boelens,R. and van Gunsteren,W. (1985) *J. Mol. Biol.*, **182**, 179.
7. Levitt,M. (1982) *Annu. Rev. Biophys. Bioeng.*, **11**, 251.
8. Jones,T.A. (1978) *J. Appl. Crystallogr.*, **11**, 268.
9. Bajaj,M. and Blundell,T. (1984) *Annu. Rev. Biophys. Bioeng.*, **13**, 453.
10. Hasegawa,M. and Yano,T. (1984) *Proc. Jap. Acad.*, **60**, 389.
11. Lanave,C., Preparata,G., Saccone,C. and Serio,G. (1984) *J. Mol. Evol.*, **20**, 86.
12. Attimonelli,M., Lanave,C., Sbisa,E., Preparata,G. and Saccone,C. (1985) *Cell Biophys.*, **7**, 239.
13. Cannings,C. and Thompson,E.A. (1981) *Genealogical and Genetic Structure*. Cambridge University Press, Cambridge.
14. Felsenstein,J. (1978) *Syst. Zool.*, **27**, 27.
15. Mortlock,R.P., ed. (1984) *Microorganisms as Model Systems for Studying Evolution*. Plenum Press, New York.
16. Kornberg,A. (1980) *DNA Replication*. Freeman, San Francisco.
17. Kornberg,A. (1982) *Supplement to DNA Replication*. Freeman, San Francisco.
18. Friedberg,E.C. (1985) *DNA Repair*. Freeman, New York.
19. Dressler,D. and Potter,H. (1982) *Annu. Rev. Biochem.*, **51**, 727.
20. Shapiro,J.A., ed. (1983) *Mobile Genetic Elements*. Academic Press, New York.
21. Campbell,A. (1981) *Annu. Rev. Microbiol.*, **35**, 55.
22. Bodmer,W.F., Weiss,R. and Wyke,J. (1985) *Oncogenes: Their Role in Normal and Malignant Growth*. The Royal Society, London.
23. Wilson,A.C., Carlson,S.S. and White,T.J. (1977) *Annu. Rev. Biochem.*, **46**, 573.
24. Dover,G.A. (1986) *Trends Genet.*, **2**, 159.
25. Felsenstein,J. (1982) *Q. Rev. Biol.*, **57**, 379.
26. Felsenstein,J. (1983) *Annu. Rev. Ecol. Syst.*, **14**, 313.
27. Sankoff,D. and Kruskal,J.B. (1983) *Time Warps, String Edits, and Macromolecules: The Theory and Practice of Sequence Comparison*. Addison-Wesley, Reading, MA.
28. Felsenstein,J. (1983) *J. R. Statist. Soc.* A, **146**, 246.
29. Kimura,M. (1983) *The Neutral Theory of Molecular Evolution*. Cambridge University Press, Cambridge.
30. Dayhoff,M.O., ed. (1972) *Atlas of Protein Sequence and Structure*. National Biomedical Research Foundation, Washington, D.C.
31. Holmquist,R. (1972) *J. Mol. Evol.*, **1**, 211.
32. Zuckerkandl,E. and Pauling,L. (1965) In *Evolving Genes and Proteins*. Bryson,V. and Vogel,H.J. (eds), Academic Press, New York, p.97.
33. Kimura,M. (1969) *Proc. Natl. Acad. Sci. USA*, **63**, 1181.
34. Neyman,J. (1971) In *Statistical Decision Theory and Related Topics*. Gupta,S.S. and Yackel,J. (eds), Academic Press, New York, p.1.
35. Holmquist,R. (1971) In *Sixth Berkeley Symposium on Mathematics and Statistics*. LeCam,L., Neyman,J. and Scott,E.L. (eds), University of California Press, Berkeley. Vol. **5**, p.315.
36. Bishop,M.J. and Friday,A.E. (1985) *Proc. R. Soc. Lond.* B, **226**, 271.
37. Li,W., Luo,C. and Wu,C. (1985) In *Molecular Evolutionary Genetics*. MacIntyre,R.J. (ed.), Plenum Press, New York.
38. Hasegawa,M. and Yano,T. (1984) *Bull. Biometric Soc. Jap.*, **5**, 1.
39. Hasegawa,M., Kishino,H. and Yano,T. (1985) *J. Mol. Evol.*, **22**, 160.
40. Hasegawa,M., Iida,Y., Yano,T., Takaiwa,F. and Iwabuchi,M. (1985) *J. Mol. Evol.*, **22**, 32.
41. Edwards,A.W.F. (1972) *Likelihood*. Cambridge University Press, Cambridge.
42. Kelly,F.P. (1979) *Reversibility and Stochastic Networks*. Wiley, New York.
43. Bishop,M.J. and Thompson,E.A. (1986) *J. Mol. Biol.*, **190**, 159.
44. Anderson,S., Bankier,A.T., Barrell,B.G., de Bruijn,M.H.L., Coulson,A.R., Drouin,J., Eperon,I.C., Nierlich,D.P., Roe,B.A., Sanger,F., Schreier,P.H., Smith,A.J.H., Staden,R. and Young,I.G. (1981) *Nature*, **290**, 457.
45. Bibb,M.J., Van Etten,R.A., Wright,C.T., Walberg,M.W. and Claydon,D.A. (1981) *Cell*, **26**, 17.
46. Anderson,S., de Bruijn,M.H.L., Coulson,A.R., Eperon,I.C., Sanger,F. and Young,I.G. (1982) *J. Mol. Biol.*, **156**, 683.
47. Anderson,S., Bankier,A.T., Barrell,B.G., de Bruijn,M.H.L., Coulson,A.R., Drouin,J., Eperon,I.C., Nierlich,D.P., Roe,B.A., Sanger,F., Schreier,P.H., Smith,A.J.H., Staden,R. and Young,I.G. (1982) In *Mitochondrial Genes*. Slonimski,P., Barst,P. and Attardi,G. (eds), Cold Spring Harbor Laboratory Press, p.5.

48. Jardine,N. and Sibson,R. (1971) *Mathematical Taxonomy*. Wiley, London.
49. Cavalli-Sforza,L.L. and Piazza,A. (1975) *Theor. Pop. Biol.*, **8**, 127.
50. Felsenstein,J. (1973) *Syst. Zool.*, **27**, 27.
51. Felsenstein,J. (1973) *Am. J. Hum. Genet.*, **25**, 471.
52. Felsenstein,J. (1981) *J. Mol. Evol.*, **17**, 368.
53. Thompson,E.A. (1975) *Human Evolutionary Trees*. Cambridge University Press.
54. Brown,W.M., Prager,E.M., Wang,A. and Wilson,A.C. (1982) *J. Mol. Evol.*, **18**, 225.
55. Bishop,M.J. and Friday,A.E. (1986) In *Major Topics in Primate and Human Evolution*. Wood,B., Martin,L. and Andrews,P. (eds), Cambridge University Press, p. 150.

Glossary

ACOUSTIC COUPLER. An attachment for a **MODEM** that allows data to be transmitted and received via a telephone handset rather than making direct electrical connections to the telephone network.

AGRENET. The computer communications network established by the Agriculture and Food Research Council (AFRC) for the agricultural and botanical research community in the United Kingdom.

ALGORITHM. A description of how to solve a problem presented in terms that could be solved using a computer.

ALPHANUMERIC. The set of symbols used to represent letters of the alphabet and the integers 0-9.

AMBIGUITY CODE. A character other than the standard codes for representing the DNA and RNA bases (A,C,G,T and U) used to indicate a base with an ambiguous identity.

ANSI. American National Standards Institute. In computing it has defined important standards such as those for programming languages (ANSI Fortran 77) and communication equipment (e.g. **ANSI TERMINAL**s).

ANSI TERMINAL. A display terminal that conforms to the **ANSI** X3.41-1974 and X3.64-1977 standards e.g. the DEC VT100 terminal.

ARCHIVE. A safe and permanent store for computer data. Sometimes refers to the migration of data from magnetic disk to magnetic tape for long term storage. At other times an archive is simply a collective noun for sets of related data.

ARPA. See **DARPA.**

ARPANET. One of the earliest national computer communications networks. It was established to foster effective communications between US government agencies, contractors and research groups funded by **DARPA** using state-of-the-art networking technology. The ARPANET is now used mainly by the academic research community.

ASCII. American Standard Code for Information Interchange. A standard method of representing **ALPHANUMERIC** characters in memory or on disk as single **INTEGER**s in the range 0-127 (i.e. 7 bits).

ASSEMBLER. The lowest level of computer programming language that can be read by humans. The operations are close to **MACHINE CODE** and are programmed using mnemonics for the processor **INSTRUCTION SET**.

ASYNCHRONOUS COMMUNICATIONS. A method of managing a dialogue between two devices such that data may be exchanged in an asynchronous manner (e.g. the **RS232C** protocol). Data may flow in both directions at once and each device can halt the flow from the other at any time in the dialogue using special flow control signals embedded in the data stream or appearing on special control lines.

BACKPLANE. The set of electrical connectors at (generally) the rear of a computer chassis, into which are plugged the component modules or printed circuit boards. The backplane provides parallel connections between all components via the data and address **BUS**es and connections to common resources such as the power supply.

BATCH PROCESSING. A method of running a program unattended. Traditionally all computing was done this way, with user input taken from punch cards. Batch systems on microcomputers and on time-sharing mainframes more commonly take user input from a file.

BAUD RATE. A measure of data transmission rate usually corresponding to the number of sent or received **BIT**s per second (**BPS**).

BBN. Abbreviation for Bolt, Beranek and Newman, the distributors of the **GENBANK** data library and contractors for the National Institutes of Health PROPHET system.

BIT. The bit (**BI**nary digi**T**) is the base unit of computer memory or data. It may have values of 1 or 0 and therefore represent on/off, present/absent etc. See **BYTE, WORD**.

BIT MAPPED. Something is bit mapped if each element can be represented by a single bit in computer memory so that direct manipulation of the memory location affects its appearance (e.g. on a graphics screen) or its interpretation i.e. there is a direct mapping between bits in memory and the representation.

BITNET. An American computer communications network sponsored by IBM for use by the academic research community. The european branch of BITNET is called **EARN**.

BPI. Abbreviation for **b**its **p**er **i**nch. A measure of data recording density on magnetic storage media such as disk or tape.

BPS. Abbreviation for **b**its **p**er **s**econd (usually the same as **BAUD**).

BUFFER. A section of computer memory allocated to accommodate the communication between two devices that transfer data at different speeds. The faster sends data to the buffer and the slower reads from it.

BUS. The electrical conductors on a printed circuit board or running along the **BACKPLANE** through which the **CPU**, memory, **DISK** and peripheral equipment communicate.

BYTE. The byte is the most commonly used measure of computer memory and 1 byte contains 8 **BITS**. A byte is typically used to represent a single **ALPHANUMERIC** character or an **INTEGER** in the range 0-255.

CARD IMAGE. An antique data format of fixed-width lines (typically 80 columns) equivalent to the format of data or programs stored on punched cards.

CARTESIAN COORDINATES. Specify a point in 2D space by its perpendicular distances from horizontal and vertical axes through a predefined point of origin. To specify a point within a 3D space a third coordinate indicating depth is given.

CCD. Abbreviation for **CHARGED COUPLE DEVICE**.

CCITT. Abbreviation for the Consultative Committee for International Telegraphs and Telephones that establishes international agreement over telecommunications standards and protocols.

CENTRAL PROCESSING UNIT. The computational heart of a computer. Where all (or most) of the comparisons, conversions and manipulations of data are carried out under the instructions provided by a computer program.

CHARGED COUPLE DEVICE. A semi-conductor device used as the photo-sensitive element in solid state cameras.

CHECK-SUM. A calculation performed on a string of data that yields an essentially unique number. When transmitted with the original data, the check-sum is re-evaluated at the receiving end and compared with the original value. If the two check-sums do not agree then the data was corrupted during transmission and must be re-sent. Check-sums are the basis of many **ERROR-CHECKING PROTOCOLS.**

CLUSTER-ANALYSIS. A statistical method used to determine whether groups of data can be segregated into clusters having one or more common properties.

COMPILER. A compiler is a computer program that converts program **SOURCE CODE** written in a high level language (approximating English) to a binary **OBJECT CODE** that can be directly executed by the computer hardware (the **CENTRAL PROCESSOR UNIT** or **MICROPROCESSOR**).

COMPUTER ARCHITECTURE. The design and methods used in the construction of a computer system. A fundamental architectural issue is whether the system uses a single processing element to serially process instructions and data, or whether many processing elements are used in a parallel (non von Neumann) architecture.

CONCEPTUAL FILE STORE/TERMINAL. In the context of computer science, the adjective *conceptual* is used to distinguish between a real physical device directly connected to a computer system and program(s) that emulate the behaviour of the device. For example, software that implements a conceptual file store enables the user of one host on a **LOCAL AREA NETWORK** to treat the physical disks and file systems of other hosts as part of his own file store.

CONSENSUS SEQUENCE. A consensus sequence is constructed by finding the mutually optimum alignment amongst a group of sequences and then deciding the base that is the most representative (most frequent) at each position.

CONTIG. A contig is a **CONSENSUS SEQUENCE** derived from a set of overlapping DNA fragments from a shotgun sequencing experiment. One or more contigs will be joined by common overlapping regions to form the complete sequence.

CONTROL CHARACTERS. In the **ASCII** character set, the first 31 are non-printing characters that are generally reserved to control the movement of the screen cursor or printer head (carriage control) or serve to signal structure within a data file. The interpretation of control characters is dependent upon the operating system and possibly also the application program.

CONVOLUTION FUNCTION. Convolution functions are generally simple calculations performed on small sections of data but applied over an entire set of measurements. The purpose of convolution functions is to increase the overall signal-to-noise ratio. A hypothetical convolution function applicable to protein sequence data might be calculated for each group of 7 adjacent amino acids (the analysis window) giving a local hydropathicity value. Moving the analysis window along the sequence and applying the convolution function at each step would complete the hydropathicity analysis.

COPROCESSOR. A coprocessor is a specialised hardware processing element that can cooperate with the main processor in order to perform certain (usually computationally intensive) computations asynchronously from the main processing stream. Specialised coprocessors are typically used when one class of computation becomes rate-limiting. For example arithmetic coprocessors for fast **FLOATING POINT ARITHMETIC** and graphics coprocessors for manipulating **BIT MAPPED** graphics images.

CP/M. A microcomputer operating system originally designed for use with computers based on Zilog Z80 microprocessors.

CPU. Abbreviation for **CENTRAL PROCESSING UNIT**.

CRASH. A serious **HARDWARE** or **SOFTWARE** failure.

CURSOR. A mobile pointer mark on a visual display screen that indicates where output will start or where text from the keyboard will be inserted. Programs that use **MENU**s may use the cursor to indicate which menu option has been selected. The cursor is moved about the screen either by special cursor keys (forward, reverse, up and down arrows), a **MOUSE** or other pointing device, or keys selected by the application program.

DAISY WHEEL PRINTER. An impact printer that uses a print-head based on a plastic disc segmented into sectors carrying the raised printing characters. The term *daisy wheel* derives from the resemblance between the disc sectors and the petals of a flower.

DAP. Abbreviation for Distributed Array Processor, a parallel **COMPUTER ARCHITECTURE** available in some ICL mainframe computers and used for fast calculations on matrices.

DARPA. The Defense and Advanced Research Projects Agency of the United States of America.

DATA BACKUP. A copy, or the process of copying data stored on magnetic storage devices (e.g. **FLOPPY DISK**) to ensure security in the event of corruption.

DATA COMPRESSION. The process of converting data to a more compact representation that occupies less disk or memory space.

DATA REDUCTION. The process of reducing the number of recorded values from an observation without significant loss of important information.

DATA STRUCTURE. Programming languages such as FORTRAN, BASIC and PASCAL provide for a small set of fundamental types of data representation such as integer and real numbers, characters, arrays logical or boolean values. These simple data types can be combined by the programmer in a variety of ways (depending on the programming language) to form data structures that that can be manipulated as single data types and used to represent the required information. In PASCAL such complex data structures are called records.

DATA TERMINAL EQUIPMENT. Any mechanical or electrical device capable of displaying or printing output from a computer system.

DATABASE. A collection of data stored on magnetic disk and structured so that a **DATABASE MANAGEMENT SYSTEM** can access, add or delete data in a flexible and efficient way.

DBMS. Abbreviation for DataBase Management System, a program used to manage, manipulate and retrieve data stored on disk in a **DATABASE**.

DCE. Abbreviation for Data Communications Equipment.

DCL. Abbreviation for **DIGITAL COMMAND LANGUAGE**.

DECNET. A communications network architecture developed by Digital Equipment Corporation.

DEC. Abbreviation for Digital Equipment Corporation, manufacturers of the VAX/11, PDP/11, DECsystem10 and DECsystem20 computer systems.

DERIVATIVE. The first derivative of a list of data describing a continuously varying observation is the first differential of the mathematical function that describes the observations. In other words, the first derivative is a measure of the local slope (the rate of change) in the observed values. It is generally calculated for several points either side of the value being considered. Values of zero in the first derivative indicate the position of peaks or troughs in the observed data.

DEVICE DRIVER. A program designed to convert a standard set of general purpose control and input/output commands into specific instructions for a particular device.

389

DIGITAL COMMAND LANGUAGE. A programming language provided with DEC VAX computers for controlling program execution under the VAX/VMS operating system.

DIGITIZER. A peripheral device for converting the position of a point in space to a set of (usually cartesian) coordinates that may be read directly by a computer, through a communications port.

DIRECTORY. A grouping of files under a common heading or name. From a users point of view a disk is a collection of directories and a directory is a collection of files.

DISK CONTROLLER. The hardware responsible for reading and writing data onto a disk memory device.

DISK DENSITY. The density of information that can be recorded on a magnetic disk.

DISKETTE. Commonly used synonym for a **FLOPPY DISK**.

DONGLE. A security device used to prevent unauthorised copying of licensed software.

DOS. Abbreviation for Disk Operating System. Originating from Microsoft, DOS has been adopted by many 16 bit personal computers as MSDOS and as PCDOS on the IBM personal computer.

DOT MATRIX. In terms of computer hardware (e.g. dot matrix printer) a dot matrix refers to a shape (character) constructed from a pattern of dots. In terms of molecular sequence analysis the dot matrix is a method for displaying the similarities between two sequences. One sequence is drawn on the horizontal axis and the other along the vertical axis of a 2D-matrix. A dot is drawn in every cell where the corresponding character in each sequence is identical. Regions of similarity appear as lines parallel to the diagonal.

DOT-MATRIX PRINTER. An impact printer with a print-head that builds characters from a matrix of dots.

DRIBBLE FILE. A file used to record all characters sent and received from the user's terminal.

DTE. Abbreviation for Data Terminal Equipment.

DUMB TERMINAL. A display terminal that has very limited screen control features and essentially reproduces the functions of an old-fashioned electromechanical teletype. See **GLASS TELETYPE**.

DYNAMIC PROGRAMMING. A technique employed in the design of **RECURSIVE** computer algorithms whereby partial solutions are stored and re-used dynamically to avoid unnecessary re-calculations.

EARN. Abbreviation for the European Academic Research Network. The European section of the IBM sponsored **BITNET** communications network.

EDGE EFFECT. Many numerical analyses of sequence data assume that the sequence is infinitely long. The possible erroneous behaviour of such methods at the beginning and end of finite sequences is called the edge effect.

EDT. A general purpose text editor.

EPROM. Abbreviation for Erasable, Programmable Read Only Memory. A non-volatile semiconductor memory device whose contents may be erased by exposure to strong u.v.-light and then re-programmed using high-voltage signals.

ERROR-CHECKING PROTOCOL. In the transmission of data in potentially noisy media it is important to check that the data received corresponds with the data that was sent. The effects of noise or corruption can thereby be detected and the data resent. A variety of methods are available that achieve this and are incorporated at many levels in a computer system (e.g. reading/writing to magnetic disk or tape, transmission of data over communications lines).

ESCAPE SEQUENCE. A sequence of characters preceded by the escape character (**ASCII** character 27). Used to control the movement of the **CURSOR** around a display screen and for altering the behaviour of a display device.

ETHERNET. A high bandwidth **LOCAL AREA NETWORK** technology based on a coaxial cable electrically sealed at both ends. Transmission protocols enabling data rates of up to 10 megabytes per second have been developed both for **DECNET** and by the Xerox Corporation (TCP/IP) used by many computer systems with the **UNIX**™ operating system.

EXPERT SYSTEM. A computer program able to mimic aspects of expert problem solving using decision rules, **HEURISTICS** and a general-purpose problem resolution strategy (e.g. logical deduction). Expert systems use the explicit representation of decision rules to enable the user to interrogate the program about why decisions were made and the reasons for asking particular questions.

FEATURE TABLE. A table within an entry in a molecular sequence database describing the location of features of interest within the sequence itself.

FILE TRANSFER PROTOCOL. An agreed method used by two or more computer systems for exchanging data files over local or a wide area computer communications networks. True file transfer methods use **ERROR-CHECKING PROTOCOLS** to ensure the accurate transmission of data.

FILTER. A program or procedure that transforms data in some way. For example, removal of all blank lines from a file or smoothing a series of data points to remove extreme or erroneous values.

FLOATING POINT ACCELERATOR. A specialised **COPROCESSOR** for fast execution of arithmetic on floating point numbers.

FLOATING POINT ARITHMETIC. Arithmetic with non-integer numbers expressed in decimal notation with positive or negative exponents.

FLOPPY DISK. A mass data-storage device using a rotating, flexible, magnetisable surface. Various sizes (3.5'' and 5.25'') are in common use. The advantages of floppy disks are that they are removable from the disk drive and can store large amounts of data (up to 1 megabyte).

FLOW DIAGRAM. A diagrammatic method for describing the flow of information and the transfer of control within a program.

FTP. Abbreviation for **FILE TRANSFER PROTOCOL**.

GATEWAY. A computer that provides a translation service between different communication protocols and links two otherwise incompatible network systems or network services (e.g. electronic mail).

GAUSSIAN APPROXIMATION. Approximating a series of observations to a mean and standard deviation describing the **GAUSSIAN CURVE** which fits closest to the data.

GAUSSIAN CURVE. The familiar bell-shaped curve predicting the expected frequency of observations when data is normally distributed.

GENBANK. The Genetic Sequence Databank collected at Los Alamos National Laboratory and distributed as a U.S. national data library by Bolt Beranek and Newman.

GKS. Abbreviation for **GRAPHICAL KERNEL SYSTEM**.

GLASS TELETYPE. A simple **DATA TERMINAL** using a video screen, but as unsophisticated as an ancient mechanical teletype.

GRAPHICAL KERNEL SYSTEM. A standard notation for writing computer graphics programs designed to make software independent of the graphical display device.

GRAPHICS PROCESSOR. A specialised **COPROCESSOR** designed to provide efficient execution of graphics operations such as drawing standard shapes, shading surfaces etc.

GREP. Abbreviation for **G**lobally find **R**egular **E**xpression and **P**rint. A program provided with the **UNIX**™ operating system for rapidly searching text files for patterns of characters or strings.

HARD DISK. A mass data-storage device using one or more inflexible disks with magnetisable surfaces in a sealed package. Hard disk drives use high rotation speeds, high data densities and multi-disk packs to store in the order of tens of millions of **BYTES** of data.

HARDWARE. The physical and electrical devices from which a computer system is constructed.

HELP INTERACTIVE/ONLINE. Written instructions and assistance available to the user whilst using the program. A comprehensive and well designed help system can significantly relieve the difficulties encountered by naive users. Online help is an important part of the design of user-friendly programs.

HEURISTIC. A heuristic is a rule of thumb or a decision based on an approximate assessment of a problem that leads closer to a solution without having to exhaustively enumerate and evaluate all possible partial solutions on the way.

HOST. A computer system on a communications network.

ICON. A small shape or pattern used in graphics-orientated programs to represent a program, file, **DIRECTORY** or other object. In **WIMP**-based systems, a **WINDOW** or program may be collapsed down to an icon to save screen space and remind the user of its existence.

INDEL. Acronym for **IN**sertions or **DEL**etions proposed by **ALGORITHMS** that determine intersequence similarities or optimal alignments.

INSTRUCTION SET. The vocabulary of instructions that can be executed by a processor or microprocessor.

INTELLIGENT TERMINAL. A display terminal with in-built processing and data storage capabilities. These may be used, for example, to make the terminal manage its communications link in an economical fashion, or to make the device easier to use. Some **TERMINAL EMULATORS** enable a microcomputer to be used as an intelligent terminal. For example, a useful feature provided in some emulators enables the user to automatically connect to selected host computers by controlling an auto-dial **MODEM**.

INTERACTIVE. A style of computing where the user is involved throughout the running of a program by answering questions or giving commands. The opposite of **BATCH PROCESSING**.

INTERNET. A network of networks being developed by **DARPA** using the **ARPANET** as its backbone.

INTERPOLATION. The estimation of an unknown value from known values that in a series of observations lie both before and after it.

INTERPRETER. A program that executes statements in a programming language directly, without using a compiler. Programs executed using an interpreter generally run much slower than when a compiler is used.

IPSS. Abbreviation for International Packet Switch Stream. A commercial, international communications network based on packet switching.

JANET. Abbreviation for Joint Academic NETwork.

JOB. A session of computer usage. Generally only applies to **MULTI-USER** or **MULTI-TASKING** computer systems. On an interactive multi-user system a job would correspond to the resources used between login and logout. Exact usage of this term varies among operating systems.

JOINT ACADEMIC NETWORK. The UK academic computer communications network administered by the Science and Engineering Research Council.

KBYTE. Abbreviation for **KILOBYTE**.

KILOBYTE. 1 kilobyte = 1024 **BYTES**.

K FLOPS. 1000 floating point operations per second. A measure of the speed of arithmetic processing of a computer or floating point accelerator.

K-TUPLE. A short string of characters (tuples) k elements long. See **WORD**.

KERMIT. A free, portable and widely available program to provide error-free file transfer between computers over serial communication lines. Technically KERMIT refers to the **FILE TRANSFER PROTOCOL**.

KERNEL. A central concept, or group of concepts, upon which can be built more complex and sophisticated tools or techniques.

LOCAL AREA NETWORK. A computer communications network designed for use over relatively short distances (e.g. less than 5 miles/km). They generally have high data transmission rate (tens of millions of **bps**) but cannot easily be extended to long distance communications.

LOGICAL CONSISTENCY (OF A DATABASE). Data in a database is said to be logically consistent if none of the values are contradictory in meaning nor contradict the definition of the database. For example, for a molecular sequence database to be consistent, the set of integers describing sequence length must actually be the correct value for each corresponding sequence.

MACHINE CODE. See **OBJECT CODE**.

MAN MACHINE INTERFACE. That part of a program that provides the communications channel between other sections of the program and the human user.

MATRIX. A two-dimensional (or more) table used for storing data.

MBYTE. Abbreviation for **MEGABYTE**.

MCKUSICK NUMBER. A numerical value used to indicate the dominant, recessive or sex-linked nature of a gene.

MEGABYTE. 1 megabyte = 1048576 **BYTES**.

MELD. See **CONTIG**.

MENU. A list of options from which the user may select one or more items.

METRIC. A measure of distance which has a geometrical interpretation. For a pair of points x and y the distance x,y is a finite real number which is 0 if and only if x=y. Three points in a metric space obey the **TRIANGLE INEQUALITY.**

MICROPROCESSOR. An integrated circuit device containing most of the fundamental elements of a computer on one chip.

MIPS. Millions of Instructions per Second. A measure of the performance of a computer system. The Sun 3 and VAX11/780 run at approximately 2 MIPS.

MISMATCH ALGORITHMS that find maximum similarities and alignments between sequences either propose insertions and deletions (**INDELS**) or allow matches between non-identical base or amino acids i.e. mismatches.

MMI. Abbreviation for **MAN MACHINE INTERFACE**.

MODEM. Acronym for **MO**dulator **DEM**odulator. A device that converts digital data to analogue signals that can be transmitted along a telephone wire. Modems enable computers to communicate over long distances using the established public telephone network.

MODULUS. The modulus of two numbers A and B is the difference between A and the multiple of B that is closest (and less than) the value of A.

MOUSE. A small hand-held device that is moved over a flat surface, and whose location is used to position the screen cursor. The mouse is the principal mode of interaction with **WIMP**-based systems. Buttons on the upper surface of the mouse enable items to be selected from **MENU**s or allow **WINDOWS** and **ICON**s to be expanded, contracted or moved.

MSDOS. An acronym for the operating system developed by Microsoft for 16 bit microcomputers based on the Intel 8086, 8088, 80286 and related microprocessors.

MULTI-TASKING. An operating system feature that allows a single user to run more than one program (task) at a time.

MULTI-USER. An operating system feature that allows more than one user at a time to use a single processor.

MULTIPLEXOR. A device that shares-out a single communications line between several users.

NBRF. Abbreviation for the National Biomedical Research Foundation, Washington; the collectors and distributors of the PIR protein sequence databases.

NETWORK. A group of computer systems (network hosts) connected via telecommunication links to provide close integration and sharing of user services such as electronic mail, file transfer, login facilities, and specialist hardware. Somewhat different approaches and services are used in **LOCAL AREA NETWORKS** and **WIDE AREA NETWORKS**.

OBJECT CODE. The binary coded machine instructions actually executed by a microprocessor. The output of a programming language **COMPILER**.

OPERATING SYSTEM. The computer program that manages system resources: the file system, running programs, input and output to the user and communication ports etc.

PACKET SWITCHING. A method of transmitting digital data over a telecommunications network. Data is split into segments or packets by a **PAD** (Packet Assembler Disassembler) and each segment is tagged with its destination address and its point of origin. Individual packets are routed towards their destination, where they are reassembled into the original data by a PAD.

PAD. Abbreviation for **P**acket **A**ssembler **D**isassembler. Used in **PACKET SWITCHING** computer networks.

PATTERN LEARNING. A technique that enables a computer program to recognize a pattern occurring in data. The program must previously have been 'trained' by presenting it with two sets of data: one where the signal is known to be present, and the other where it is absent. The **PERCEPTRON** algorithm is a pattern learning algorithm.

PCDOS. A variant of the Microsoft operating system **MSDOS** for the IBM personal computer.

PERCEPTRON. A **PATTERN LEARNING** algorithm.

PIPE. A feature provided by some operating systems (e.g. **MSDOS** and **UNIX**™ for redirecting the output of a program to the input stream of another program.

PIR. The Protein Identification Resource. A protein sequence data library collected at the National Biomedical Research Foundation (**NBRF**) Washington.

PIXEL. The smallest picture element that can be plotted on a display device.

PORT. A port is a connector that enables a computer to be connected to another digital device (e.g. a printer, digitiser etc).

PLOTSTREAM. A stream of ASCII characters which are interpreted into vectors or characters and displayed by Tektronix graphics devices.

POSTSCRIPT. A command language for graphics display devices with features particularly suited to production of mixed high quality text and graphics as required in photo-typesetting. A number of laser-printer manufacturers have adopted Postscript as their control language.

PROGRAM COUNTER. A memory location (register) that indicates the current state of program execution by the position reached in the program.

PROGRAM PORTABILITY. The ease with which it is possible to take a program written on one machine or under one operating system and move it (port it) to a different computer system.

PROTEIN TOPOLOGY. A description of protein structure based on the relative spatial arrangement of protein secondary structures.

PSS. Abbreviation for Packet Switch Stream, a commercial UK communication network that uses **PACKET SWITCHING**.

PSTN. Abbreviation for **P**ublic **S**witched **T**elephone **N**etwork.

PTT. Postal, Telegraph and Telephone authorities. In Europe these are often government bodies managing the PSTN.

RAM. Abbreviation for **R**andom **A**ccess **M**emory.

RANDOM ACCESS MEMORY. Fast semiconductor memory in which the program runs and where the primary data is stored. Data may be read and written in any order and at any time without any overheads.

RASTER. The scanning pattern of the electron beam that forms the display area in a cathode ray tube visual display device.

RECORD. A single line in a text file or a single entry in a **DATABASE**.

RECURRENCE EQUATION. Evaluation of an expression by means of a series whereby the previous value is used in the calculation of the next until the required degree of accuracy is obtained.

RECURSION/RECURSIVE. A method of describing a mathematical or computational function in terms of itself. The classical mathematical example is the definition of the function that evaluates the factorial of a positive integer *factorial*(N) i.e.

i. $factorial(1) -> 1$
   i.e. the *factorial* of 1 is 1
ii. $factorial \rightarrow N \times [factorial\ (N-1)]$
   for $N > 1$

equation (ii) defines factorial recursively.

REGIS GRAPHICS. A set of commands for drawing and controlling graphics display devices developed by **DEC**.

REGISTER. A memory location on the processor or microprocessor itself where important values are placed during the execution of a machine instruction.

RELATIONAL DATABASE. A model for storing data in a database based on 2D-tables (relations). The query language for relational databases has features that enable tables to be joined or data selected from individual rows and columns.

RELATIONAL OPERATOR. An operator or function that tests for a relationship between two (or more) values e.g. greater-than, equal-to etc.

RFLP. Abbreviation for **R**estriction **F**ragment **L**ength **P**olymorphism.

RISC. Abbreviation for **R**educed **INSTRUCTION SET** Computer. A computer **ARCHITECTURE** in which performance is determined by a small number of very efficiently encoded instructions. This contrasts with the 'traditional' approach of providing an increasing number of complex instructions in order to gain computing power.

ROM. Abbreviation for **R**ead **O**nly **M**emory. Non-volatile memory from which data may only be read or programs executed. Most personal computer systems have the BASIC language interpreter available as a ROM.

RS232C. A serial asynchronous communications protocol.

SECTOR HARD/SOFT. Magnetic disks are segmented into sectors where data may be recorded. If the type and number of sectors available is reconfigurable by the operating system software then the sectors are said to be "soft". Sectors with boundaries permanently defined are called "hard".

SERIAL INTERFACE. A piece of **SOFTWARE** or **HARDWARE** that converts data presented in digital form (e.g. on a data **BUS**) to a series of electronic pulses that may be transmitted serially or *vice versa*.

SERIAL PORT. A serial **PORT** is one that enables communications using a serial asynchronous communications protocol (usually **RS232C**) via the **SERIAL INTERFACE**.

SHELL. The name given to programmable command interpreters for the **UNIX**™ operating system.

SOFTWARE LOOKUP TABLE. When frequent evaluations of a complex expression are needed it is often more efficient to pre-compute and store the results in a lookup table. Subsequent calculations may then refer to the table rather than the original expression.

SOFTWARE. Instructions provided to a general purpose processor in a programming language.

SOURCE CODE. The original program text written in a computer programming language. Source code is presented to a **COMPILER** or an **INTERPRETER** in order to convert the program into executable processor instructions.

SPOOLING A commonly used process in **MULTI-TASKING** operating systems for collecting **JOBS** or users requests in a queue that do not require interactive processing or that have a low priority and can thus be run independently when resources become available. Spooling is the act of sending a job to a queue (spool) that is managed by a program called a spooler (queue manager).

STACK. A programming language feature that allows **DATA STRUCTURES** to be stored and accessed in a metaphor of a canteen dinner plate stack. Each new data structure pushes down the previous one on the stack. Accessing the data can only occur from the top of the stack and as each structure is taken, its predecessor becomes the next one available.

STEM. A double stranded helical region of DNA or RNA.

STOCHASTIC PROCESS. A process determined by probabilistic events in time (e.g. radioactive decay).

SWAPPER. The process within an **OPERATING SYSTEM** that dynamically exchanges sections of memory on disk with that resident in the systems **RAM**. This is the machinery that enables programs and data to be used in a **VIRTUAL ADDRESS SPACE** greater than the available physical random access memory.

SYNCHRONOUS COMMUNICATIONS. A method of managing a dialogue between digital devices in which there is an agreed synchronisation between the transmitter and the receiver of signals.

TCP. Abbreviation for Transmission Control Protocol, a communications protocol between host computers on **ARPANET** and the **INTERNET**.

TELNET. The display terminal control protocol used in the **ARPANET** and the **INTERNET**.

TERMINAL EMULATOR. A program that reproduces the functions of a display terminal.

TIP. Abbreviation for Terminal Interface Processor, the equivalent of a **PAD** on the **INTERNET** network.

TRANSPAC. The French national **PACKET SWITCHING** communications network equivalent to the UK PSS.

TRANSPUTER. A 32 bit computer on a single chip developed by Inmos for applications in parallel computing and experiments in parallel **COMPUTER ARCHITECTURES**.

TREE SEARCH. An **ALGORITHM** or program that searches through a hierarchical, tree-structured organisation of data in order to quickly locate an individual datum.

TRIANGLE INEQUALITY. If x,y,z are three points in a METRIC space then the distance from y to z must be less than or equal to the sum of the distances from x to y and x to z.

TROFF. A text processing language available under the **UNIX**™ operating system.

TYMNET. An American national computer communications network linked to **TRANSPAC** and **PSS**.

ULTRIX. An implementation of the **UNIX**™ operating system available for DEC VAX computers.

UNCERTAINTY CODE. See **AMBIGUITY CODE**.

UNIX™. A popular **MULTI-USER, MULTI-PROCESS** operating system developed and copyrighted by Bell Laboratories.

UWCGC. Abbreviation for the University of Wisconsin Genetics Computer Group.

VIRTUAL ADDRESS SPACE. Operating systems such as VAX/VMS and **UNIX**™ extend the memory that a program has available beyond real memory (physical solid state memory installed in the computer) by utilising disk storage. The total amount of memory that an operating system will allow a program to use is called the virtual address space.

VLSI. Abbreviation for Very Large Scale Integration.

WIDE AREA NETWORK. A computer communications network designed to operate over long distances. Data transmission rate is usually sacrificed for distance except when cost is no object.

WIMP. Abbreviation for Windows Icons Mouse and Pointers. A style of interactive computing popularised by the Apple Macintosh that takes advantage of high resolution graphics and the mouse pointing device.

WINCHESTER DISK. A **HARD DISK** technology used by microcomputer and minicomputers.

WINDOW. A window is a region of the display screen that can be used as if it were a complete screen i.e. an emulation of a display device that can be positioned on any part of the physical display.

WORD. Word may have 2 related meanings. In the context of computing it is a unit of memory corresponding to the bit-width of a single microprocessor or processor instruction e.g. the Motorola 68000 series uses 32 bit words. In the context of molecular sequence analysis, word has the some meaning as **K-TUPLE** i.e. a short character string.

WORKSTATION. A high-performance (16 or 32 bit) computer system designed for use by one or perhaps a few users (e.g. the Sun Microsystems SUN-3).

XENIX. An implementation of the **UNIX**™ operating system available for the IBM personal computer.

INDEX